ALSO BY KEVIN WEST

Saving the Season

THE COOK'S GARDEN

THE COOK'S GARDEN

A GARDENER'S GUIDE TO SELECTING,
GROWING, AND SAVORING
THE TASTIEST VEGETABLES OF EACH SEASON

KEVIN WEST

Alfred A. Knopf
2025

A BORZOI BOOK
FIRST HARDCOVER EDITION PUBLISHED BY ALFRED A. KNOPF 2025

Published by Alfred A. Knopf, a division of Penguin Random House LLC, 1745 Broadway, New York, NY 10019.

Library of Congress Cataloging-in-Publication Data
Names: West, Kevin, 1970– author.
Title: The cook's garden : a gardener's guide to selecting, growing, and savoring the tastiest vegetables of each season / Kevin West.
Description: First edition. | New York: Alfred A. Knopf, 2025. | Includes index.
Identifiers: LCCN 2024023317 | ISBN 9780593319321 (hardcover) | ISBN 9780593319338 (ebook)
Subjects: LCSH: Kitchen gardens. | Vegetable gardening. | Cooking (Vegetables)
Classification: LCC SB321 .W468 2025 | DDC 635—dc23/eng/20241029
LC record available at https://lccn.loc.gov/2024023317

penguinrandomhouse.com | aaknopf.com

Book design by Shubhani Sarkar, sarkardesignstudio.com

Manufactured in China

The authorized representative in the EU for product safety and compliance is Penguin Random House Ireland, Morrison Chambers, 32 Nassau Street, Dublin D02 YH68, Ireland, https://eu-contact.penguin.ie.

FOR MY MOTHER

CAROL MARTIN SATTERFIELD

1949–2024

“PRESERVE THE LEGACY”

I believe a leaf of grass is no less than the journey-work of the stars,
And the pismire is equally perfect, and a grain of sand, and the egg of the wren,
And the tree-toad is a chef-d'oeuvre for the highest,
And the running blackberry would adorn the parlors of heaven.

—WALT WHITMAN, "Song of Myself"

CONTENTS

PART

02

VEGETABLE KNOWLEDGE

JANUARY 14: Baked Honeypatch Squash from winter storage

PREFACE

This is a book about flavor. It is a book about how to become a better cook by stepping into the garden. It's about a way of cooking that is seasonal, spontaneous, improvisational, simple, varied, full of life, and deeply, unexpectedly flavorful. How could a sliced cucumber taste so good? Or plain baked squash? How can braised cabbage, that unjustly maligned vegetable, produce a broth so soulful from only salt, pepper, and butter? Could broccoli really taste so sweet?

The answer to those questions is in the garden.

As the grandson of East Tennessee farmers and Smoky Mountain hillbillies, I grew up with my ears full of the old-timey phrase "cooking from a garden." It didn't necessarily mean you grew everything you ate. It meant you brought to the table the flavors of the season. You ate what was best at that moment. You looked forward to when a favorite crop came in and you appreciated it during that month of the year. There was nothing so good as what had just been picked. The second-best was whatever you had frozen or canned yourself.

Often the quality we anticipated wasn't simply a matter of freshness. Everyone knew certain varieties of snap beans made for better eating—Kentucky Wonder for the table and Blue Lake for canning. Even potatoes had their qualities and uses—little red-skinned "new potatoes" dug in June and eaten that same day versus leather-skinned russet "Irish potatoes" forked up at the end of the season and stored for midwinter use.

The garden was full of variety, contrasts, alliances, whole meals. "What grows together, goes together," the folk wisdom held. Making supper didn't require a cookbook. You walked out to see what was ready to pick. There were your recipes. On the table, flavors would sing.

Good country cooks knew how to extract the most from each season without too much fuss; there were other things to do besides getting supper together. Skillful farmers and gardeners planted crops to suit their own taste and that of their families and customers. Happy eaters could tell the difference between homegrown vegetables and the lesser quality of store-bought produce trucked in from Florida or California.

At its most bountiful, such as at my West grandparents' farm, cooking from a garden meant a table covered with bowls of cream-style Silver Queen corn, stewed white half-runner beans, fried Clemson Spineless okra, Green Mountain potatoes mashed with their cooking stock, and a platter of trimmed spring onions—the last of which Pappaw ate raw in a show of awesome bravery, or so it seemed to me at the time. Sugar Snap peas with their edible pod were as good raw as steamed and buttered. Early Girl tomatoes or the late-season Big Boy would be sliced on a platter. Looseleaf Black-Seeded Simpson lettuce was "killed" by tossing it with a hot bacon-grease vinaigrette. Kirby cucumbers went into Gran's pickling crocks. We recognized these names and others in the garden. Vegetables were not generic.

It's worth repeating: You weren't expected to grow everything yourself. We gave full credit to produce bought at farm stands, family-owned local markets, front-yard honor-system tables, farmers' markets, and off the tailgates of pickup trucks parked along the roadside. Maybe you'd be moved to bring home a sack of yellow crookneck summer squash at their prime—5 inches long, shiny, taut-skinned, tender, the green stem still moist and fragrant of the living plant. Buying vegetables from the grower was a form of sociability, an excuse to talk and find out where such fine squash came from. The culminating purchase was a compliment.

And then other gardeners shared what they had. In August you might come home from church or a neighbor's house with a dozen ears of Hickory King corn to shuck—a rare variety favored by older growers with long memories. Everybody had zucchini and would eagerly contribute if you didn't have your own, which was unlikely. "Lock your car doors," people joked. A healthy garden creates a communal quantity of food, out of keeping with the individual effort expended to grow it. To give away freely what is generously provided by the soil is a gardener's unique pleasure.

In short, "homegrown" cooking never pretended that the zucchini or tomatoes or beans or whatever came from *your* backyard. Instead, it meant that you knew, at least in some general way, where it did come from—*whose* backyard or garden or farm. The grower had a name that could be found out if you didn't already know it. There existed a place on the map where his or her vegetables grew. A smear of damp red clay on the underside of a Crimson Sweet watermelon was proof of its local authenticity, even if you did buy it.

In today's lingo, our mindset was "locavore," without being self-consciously so. On each side of my family—the Martins and the Wests—we maintained in our collective imaginations a shared map of the local foodscape. It was subject to constant update and revision. We could tell you where to buy the first picking of Peaches and Cream bicolor corn and which markets carried Beauregard sweet potatoes. We also held in mind a broader overview of the regional foodshed. We'd pick out Vidalia onions from Georgia or O'Henry peaches from South Carolina because each had a flavor that couldn't be beat.

Both of my parents were raised in families fed by large gardens. The gardening I learned had a recreational bent. My mother, who died in 2024, as I finalized this book, taught me to put out cherry tomatoes called Tommy Toes, fun to say and eat. My father, who lives in the country, has always kept a pleasure garden for his summertime cooking—corn, tomatoes, okra, peppers, squash, eggplant. Once we dug potatoes at Christmas, late even by his lackadaisical standards. Growing a little something was for them the habit of a lifetime, a better way to eat. In recent years, my mother limited herself to potted thyme, dill, and rosemary. My father stopped putting in a garden and instead now tucks a few vegetables among his flower beds. I'm the one with the big garden now. I grow Carola potatoes because they were my mother's favorite, and I take shallots to Tennessee for my stepfather, Don. My father and Bob, his partner of forty-some years, get garlic and Green Mountain potatoes. This book is a tribute to the farmers and gardeners who raised me.

But legacy isn't what inspires me to put out a garden every year, much less nostalgia or filial obligation.

I do it for appetite—for flavor. Where else would I get chervil, Country Gentleman corn, Spanish Roja garlic, Belle Isle cress, green coriander seed, Amish Paste tomatoes, Perpetual Spinach chard, Golden purslane, or my favorite potatoes if I didn't grow them myself?

Each of these crops has a clarity of flavor that once startled me but now strikes me as the hallmark of homegrown food. I sometimes

OCTOBER 10: Scarlett Emperor beans

wonder if the vigor of just-pulled carrots is a verifiable taste. Or if the life force of green peas really can be sensed when you eat them mere seconds off the vine.

Probably not, but the deliciousness of homegrown vegetables and herbs grown in healthy soil does have a unique charge, a jolt of vitality and well-being, the essence of life and health.

Put simply, it is the flavor of a garden.

JULY 14: Green Arrow peas

INTRODUCTION

I was worried.

May had been unusually cold and dry in the Berkshires. There were snow flurries at Mother's Day and a frost on May 20. Rows of seedlings huddled close to the ground, half smothered by mulch. How would this ever make a garden?

"Don't worry," my mother said when I called to fret. "A **SEED** wants to grow."*

I reached for a pencil to take notes.

"You always see volunteers at the edge of the compost pile or in last year's beds," she continued, using the gardener's term for self-sown plants. "Lord knows, I have a million cherry tomatoes out there. Seeds want to grow. Nature wants to bring them forth."

In the weeks after our call, chilly May gave way to sunny June. The garden leapt up.

What I've come to understand with more time is that my mother was right, as well, about a more fundamental truth. Our role among the plants is a modest one: We merely encourage what nature already intends to do.

* Here and throughout the coming chapters, bolded words indicate useful vocabulary and key concepts. Refer to the Glossary, page 96, for definitions and further information.

THE VICTORY GARDEN

Americans first planted Victory Gardens over a century ago to defend and protect freedom abroad. The very notion of growing **VEGETABLES** to boost the war effort sounds quaint today, but during World War I the federal government exhorted citizens to take up gardening at home and in public parks as a patriotic duty. Millions of average Americans dug up their yards. "Food will win the war" became a national slogan. During World War II, the campaign was even more effective: As much as one-third of all vegetables grown nationwide came from home-based Victory Gardens.

The fighting spirit of the Victory Garden caught on, I suspect, because it resonated with the foundational American story of self-reliance in defense of liberty. The term *Victory Garden* and the idea of food independence spread most widely in the postwar era, merging at the periphery of American society with Back to the Land idealism and homesteader movements of all political stripes.

One measure of its lasting cultural influence was *The Victory Garden,* launched in 1975 on Boston's public television station, WGBH, the broadcast home of Julia Child. The show ran until 2015 and spun off popular books, among them 1982's *Victory Garden Cookbook* (Knopf). My well-thumbed copy was an inspiration for this book, which is meant to be, like its precursor, an up-to-date guide for the cook-gardener. *The Cook's Garden* grew out of my personal experience, and I offer it with the hope it will be a resource for anyone who wants to raise and prepare healthy, flavorful vegetables at home.

THE FLAVOR OF A GARDEN

As a lifelong cook from a family of farmers and backyard growers, I've learned that the excellent flavor of homegrown vegetables comes from healthy plants, and healthy plants spring from a benevolent conspiracy of gardener and nature. Everyone knows that a vine-ripened tomato in August tastes better than a hydroponic greenhouse tomato in February. But the same holds true for green beans and peas and cucumbers. Even zucchini and humble potatoes taste better when homegrown. A garden is a collection of flavors.

A book like this gives the unfortunate impression that gardening is a skill learned by applying set rules to known situations. Not so. Learning to grow food isn't a mechanical craft, like plumbing. It is more akin to learning to ski or surf. Theory doesn't fully prepare you for how the real world hits, and who knows how things will go the first time you commit to a slope, a wave, a garden. Practice is the teacher.

I hope that doesn't sound discouraging, like you're bound to fail. What I mean is the opposite: You're bound to figure it out. And to have fun doing so. You get the hang of things as you go and quickly progress from learning to doing. You pick up skills and take encouragement from gardeners who have been at it longer. Throughout these pages, I repeat useful tips I've collected from master growers along the way.

SOIL, SUN, WATER, AND TIME

The essentials for growing tasty vegetables are few: **SOIL**, sun, water, and time.

Above all, healthy soil. I'm inclined to say *living* soil. Indeed, many of the growers I spoke to while researching this book—and other authorities I read during the long annual hiatus known as the New England winter—rally to this common theme. The chief project of gardening is tending the soil itself. "I grow soil and the soil grows the plants," said Jan Johnson of Mill River Farm, as we walked out toward vegetable beds and hoop houses at her small, diversified commercial vegetable farm a few miles away from my house.

The truism holds for home gardeners as well, even if your garden is as small as a window box of herbs. The topic of soil will receive considerable attention in the pages ahead—how to assess it, enrich it, sustain it—because healthy soil is the mother of flavor.

The next essentials are plentiful sunshine and adequate water. We'll soon enough get into the specifics of what that means. Or, to be honest about how gardening really works, we'll discuss the *generalities* of each topic. Not only is gardening a flexible, improvisational, repetitive practice, but vegetables prove to be adaptable partners. To be sure, all plants have ideal **GROWING CONDITIONS**, the set of environmental factors that favor their development, but they tolerate less-than-ideal reality. Otherwise the common bean, a species native to the New World tropics, would not have thrived in Henry David Thoreau's New England garden. Vegetable crops have survived thousands of years of unpredictable weather, droughts, downpours, cold snaps, and heat waves. They'll get by when humans lapse. Your garden will survive a few days of neglect when you get busy with something else.

The final essential for growing food at home—after soil, sun, and water—is less tangible but no less important. A garden takes time, in both senses of the phrase. You will need to check your plants frequently—giving them a few minutes of your time—to see that they are watered and happy. But you will also have to allow them time to grow at their own pace. Baby lettuce, arugula, and radishes could be ready to harvest in as little as three weeks. But some types of corn and winter squash need three months to mature. Cultivate patience.

The gardener's baseline commitment is to keep up with a few regular and recurring tasks during the **GROWING SEASON**, a span of five to seven months in most parts of the country.

GARDEN ECOLOGY

Vegetables are alive, of course, and the garden in its totality is a diverse living community of plants, animals, microorganisms, and fungi. In a nutshell, that is the worldview of garden ecology, or an understanding of the garden as a biodiverse habitat within the all-encompassing context of the natural world.

Ecological studies have shown that complex biological systems are resilient and self-regulating. The ecological gardener guides nature and encourages biodiversity but does not attempt to control the system. You strive to become a genial conductor.

The garden's ability to reach dynamic equilibrium within changeable natural conditions—one definition of health—doesn't rely on artificial inputs, by which I mean chemical fertilizers, pesticides, and all that mess. It is achieved instead through the measured development of balanced **FERTILITY** and the subtle influences plants exert on each other, the basis of **COMPANION PLANTING**. It also includes interactions between plants and beneficial garden creatures, such as when predator insects limit the spread of munching, grazing, and sap-sucking pests. A gardener tips the balance in favor of health by selectively removing nemesis species. In return for that modest task, earthworms and soil microorganisms labor mightily to build soil fertility and increase the overall vitality of the garden. Soil mycorrhizae, a microscopic fungal network, operate at a still more mysterious level beyond our control, but not beyond destruction through mismanagement. A holistic understanding of garden ecology embraces all these essential collaborators.

Things get out of balance when ecological thinking is displaced by the economic logic of mechanical thinking. Attempts to force natural systems lead to unpredictable consequences. Chemical fertilizers boost short-term yields but undermine long-term systemic health. My garden is entirely organic—no artificial fertilizers, pesticides, herbicides, or fungicides—and this book is intended to offer both practical advice and moral encouragement for those who would like to garden without artificial inputs.

AUGUST 22: Brandywine

A final thought on how garden health ripples out: Homegrown vegetables benefit the body and gut, a prescription for long-term wellness. They bring together families who learn that if a kid grows it, he'll eat it. Local food production builds local economies and social networks. Homegrown food is supreme from an environmental perspective. It generates hardly any carbon footprint, apart from the emissions created by shipping a few ounces of seeds

QUICK START GUIDE

Gardening is a gateway to a world of learning. But you don't need to absorb this entire book before you start. Below are the essential steps at a glance. The key decision to make is how much time you can commit. A few minutes a day for potted herbs or a few hours a week for a raised bed? Or more time for a larger in-ground garden? Better (and Worse) Backyard Crops on page 14 provides recommended vegetable varieties to match your ambitions.

HOW MUCH TIME?

A few minutes a day	3 containers
A few hours a week	4 × 8-foot raised bed
4 to 6 hours a week	10 × 10-foot in-ground garden

HOW MUCH SPACE?

1/2 gallon container	1 thyme
1 gallon container	1 basil
5 gallon container	1 cherry tomato
Window box	1 each thyme, basil, cilantro, chives, and parsley
4 × 8-foot raised bed	Mixed lettuces and arugula or interplanted radishes and carrots
	1 cherry tomato and 1 slicing tomato
	1 zucchini and 1 cucumber (to run over the side)
	2 chiles/peppers and 1 eggplant
	4 chard or kale
10 × 10-foot in-ground garden	1 each thyme, basil, cilantro, chives, and parsley
	Peas/beans on a trellis
	Mixed lettuces, arugula, radishes
	Carrots, beets, or turnips
	2 cherry tomatoes and 2 slicing tomatoes
	2 zucchini and 2 cucumbers
	3 chiles/peppers and 2 eggplants
	4 chard or kale

CONTAINERS OR RAISED BED

Place in a sunny spot	page 26
Fill with potting soil or compost	page 51
Plant transplants, or	page 66
Sow seeds	page 64
Water frequently	page 69
Harvest herbs early, other vegetables at maturity	page 78

IN-GROUND (TURNING THE SOIL)

Choose a 10 × 10-foot site on level, well-drained, sunny ground	pages 26–29
Dig and turn the ground with a shovel	page 53
Lay down 2 to 3 inches of compost	page 53
Plant transplants, or	page 66
Sow seeds	page 63
Water regularly	page 69
Weed regularly	page 71
Watch for pests	page 75
Fertilize with more compost as needed	page 35
Harvest daily in season	page 78

IN-GROUND (NO-TILL)

Choose a 10 × 10-foot site on level, well-drained, sunny ground	pages 26–29
Smother grass with a tarp for 2 to 4 weeks*	page 54
Aerate the soil with a garden fork	page 54
Lay down 2 to 3 inches of compost	page 54
Plant transplants, or	page 66
Sow seeds	page 63
Water regularly	page 69
Mulch transplants/seedlings when 2 to 3 inches tall	page 73
Weed regularly	page 71
Watch for pests	page 75
Fertilize with more compost as needed	page 35
Harvest daily in season	page 78

* An alternative is to aerate first, put down a layer of cardboard to smother, cover with compost, and plant—all on the same day. See page 54.

through the mail. Indeed, the gardening practices I recommend—less tilling and turning of the ground, more **COMPOST** and mulch—will boost organic matter in the soil, sequestering carbon.

To me, the biology and chemistry of healthy soil is fascinating. But the beauty of growing vegetables is that we don't need to master the science to reap the benefits. Nature has worked out the details. Our task is much easier: to use and enjoy the harvest.

HOW TO USE THIS BOOK

This book came about after I started gardening seriously enough to raise most of my own food. I became attuned to the differences among named **VARIETALS**, or "cultivated varieties." Some thrived in my garden—on a well-drained southerly slope at an elevation of 1,300 feet in southern New England—while others underperformed. The bigger surprise for me, even as a lifelong home cook and thirty-year habitué of great farmers' markets in Berkeley, New York, Paris, San Francisco, and Los Angeles, was that some varieties tasted so much better. Through several pleasurable years of cooking from my garden, I developed what I came to think of as a varietal-specific approach to selecting, growing, and cooking the tastiest vegetables.

Growing a garden and cooking from a garden, it turned out, were two aspects of a single activity, two stages of a life cycle. "I'm a farmer and a cook," Tracy Hayhurst said to me while walking me around Husky Meadows Farm in Norfolk, Connecticut. "They're the same thing." Gardening is now a facet of what it means to me to eat well.

This book is a cook's guide to the garden, intended for home cooks who already enjoy shopping at farmers' markets and would like to learn more about seasonal produce. Cooking with homegrown food is a shortcut to becoming a more creative and more instinctive home cook, a way to enjoy the kitchen more and produce tastier results with less effort. A large garden is not required. In fact, I'd strongly advise starting small. Success with a single potted basil and a Sweet 100 cherry tomato in a planter box will transform a dozen summertime meals and perhaps unleash a lifelong passion for homegrown food, as it has for me.

This book is equally a gardener's guide to the kitchen, intended for backyard growers who already enjoy gardening but would like to learn more about choosing and preparing the most flavorful varietals. Many gardeners will already grasp the significance of varietal performance—how the first picking of little French filet beans differs from the high-summer splendor of meaty Romano beans. My invitation is to expand that way of thinking into a varietal-specific approach to cooking.

My philosophy derives from one part upbringing, two parts reading, and three parts lived experience. Garden authority Margaret Roach uses the title *A Way to Garden* for her book and newspaper column, and I love that she acknowledges right up front there is no such thing as *the* way to garden. This introduction and the other narrative passages throughout, such as chapter introductions and recipe headings, explain how I got to *my* way.

The next eight chapters will lay out my Garden Basics: an introduction on how to establish and tend a garden of any size, whether in containers, a **RAISED BED**, or in-ground.

The Garden Basics are meant to apply broadly to all vegetables, and they are pitched to curious gardeners, whether novices or those with with some experience, who want to know more, not just the how but also the why of successful gardening practices.

First-time beginners or anyone daunted by the sheer volume of information in the pages ahead can refer to the Quick Start Guide on page xviii. This one-page cheat sheet provides an initial orientation and also serves as a handy reference to return to throughout the gardening year.

AUGUST 13: My garden in high summer: flowering cosmos, pumpkin vines, collards, Tuscan kale

Each of the last thirteen chapters is dedicated to a single vegetable or vegetable cohort, which is to say a group of crops with similar growing requirements and kitchen uses. The cohort might represent a family gathering, such as onions and garlic, squash in its multiplicity, or the legumes. Less obviously, perhaps, the tomato chapter also covers peppers, chiles, eggplants, and tomatillos because, as closely related nightshades, all have similar needs.

This book is not meant to be an encyclopedia, and I skip some vegetables less suited to backyard cultivation. No artichokes. No melons, which might surprise gardeners in the warmest regions. See Better (and Worse) Backyard Crops, page 14, for an explanation. Sorry on all counts! I also know I will likely confound some Southerners by including okra in the chapter on corn, but their planting schedules overlap, as do their **SPACING** needs. And okra had to go somewhere. A botanist might roll her eyes. Other readers might sort the vegetable panoply differently than I do. And so you should. Organize your garden according to your habits, just as you design a kitchen to suit your needs. If you can't find a favorite vegetable in the chapter where you think it should be, please check the index before giving up.

GARDENS OF DEFIANCE

Wendell Berry's famous reminder that "eating is an agricultural act" converges with a corollary truth. Agriculture is a political act. Growing food is activism. It is self-reliant, progressive, dissident. Guerrilla gardener Ron Finley calls it gangster. The extraordinary vegetable gardens planted by US citizens of Japanese descent imprisoned at Manzanar internment camp near Lone Pine, California, during World War II were nothing less than an organized resistance movement. I did archival research at Manzanar for Los Angeles artist Lauren Bon, who described them as "gardens of defiance."

Imprisoned master gardeners supplied the camp's prisoner-run kitchens with homegrown vegetables throughout the year, including specialty crops such as Asian greens and daikon radish for traditional Japanese meals. The productivity and sheer beauty of the camp gardens expressed a human dignity that would not be debased through physical coercion.

Manzanar's Gardens of Defiance were a poignant mirror image of the backyard Victory Gardens planted nationwide to aid the war effort. As much as I embrace the Victory Gardens' spirit of food independence, there is also a great lesson for our time in the legacy of Manzanar. Let us all plant a Garden of Defiance.

The reasons to do so are many. A garden establishes one's practical and spiritual independence from industrialized consumer capitalism. It embodies both resistance and resilience, the same activity at two different moments in the struggle, to borrow a phrase from Adam Gopnik. In this case, the struggle is against the industrial food system and its bleak twin, industrialized medicine.

A garden expands the basic human right to wholesome food and enlarges the future for anyone who takes it up. "A garden is made of hope," wrote M. S. Merwin. Cooking from the garden brings that hope into the kitchen, the very center of domestic stability and human renewal.

Gardening is an ancestral practice, an apprenticeship in cultural wisdom. It remembers women's experience and indigenous knowledge elsewhere lost. A garden takes in the ages. The earliest cultivated plots, sown perhaps ten thousand years ago, gave rise to human settlement and its ornaments, what we loosely call "civilization." Sowing seeds—among them wheat, rice, beans, corn, squash, tomatoes, chiles, and quinoa—is what kept us alive, as individuals and as a species. Small and slight, seeds would seem a weak foundation for human existence, a terribly provisional safeguard against hunger. And yet a seed wants to grow. Vegetables in healthy soil grow like **WEEDS**. "What is a weed?" asked Emerson. "A plant whose virtues have not yet been discovered." Vegetables are weeds whose virtues are known.

In a garden, the human imagination enters a duet with the nature world. The word *horticulture* derives from *hortus,* the Latin for garden, and *colere,* to attend to or to honor. *Agriculture* honors the field, *agros*. Both words embed, however deeply, a sense that growing food pays homage to something larger: the setting within the surrounding habitat, the global ecosystem, human memory, the all-encompassing cosmos. A gardener enters communion with that greater realm, and the garden, in return, bestows favor. In a garden, the natural truth that a seed wants to grow meets the biological reality that people need to eat. Cooking from the garden closes the circle.

When I plant in the spring, by habit I am apt to say, "I'm growing a garden." Later in the year—when I tuck into a bowl of buttered Champion of England peas or catch a whiff of baking Honeypatch squash as it rises to the upstairs room where I write—I realize I was wrong to think that I grew a garden. It is the garden that has grown me.

MY GARDEN

I grow vegetables on land owned by my neighbors Del and Christine Martin. The plot had been their garden. After I moved to town in 2016, Del invited me to help harvest potatoes in return for all I could eat. Over several seasons, as we became friends, I weaseled my way into his garden and progressively took it over. Now, in return for a share of the crops, the garden is "mine." Del teases me that I'm a gardener who doesn't have a garden.

We are an odd duo out there. Del loves hijinks, machinery, efficiencies of scale, and high-speed work. I am earnest and enjoy hand tools, fussy small projects, and interrupting my work to admire the day. I've shifted our/my garden away from diesel-driven horsepower to beer-fueled manpower and have diversified the crops we grow. For his part, Del handles vital infrastructure. He put up an 8-foot deer fence to protect our main growing plot, 60 feet by 78 feet, and devised an elaborate irrigation system. Across the driveway from the main plot, an unfenced parcel provides space for the few crops deer like less than we do—the witchy nightshades (potatoes, tomatoes, and peppers), stinking alliums (onions, garlic, shallots), and prickly cucurbits (summer and winter squash).

Every garden exists within a regional climate and a local microclimate. I garden in **USDA PLANT HARDINESS ZONE** 5b—see Climate in chapter 2, page 21. Winter lows here plunge to –15°F. Summer highs peak around 90°F. The average last frost, based on 1991–2020 climate normals from the National Oceanic and Atmospheric Administration as recorded at the official reporting station nearest my house, is May 13. Average first frost is September 29, making for an official growing season of 138 days. But because my garden is some 300 feet higher in elevation than the reporting station, my actual growing season will be a bit shorter and likelier to be clipped at either end by a rogue frost. In my observation, roughly May 20 to September 20 is reliably frost free. Annual rainfall is over 40 inches.

All these climate averages are made less dependable by the destabilizing effects of climate change. In addition to warmer temperatures in every season, aberrant temperature swings and extreme weather events have become more frequent here and everywhere. In my corner of southern New England, there are fewer weeks of winter snow cover and more summer days above 90°F. Finding ways to make our gardens and farms more resilient to withstand the extremes of climate change is an urgent priority and has reshaped my gardening practices in ways we'll discuss ahead.

What do I grow? The seasons of my garden always include herbs. The spring garden has radishes by the handful, lettuces and other salad greens, spring onions, green garlic, and trellised peas. Summer brings pole beans, zucchini, and tomatoes. Fall means winter squash, Brussels sprouts, and leafy collards, green as dollar bills, which will hold steady until the onset of deep cold in the new year.

In the winter months, I go to the freezer for homegrown peas, beans, corn, okra, and chard. The pantry holds canned tomatoes, pickles, and sauerkraut, as well as dried herbs and dried chiles. I store potatoes in the basement, winter squash in the chilly front hall, and all-important shallots and onions in a cold back room. By April, the potatoes get sprouty and the garlic has withered within its papery wrappers, but outside spring is arriving. Self-sown chervil volunteers where wind blew away mulch to expose bare ground and cress will soon be big enough for salads. I'll sow English peas, pushing the dimpled seeds into the soil, one by one, to the depth of an inch, or to the first joint of your index finger.

The rhythm of the gardener's year is four-beat and repeating, an annual cycle. The melody is composed of vegetables, herbs, and flowers.

"LA CÔTE BASQUE"

by TRUMAN CAPOTE (1924–1984)

The short story "La Côte Basque," published in *Esquire* in 1975, ended the social ambitions of its author, Truman Capote, because he put into print thinly veiled gossip entrusted to him by prominent New York City hostesses. It all feels very dated now, and rather nasty, although one passage sticks in mind as a catalogue of delights.

Over lunch at La Côte Basque, a stylish restaurant, the jaded character Lady Ina perks up as she recalls dining at the homes of the very rich on . . . homegrown vegetables. Her tipsy rhapsody is a paean to vegetable delights, any of which might be familiar to backyard growers. It's a reminder that a vegetable garden is not like a polo pitch, the luxury preserve of aristocrats, but an open-gated democratic conclave.

> *Ina selected from her salad a leaf of Bibb lettuce, pinned it to a fork, studied it through her black spectacles. "There is at least one respect in which the rich, the really very rich, are different from . . . other people. They understand* vegetables. *Other people—well, anyone can manage roast beef, a great steak, lobsters. But have you ever noticed how, in the homes of the very rich, at the Wrightsmans' or Dillons', at Bunny's and Babe's, they always serve only the most beautiful vegetables, and the greatest variety? The greenest petit pois, infinitesimal carrots, corn so baby-kerneled and tender it seems almost unborn, lima beans tinier than mice eyes, and the young asparagus! the limestone lettuce? the raw red mushrooms! zucchini . . ." Lady Ina was feeling her Champagne.*

PART 01 GARDENING BASICS

No matter how the harvest will turn out, whether or not there will be enough food to eat, in simply sowing seed and caring tenderly for plants under nature's guidance there is joy.

—**MASANOBU FUKUOKA,**
The One-Straw Revolution

CH.

01

WHAT TO GROW

SEPTEMBER 12: Annina eggplant

The summer I began this book, I visited gardens and farms in the Berkshires, where I live, and in East Tennessee, where I come from. My purpose was to discuss with wise growers the question presently at hand. What to plant if you're a new gardener? Where to start if you're someone who appreciates vegetables, cooks seasonally, and follows the rhythms of a farmers' market, and now would like to try growing a few things at home? Our conversations turned up much prudent advice, which is layered into the pages ahead. One key theme recurred: "Start small."

In setting out to make a garden, resist the temptation to overdo. A small vegetable patch builds confidence and culminates in the desire to try again. Overambition is fatal to second gardens. Consider your plans, then scale back. A garden will take twice as much time and half as much space as you think. Account for the work involved. Here's a formula:

garden size ÷ work hours = pleasure

The ideal garden—whether it's your first or your twentieth—spreads light effort over narrow ground, leaving you the energy and spirit to enjoy being among your plants. The fragrance of tomato leaves, the bouquet beauty of Swiss chard, the reach of a pea tendril as it stretches toward the trellis with a toddler's confidence. Those are the pleasures of a garden. Allow yourself the serenity to enjoy them. Grant yourself permission to plant less than you could.

The impulse to create a first garden might be satisfied by a few terra-cotta pots or a single four-by-eight-foot raised bed. At most, stake out a modest piece of yard. A ten-foot-square garden won't let you grow everything, but it will let you grow a lot, certainly enough to improve many meals.

Elizabeth Keen of Indian Line Farm crystallized a new and useful perspective for me.

Do not plant a garden based on *space,* she advised. The limiting factor in every gardener's life is *time,* so create a garden that respects your most precious commodity. "How many hours do you have to dedicate to your garden?" she said as we sat on upturned buckets beneath a shade tree. "Putting a few plants in the ground is the easy part."

For a sense of scale: A container garden of four or five pots will take an hour or two to put together and maybe ten minutes a day to keep going. A raised bed of 4 feet by 8 feet or a 10 × 10-foot garden plot can be installed in a day and tended in 20 to 40 minutes daily. A larger in-ground garden of 400 square feet will produce a substantial and varied harvest for a family throughout the growing season, with surplus for the freezer or root cellar, but it will soak up entire weekends and require an hour or more of daily tending and watering. Beyond 1,000 square feet lies the demands of self-sufficiency, a topic for another book.

If in doubt, make it smaller, and still it will be large enough to contain the great question—what to grow.

I called my friend Samin Nosrat one day in May and caught her outside among the fava beans. We gossiped: about the weather (her foggy Bay Area summer would ripen cherry tomatoes but not big **HEIRLOOMS**), our favorite matinee-idol pole beans and how to stake them, what to do with a single spear of asparagus, and the status of our respective compost piles. As a source of pride and worry, compost ranks somewhere between sourdough starter and children, placing it near pets in the upper-middle ranks of a gardener's affection.

Samin's garden, her first, brought home ideas she already knew in a general way, having learned them early in her career as a chef: a vegetable's taste comes from the soil, and good soil grows good vegetables. Now a new idea captivated her.

"You can grow what you like to eat!" she said.

The insight was sharp and true. A garden becomes a self-portrait. More broadly, it also represents a culture. It both reflects a cuisine, which I would define as the full expression of a culture's culinary imagination within the limits of its climate, and makes that cuisine possible.

In my garden, growing what I like to eat means growing to suit my style of cooking—or, rather, the styles of cooking I like. Because a garden is not defined by place (other than the natural limits of climate), you can plant an Italian garden in Idaho, a Thai garden in Maine, or a French garden in Kentucky. My approach is to catalogue a cuisine's garden flavors—its signature herbs, chiles, alliums, and vegetables. For example, to cook Italian food—or let's say "Italianate," to avoid the fraught question of authenticity—I'll need basil, oregano, rosemary, fennel seed, garlic, and dried chile flakes. For vegetables, I'd select broccoli rabe, zucchini, fennel bulb, and paste tomatoes. And I'd want arugula and radicchio for salad greens. That's a good-sized garden and well suited to someone inspired by Italian cooking.

Another example: for Mexican (Mexican-ish) cooking, I'd plant cilantro, jalapeños, white onions, and tomatoes for salsas, with vegetables such as grey zucchini (calabacitas), corn, and tomatillos. (Purslane, in New England regarded as a weed, is in Mexico the esteemed vegetable verdolagas.) See Growing Cuisine (page 7) for further examples.

What's thrilling about the American garden is how, from the start, it has been eclectic, expansive, cosmopolitan. Indigenous farmers had established corn, beans, and squash as staple crops two hundred generations or more before Columbus. Enslaved Africans brought seeds for okra, rice, and watermelon across the terrible Middle Passage. Tomatoes and potatoes came north from Mexico and South America. Colonial gardeners planted them all alongside Old World cabbage, lettuce, onions, and herbs.

The American garden has never stopped expanding. Cilantro and chiles are as commonplace today as garlic and broccoli, which in earlier times were also considered suspiciously "foreign" before they became garden essentials. Food moves with people. Gardens travel. Planting a garden even can be a tribute to those who came before us—a cultural legacy.

For example:

The Three Sisters: First grown together perhaps five thousand years ago in Mesoamerica, the milpa system of interplanted corn, beans, and squash—known in North America as the Three Sisters—represents one of civilization's greatest agricultural inventions. The trio of crops supplies complete human nutrition and also cooperates in the garden. Heavy-cropping corn supports climbing beans. Big-leafed squash shades the ground, conserving moisture. Beans restore soil fertility by drawing nitrogen from the air and "fixing" it in the soil. The milpa system is sustainable, resilient, and enduring.

A Pre-Columbian Garden: Crops that originated in the Americas today supply approximately 60 percent of global food consumption. In addition to corn, beans, and squash, the New World produced tomatoes, potatoes, sweet potatoes, and chiles, which spread as far as China within a century of the Columbian Exchange, or the transfer of species between the two hemispheres. Other pre-Columbian crops include peanuts, amaranth, the Andean tubers oca and olluca, quinoa, Jerusalem artichokes, and sunflowers. The indigenous herb epazote flavors beans and supposedly makes them less windy.

An African Diaspora Garden: Enslaved Africans and African Americans made incalculable improvements to American agriculture and cuisine. African crops spread with diaspora gardens: okra, cowpeas/black-eyed peas, benne/sesame seeds, eggplant, West Indian gherkins, watermelon, African rice (the source of Carolina Gold), and millet. African Americans expanded the range and uses of cymling (pattypan summer squash), collards, peanuts, tomatoes, and sweet potatoes.

A Thanksgiving Garden: Plant in spring, serve on the fourth Thursday in November: winter squash and pumpkins, Brussels sprouts, chard, kale, turnips, sweet potatoes, and sage.

A Forager's Garden: Every garden is blessed with vegetables the gardener didn't plant—weeds. The most flavorful include lamb's-quarter (*Chenopodium alba*), purslane, dandelion, garden cress, and nettles.

In a similar spirit, you can also plan a backyard garden to serve a specific use, such as:

Kids' Gardens: A literal kindergarten, or children's garden, might focus on radishes, green peas, cucumbers, and zucchini—jump-up vegetables for impatient young growers. A potted cherry tomato and a windowsill basil plant yield spaghetti sauce. A tiny taco garden needs only cherry tomato, chile, scallions, and cilantro for simplified pico de gallo.

Other Special-Use Gardens: A preserver's garden would feature Kirby cucumbers for pickling, paste tomatoes, cabbage for sauerkraut, and dill, the essential pickling herb. To make fall kimchi, plant napa cabbage and daikon, along with garlic, chives, scallions, chiles, and even ginger.

For freezing summer produce, peas, green beans, corn, okra, chard, and kale are best. Crops for long winter storage include potatoes, winter squash/pumpkins, onions, garlic, and shallots. Herbs for drying include thyme, savory, rosemary, sage, marjoram, and oregano. Dried chiles keep their heat and flavor for a full year.

AUGUST 5: Gold Rush wax bean, Velour filet bean, and Blue Lake 274 snap bean with marjoram

GROWING CUISINE

A cuisine contains cultural wisdom. It is a body of knowledge bound by place and time that reflects how a group of people come together to grow, prepare, and share food within larger social, political, and spiritual contexts. Individual recipes and even signature flavor pairings are cultural artifacts to acknowledge and honor.

And yet a cuisine is not a museum, either. Cuisines evolve. My own way of cooking has changed dramatically over the years as I've eaten more broadly, traveled farther from home, and learned from talented cooks and gardeners along the way.

A garden reflects all of that: it is a self-portrait that draws from the world.

CUISINE	SIGNATURE FLAVORS	SIGNATURE VEGETABLES	ADDITIONAL VEGETABLES
Chinese	Ginger, scallions, garlic, garlic chives, chiles	Bok choi, napa cabbage, daikon, Chinese eggplant, snow peas, leaf celery	Gai lan (Chinese broccoli) Sweet potato leaves Celtuce
French	Chervil, chives, tarragon, parsley, savory, shallots, thyme	Celeriac, leeks, mâche, lettuces, sorrel	Fingerling potatoes, such as La Ratte Très Fine Maraîchère frisée Tonda di Parigi carrots French Breakfast radish
Italian	Basil, fennel, oregano, parsley, rosemary, garlic, cayenne chiles (for drying)	Arugula, broccoli rabe, eggplant, garlic, radicchio, tomatoes, zucchini	Nero di Toscana kale San Marzano tomatoes Costata Romanesca zucchini Jimmy Nardello peppers Rosa Bianca eggplant
Mexican	Chiles, cilantro, garlic, onions (white), oregano	Calabacita, corn, tomatillos, tomatoes	Epazote Mexican oregano (*Lippia graveolens*) Chayote
Southern	Dill, sweet onions, thyme, scallions	Cabbage, beans, cucumbers, collards, corn, okra, tomatoes	Greasy-type pole beans Seven Top turnip greens Southern Giant Curled mustard greens Cushaw

PLANT YOUR RECIPES

It's easy to focus too narrowly on planting vegetables as if making a shopping list—broccoli, zucchini, kale, potatoes—somewhat in the same way you look up a recipe by its main ingredient or search online with a keyword: What to do with *kohlrabi*? The gardener-cook pursues a different strategy. You don't just plant the vegetable. You plant the recipe.

As you go about choosing favorite vegetables to grow, think as well about the secondary elements that complement them in cooking, the building blocks of texture and color, as well as the herbs, alliums, and counterpoint flavors, such as bitterness and spiciness, that make a dish sing. When you plant recipes in this way, you grow flavor and gather in meals.

For example, one of my standby summer recipes for groups of any size is blanched yellow wax beans tossed in a vinaigrette with herbs. Newly pulled shallots and blooming marjoram make the dish, so I plant them both, and to take things one step further, I might plant more beans of different colors. Pretty is an ingredient, too. (See the picture on page 6.)

A more layered example is salad. Why not grow several types of lettuce, combining old friends—Little Gem and oakleaf—with an attractive stranger, such as Speckled Trout romaine? Arugula and other minor greens round out the ensemble. An afterthought of edible blossoms—borage, nasturtium—interrupts the chlorophyll monopoly. Sliced scallions add neat circularity to lettuce's free-form billows. Crisp radish and snapping cucumber add contrast. Tie the bowl together with ribbons of shaved carrot.

By the inverse logic, the recipe for ratatouille makes perfect sense when you stand in an August garden surrounded by tomatoes, zucchini, eggplant, and peppers, all of which plead for stewing with new onions and garlic. I believe the symmetries between the garden and the kitchen are cultural artifacts, not chance occurrences. At some point in the past, a gardener and a cook must have found common purpose—one sowed, the other cooked, both were happy with the outcome. They repeated the experiment the following year. They kept it up, and others followed suit until their great-great-great grandchildren called it ratatouille. Over time, a body of recipes coheres into a cuisine.

FAMILY NAMES, LATIN NAMES, COMMON NAMES

Gardeners and farmers almost always prefer a vegetable's common name to formal botanical nomenclature. Silver Queen is corn, not *Zea mays*. That said, gardeners (and cooks) automatically sort vegetables into botanical families for convenience. We talk unselfconsciously about brassicas, alliums, legumes, and nightshades. So it's worth taking a detour here to consider scientific naming conventions and what they reveal about plant kinship. Understanding the garden as a gathering of families with shared tendencies rather than an anarchic gang of individuals simplifies our work.

A vegetable's scientific name, usually written in italics, comes in two parts, a binomial. The capitalized first word is the generic epithet, from *genus,* the narrowest taxonomic category to group closely related species. By rough analogy, the generic epithet is akin to a person's surname; it indicates kinship. The binomial's second word, or specific epithet, indicates individuality akin to a person's given name. The combination of the two is unique. Cabbage is *Brassica oleracea,* so no other species will ever be assigned that name. The binomial is often called the Latin name, but inaccurately, inasmuch as Greek and many other languages have been raided since eighteenth-century Swedish botanist Karl Linnaeus codified the naming convention, a permanent legacy of the Enlightenment's grand project to describe and catalogue the world. In writing, the generic epithet is abbreviated on second mention, hence *B. oleracea*.

Confusingly, some species take on multiple forms in the garden. Cabbage and broccoli, despite their different appearance, are both *B. oleracea*. To distinguish between them as subspecies, each is given a varietal epithet. Cabbage is of the variety *capitata,* a Latin reference to its headlike shape, written as *B. oleracea* var. *capitata*. Another common vegetable species, *Beta vulgaris,* gives us both chard and beets. What's more, both forms belong to a single subspecies, *B. vulgaris vulgaris*. At this point, botanists make a still-narrower distinction between cultivar groups within the same subspecies, but that's splitting the hairs pretty fine.

To move back up the taxonomic ladder, multiple species can belong to a single genus and multiple genera can belong to a botanical family, which is an allied group of species with notable genetic similarities. Tomato shares the genus *Solanum* with eggplants and potatoes, while the chiles and peppers sort into *Capsicum,* but all are *Solanaceae,* commonly called nightshades.

Latin names are universally understood among botanists, whatever their native tongues, because each binomial is unique and each species receives but a single binomial. Common names swarm like gnats. One ubiquitous self-sown garden visitor, *Chenopodium album,* is commonly called goosefoot, lamb's-quarter, fat-hen, and pigweed. The poetry of vernacular speech misses an important fact: The weed is related to spinach, another *Chenopodium,* and equally delicious to eat.

See the chart on pages 10–12 for a breakdown of the main crop families.

VEGETABLES BY FAMILY

BOTANICAL NAME	ALSO KNOWN AS	COMMON FAMILY NAME	VEGETABLE NAME	VEGETABLE BINOMIAL
Apiaceae or Umbelliferae	Umbellifers	Carrot family	Carrots	*Daucus carota*
			Celery	*Apium graveolens*
			Fennel	*Foeniculum vulgare*
			Parsnips	*Pastinaca sativa*
			Chervil	*Anthriscus cerefolium*
			Cilantro	*Coriandrum sativum*
			Dill	*Anethum graveolens*
			Parsley	*Petroselinum crispum*
Amaranthaceae (subfamily *Chenopodioideae*)		Amaranth family (Goosefoot family)	Spinach	*Spinacia oleracea*
			Chard	*Beta vulgaris*
			Beets	*Beta vulgaris*
			Amaranth	*Amaranthus* spp.
			Lamb's quarters	*Chenopodium album*
Compositae or *Asteraceae*	Composites	Sunflower family	Jerusalem artichokes	*Helianthus tuberosus*
			Artichokes	*Cynara cardunculus* var. *scolymus*
			Cardoons	*Cynara cardunculus*
			Lettuce	*Lactuca sativa*
			Endive, frisée	*Cichorium endivia*
			Radicchio	*Cichorium intybus* var. *foliosum*
			Salsify	*Tragopogon porrifolius*
			Tarragon	*Artemisia dracunculus*
			Witloof (Belgian endive)	*Cichorium intybus*

BOTANICAL NAME	ALSO KNOWN AS	COMMON FAMILY NAME	VEGETABLE NAME	VEGETABLE BINOMIAL
Convolvulaceae		Morning glory family	Sweet potato	*Ipomoea batatas*
***Brassicaceae* (formerly *Cruciferae*)**	Brassicas, crucifers	Cabbage family, mustard family, cole crops	Cabbage	*Brassica oleraceae*
			Collards	*Brassica oleraceae*
			Kale	*Brassica oleraceae*
			Broccoli	*Brassica oleraceae*
			Cauliflower	*Brassica oleraceae*
			Brussels sprouts	*Brassica oleraceae*
			Kohlrabi	*Brassica oleracea*
		Asian greens	Napa cabbage	*Brassica rapa pekinensis*
			Bok choy	*Brassica oleraceae*
		Spicy greens	Mustard greens	*Brassica juncea*
			Arugula	*Eruca sativa*
			Radish	*Raphanus sativus*
			Watercress	*Nasturtium officinale*
			Wild arugula	*Diplotaxis tenuifolia*
		Turnips	Turnip	*Brassica rapa*
			Rutabaga	*Brassica napus*
Cucuritaceae	Cucurbits	Cucumber family	Cucumber	*Cucumis sativus*
			Gourd	*Lagenaria siceraria*
			Melon	*Cucumis melo*
			Squash—summer	*Cucurbita pepo*
			Squash—winter	*Cucurbita maxima, C. mixta, C. moschata, C. pepo*
			Watermelon	*Citrullus lanatus*
			West Indian gherkin	*Cucumis anguria*

(CONTINUED)

BOTANICAL NAME	ALSO KNOWN AS	COMMON FAMILY NAME	VEGETABLE NAME	VEGETABLE BINOMIAL
Lamiaceae		Mint family	Basil	*Ocimum basilicum*
			Marjoram	*Origanum majorana*
			Mint	*Mentha* spp.
			Oregano	*Origanum vulgare*
			Sage	*Salvia officinalis*
			Thyme	*Thymus vulgaris*
Leguminosae* or *Fabaceae	Legumes	Pea family	Pea (green)	*Pisum sativum*
			Bean (green, dried)	*Phaseolus vulgaris*
			Cowpea or field pea	*Vigna unguiculata*
			Peanut	*Arachis hypogaea*
		Nitrogen-fixing cover crops	Clover	*Trifolium* spp.
			Vetch	*Vicia* spp.
Liliaceae		Lily family	Asparagus	*Asparagus officinalis*
***Amaryllidaceae* (subfamily *Allioideae*)**		Amaryllis family (onion subfamily)	Onions	*Allium cepa*
			Garlic	*Allium sativum*
			Shallots	*Allium cepa* var. *aggregatum*
			Leeks	*Allium porrum*
Malvaceae		Mallow family	Okra	*Abelmoschus esculentus*
Poaceae		Grass family	Corn	*Zea mays*
Polygonacea		Knotweed family	Rhubarb	*Rheum* spp.
			Sorrel	*Rumex acetosa*
Solanaceae		Nightshade family	Chiles, peppers	*Capsicum annuum*
			Ground cherries	*Physalis pruinosa*
			Eggplant	*Solanum melongena*
			Potato	*Solanum tuberosum*
			Tomatoes	*Solanum lycopersicum*
			Tomatillos	*Physalis philadelphica*

ANNUAL AND PERENNIAL CROPS

Most vegetable crops are **ANNUALS**, plants that complete their life cycle in one growing season. A handful are **PERENNIALS** and return yearly—asparagus, artichokes, sunchokes, horseradish, rhubarb, and a few obscure greens beloved by permaculturalists. **BIENNIALS** live two years, the first for vegetative growth, the second to flowering. Cabbage, onions, carrots, parsnips, beets, and chard are biennials but grown as annuals for harvest in their first season. In mild climates, they will overwinter in the garden. Once harvested, they also keep well in the refrigerator or root cellar because, in effect, nature designed them for long-term storage.

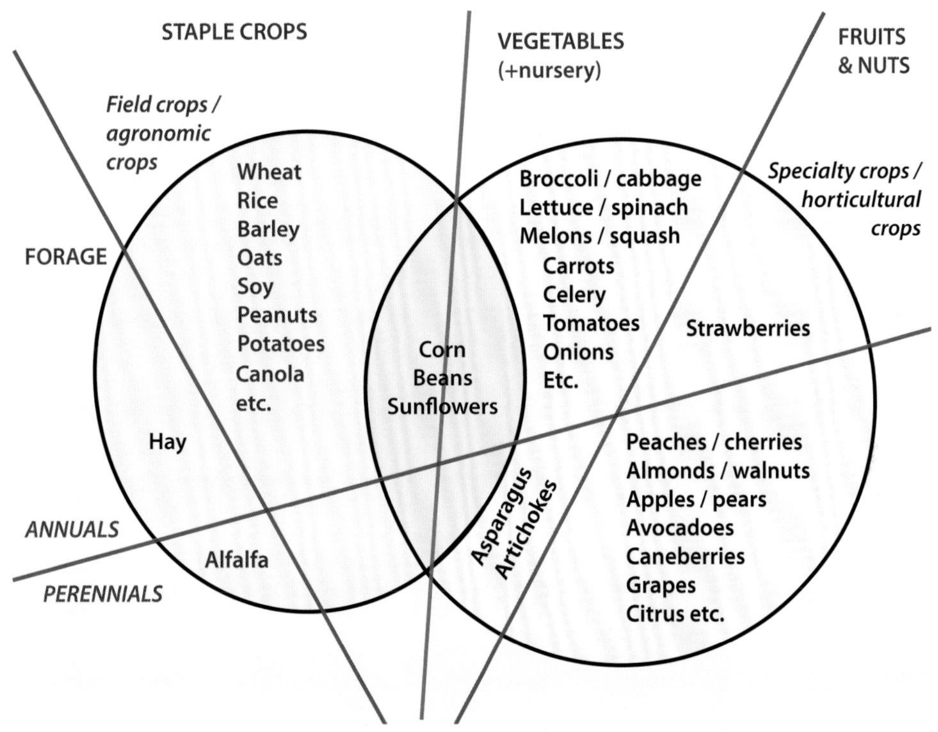

Created by Amber Kerr of the USDA California Climate Hub. Used by permission.

BETTER (AND WORSE) BACKYARD CROPS

Most vegetables are worth a try. If you love it, go for it. Everything in this book is within reach of home gardeners, apart from the limits of space. A window box won't suit pumpkins. Not all vegetables are equally satisfying to grow, however. Some are too slow, too big, too attractive to pests, too temperamental. Focus elsewhere or brace for trials and heartache.

The best self-starters for novice gardeners include:

- **Herbs** Easy to grow, they improve everything they touch in the kitchen. Most are sturdy, even rugged, and many tolerate poor growing conditions.

- **Tender Greens** The salad bowl fixings, from lettuce to arugula to your favorite small leafy green, grow quickly in compact spaces and pretty up their surroundings.

- **Hearty Greens** Robust, productive, beautiful, and almost self-sufficient, the hearty greens unfurl new leaves as quickly as you harvest the outer ones. In return for modest space and basic care, kale, chard, collards, mustard greens, and other cooking greens thrive over a long growing season.

- **Radishes** They come up fast and reach an ideal marble size in three weeks. Their haste, color, and elfin cuteness make them a top choice for children to grow.

- **Tomatoes** While somewhat needier than the above, tomatoes reward care. A vine-ripened tomato is the trophy of the backyard garden. Cherry tomatoes and small-fruited patio tomatoes grow well in containers and fruit generously.

Once you've gained confidence, expand your vegetable ambitions.

- **Summer Squash** In the garden, zucchini rhymes with abundance. Plant them and stand back. The flavor and succulence of homegrown summer squash far surpass grocery-store types.

- **Green Beans** The many types of green beans—some are yellow and purple—have beautiful big seeds that are a pleasure to handle, to sow, to collect. The seedlings leap up with bushy confidence. Some dash up a trellis and provide multiple pickings.

- **Peas** Sow peas thickly and pray for consistently cool weather. You'll learn the pleasure of watching pretty vines, pretty blossoms, and pretty pods grow. It is a pity that not everyone has tasted a just-picked pea.

- **Cucumbers** A cucumber on the vine is full of flavor—a surprise if you know only the bland, plastic-wrapped things sold in grocery stores. Freshly picked, it will taste like chlorophyll, sunshine, and well-being—a hydrating treat. Suited to containers and will climb a trellis.

- **Chiles** The varied members of the pepper family want heat, sun, and, preferably, more heat and sun. Are bell peppers worth the bother? No one ever asked that about chiles. Their abundant fruit is small but mighty, even when green. Dried red chiles last an entire year in the spice cupboard.

The next group of backyard favorites come with modest caveats.

- **Cabbages** One species, many forms. European head cabbage looks wonderful while growing and many varieties are delicious. But it is slow to mature and a space hog. The Asian cabbages, both tightly wrapped napa and shaggy headed bok choy (also spelled pak choi or pok choi) are quicker to harvest, smaller, and milder in flavor.

- **Winter Squash** Sweet and orange on the table. If you have room for the vines to sprawl and patience to wait them out (some varieties require nearly four months to mature), winter

SEPTEMBER 12: Dried and fresh cayenne chiles

AUGUST 6: Killarney Red, an extra-delicious varietal of hardneck garlic, one of the rocambole group

squash and pumpkins deliver a heavy fall harvest. The long-storing fruit comes in small, medium, large, and truly gigantic sizes.

- **Root Vegetables** Beets, carrots, turnips, and parsnips are underground sensations. They grow best in cool weather when sown in deep soil. Put out early for a first-of-summer harvest or in July for a fall harvest. Larger varieties mature slowly but keep well in storage. Carrots take forever to germinate.

- **Onions, Garlic, Shallots** Ubiquitous but underappreciated. Their slender, upright habit allows for dense plantings—a good return on space—but they are slow-maturing, long-season crops, not well suited to small gardens or impatient gardeners. Scallions and chives deliver more quickly.

- **Potatoes** What others feel for their prize tomatoes I feel for my potatoes. They take extra work and a long season to mature, but potatoes grow surprisingly well in containers. I once used a burlap sack filled with compost.

- **Corn** The queen of summer demands space and draws pests of all sizes. Some varieties take three months or more. Corn sucks nitrogen from the soil. But what flavor!

- **Okra** Okra wants as much heat as peppers and as much space as corn.

- **Celery** Not everyone will see the point of growing a slow, thirsty vegetable known more for its crunch than its flavor. Those people have not tried homegrown celery. Newly harvested, it is remarkably fragrant. Still, a minor vegetable in the eyes of many.

- **Broccoli, Cauliflower, Brussels Sprouts** Broccoli plants are bulky and provide a single harvest. Cauliflower is heat sensitive. Brussels sprouts grow so slowly. The same amount of space planted to kale or chard will keep a family in greens all season. The brassicas are nonetheless stately in the garden and delicious in the kitchen. It's a matter of space.

A first-time gardener should steer clear of a handful of crops.

- **Melons** Wonderful to eat, watermelons and cantaloupes will overrun a garden like Caesar took Gaul. They require long stretches of hot, sunny weather and a deft touch to bring them to perfection. Grand English country estates once maintained heated, glass-clad "melon houses" to ripen them. When I asked my uncle David, a lifelong farmer who found placid retirement among the rows of his magnificent backyard garden, for tips on melons, he squinted his eyes and rocked back on his heels, as if to distance himself from the advice he reluctantly delivered. "Most people won't want to bother with watermelons," he said in his gentle way, which I took to be strong counsel against them. Melons are for black-belt gardeners.
- **Dried Beans** A ten-foot row of green beans will pick out at about five pounds, enough for several meals and a little extra for the freezer. Ten feet of dried beans, which you will weed and water for months, then harvest one pod at a time and shell out by hand, will yield something like ten ounces, a scant cup and a half. Or not quite enough to make a full pot of soup beans.
- **Sweet Potatoes** Grown from slips, or sprouts off a mother tuber, chill-averse sweet potatoes get a slow start in spring and take three to four months to mature. Novices in northern climates should avoid them. Deer are drawn to sweet potatoes in every climate. Also note that the tubers bruise and crack during harvest and sweeten only with curing.
- **Asparagus** One of a handful of perennial vegetables, asparagus takes three years to mature and, even then, the yield is scanty compared to earthier crops. Asparagus beds want annual manuring, regular mulching, and weeding in perpetuity. As with other passion projects, growing asparagus is a hobby pursued in defiance of reason or economic rationale, like fly fishing.
- **Artichokes** Another perennial, the artichoke is harvested commercially only in California, which tells you its preferred growing conditions. Jefferson's enslaved gardeners labored mightily to overwinter them beneath mulch and covers. When I lived in California, I often ate artichokes. Here in New England, I never do. Life choices come with unintended consequences.

FLAVORFUL VARIETIES

You will see many caveats in the pages ahead about the specifics of region and climate. The garden I worked in as a student at Deep Springs College, at over 5,000 feet of elevation in the high desert of California's Eastern Sierra region, was very different from my current garden in New England. A vegetable that tastes best in *my* garden might not produce the best flavor in *your* garden. Still, the following recommendations provide a starting point. More options are given in the vegetable chapters that make up the second half of the book. Also consult Slow Food's Ark of Taste, an international catalogue of heirloom varieties of exceptional culinary quality. See pages 90–91 for recommended seed sources.

VEGETABLE	SUBTYPE	VARIETY
Broccoli	Heading	Green Magic
	Sprouting	Happy Rich
Cabbage		Early Jersey Wakefield, Tendersweet
Carrots		Red Cored Chantenay, Tonda di Parigi
Chard		Perpetual Spinach, Silverado
Corn		Silver Queen (white), Peaches and Cream (bicolor)
Cucumbers		Experimental 7082
Green Beans	Bush	Blue Lake
	Filet	Velour (purple when raw)
	Pole	Kentucky Wonder
	Romano	Northeaster (a pole bean)
	Wax	Dragon's Tongue
Green Peas	Short trellis	Laxton's Progress
	Tall trellis	Sugar Snap, Champion of England
Kale		Red Russian, Tuscan Black
Lettuce		Tennis Ball, Little Gem, Deer Tongue
Okra		Clemson Spineless, Aunt Hettie's Red
Onions	Fresh eating	Elisa Craig, Italian torpedo types
	Storage	Patterson, Blush, Rossa di Milano
Peppers	Pimento	Lipstick
	Sweet "frying"	Jimmy Nardello
Potatoes	All-purpose	Yukon Gem
	New	Upstate Abundance
	Starchy/baking	Green Mountain
	Waxy/fingerling	Pinto Gold
Pumpkin	Cooking	Winter Luxury, Long Island Cheese
	Carving	Howden
Squash	Summer, yellow	Yellow crookneck
	Winter	Honeynut, Robin's Koginut
	Zucchini	Costata Romanesco

SEED SOURCES AND "STARTS"

Where to find seeds and plants? Between late winter and midsummer, garden centers, hardware stores, and even some grocery stores stock seed packets for home gardeners. They are often eye-catching—printed with enticing photos of vegetable pinups. The back panel will be printed with growing instructions and notes on the variety, including its culinary qualities. Some seed packets also relate history, anecdote, and garden wisdom. Each seed company has its own character. Botanical Interests uses fine hand-drawn vegetable portraits for its packets and includes reams of information on the inside of the packet: you open it along a seam to read like a pamphlet.

A good nursery will sell seeds to suit local growing conditions and regional taste. For a wider selection, consult seed catalogues, including those found online. They offer a vast and even daunting range of options. Seed catalogues also contain volumes of free reference material, such as growing guides, varietal "biographies," planning tools, and yield calculators. Look ahead to chapter 9, Planning the Next Garden, for a selection of trusted seed companies, on pages 90–91.

Even more enticing than seeds are the six-packs of vegetable seedlings and the little potted herbs sold widely in spring, including at farmers' markets. "**STARTS**," or seedlings for **TRANSPLANT**, are cute as puppies and the easiest way to create a container garden or fill a raised bed.

Look ahead to chapter 6, Sowing and Transplanting (page 56), for a full discussion of what to do with seeds and starts once you get them home.

GARDEN RECORDS

I encourage you to keep records of what you plant in the garden and what happens over the course of the summer. You'll want to remember your sources for seeds and starts, as well as quantities bought, planting dates, and dates of first harvest. Earlier in this chapter, I advised you to consult your taste to know what to plant. For your next garden, you'll want to consult your garden records to know what to replant.

I carry a pocket-sized notebook and pencil with me to jot notes. A simple "plant more" or "too much!" will settle the question of how much to put out next time. I also record observations on a variety's performance and flavor. If I'm growing out multiple varieties of a single vegetable, I'll track yields as well, to learn not just the total weight of my potato crop but also the average yield per hill or per pound of seed for each variety. Without notes, I would forget, in the interval between July, when I dig new potatoes, and January, when I reorder, that Dark Red Norland is bountiful but bland and that buttery Upstate Abundance is more susceptible to black scab than nutty Pinto Gold. I just now looked at my notes from two years ago and saw that Copenhagen Market cabbage produced handsome heads—the word *market* in the name is always a tell—and was also less attractive to caterpillars than other varieties. I had forgotten.

I try, too, to note rainfall and unusual weather conditions—droughts, downpours—as well as any lovely natural occurrence, anything my mother would have called "a daily moment of joy." The first crocus, for instance, and then the first daffodil, blooming wildflowers, the first flash of autumn color. I'll put down the start date of April's tree-frog symphony and July's lightning-bug display. Natural indicators remind me to get things done before the moment slips past. Over years of watching, one becomes sensitive to natural time, and it can guide you. New England folk wisdom says to plant potatoes when dandelions bloom. My experience bears out the advice. Calendrical timekeeping is not the only way to understand the flow of the year, and not even the best.

FINAL THOUGHTS

A few last notes before we get to the hands-dirty part of creating a garden:

- Because diversity is an aspect of beauty, a garden should be as varied as the space allows. Two patio containers give you a cherry tomato and a leafy basil plant. If you have a window box, grow parsley *and* chives *and* tarragon. In the corner of a raised bed, there will be room for three types of salad greens, so let one of them be pert or new to you—mâche or garden cress. The variations you cultivate in a garden make it intimately yours, fitted to your way of cooking, shaped by your creativity.
- Growing a garden is simple in theory, but in practice there are a million ways to fail, and I discover new ones each year. "But tho' an old man, I am but a young gardener," wrote a sixty-eight-year-old Thomas Jefferson to Charles Wilson Peale on August 20, 1811, the peak of the growing season in Virginia's Piedmont. What Jefferson possessed were the three attributes a successful gardener needs: curiosity, the power of observation, and patience. Jefferson also had enslaved gardeners. To be frank about historic realities, the burden of work on Jefferson's estates fell on Wormley Hughes and others whose names have been lost. Their unheralded skill accrued wealth and prestige to Jefferson in his day and contributed to his stupendous horticultural legacy in ours.
- A garden requires work, but the work is outdoors, among living things. It is work leavened with joy.
- Consider every phase of a plant's life cycle and every part of the plant. Turnip greens and turnips roots. Fennel frond, fennel bulb, fennel pollen, fennel seed. The Swiss chard's thick stem, usually stripped from the leaf and thrown away, is practically a second vegetable. Every bit that's left over—the kitchen scraps from making dinner and the spent plant residues at season's end—goes into compost, another phase of the growing cycle.
- For further inspiration, the vegetarian cookbooks that expanded my garden ambitions include *The Greens Cookbook* and anything else by Deborah Madison, but especially *Vegetable Literacy,* which addresses vegetables by family. Other gold mines for vegetable-forward cooking include *David Tanis Market Cooking,* Alice Waters's *Chez Panisse Vegetables* and *The Art of Simple Food,* Amanda Hesser's *The Cook and the Gardener,* Jane Grigson's *Vegetables Book,* Travis Lett's *Gjelina,* Niloufer King's *My Bombay Kitchen,* and two canonical volumes from Edna Lewis, *The Taste of Country Cooking* and *In Pursuit of Flavor.*

CH.

02

CLIMATE, WEATHER, LOCALE

MAY 19: Dandelions bloom in the Berkshires when it's time to plant potatoes. For an explanation of so-called indicator species, see page 22.

Every gardener works within the limits of climate and weather, so we'll begin with the broadest considerations: In which parts of the country can a person reasonably hope to establish a vegetable garden, and when during the year will it grow? Next comes the crucial matter of choosing a site. Few are perfect. "You work with what you got," said Steve Cunningham of Berkshire Bounty Farm, as he described transforming a rocky hillside into the lush garden I visited on a June afternoon. He chose the site because it received a full day of sun, unlike the rest of his wooded property. Then he went about building up the scant topsoil, one bed at a time. What looked to me like a natural garden spot had, in fact, been made from scratch. It's what he had to work with.

CLIMATE

The continental United States, apart from the extremes of altitude and latitude, is well suited to vegetable gardening during at least part of the year. At the northern tier, brief summers will shortchange heat-loving crops, such as sweet potatoes, beefsteak tomatoes, and melons. And at the extreme southern margins, fierce summers will scorch tender spring vegetable—peas, radishes, and butter lettuce. But most backyard gardeners will find plenty to grow, including selected varieties suited to regional temperature variations. Parris Island romaine, developed in muggy South Carolina, takes the heat better than most lettuce varieties, and Anna Russian tomato is a northern **CULTIVAR**, or cultivated variety, of a sun-loving vegetable.

A touchstone reference for all gardeners is the USDA Plant Hardiness Zone Map, which can be accessed at planthardiness.ars.usda.gov. The map divides the country into numbered zones based on average annual winter minimums. I garden in zone 5b, where lows can be expected to dip to between –10°F and –15°F on the coldest nights. Where my father lives in east Tennessee, zone 7a, winter lows rarely drop below zero. In his region, Brussels sprouts and collards will overwinter, and a gardener can continue to harvest them straight through to spring. The current USDA map, published in 2023, tracks the rapid pace of climate change; most regions of the country have warmed by a half-zone since the 2012 map.

GROWING SEASON, FROST DATES, MICROCLIMATE

Winter is the season of garden dormancy. The remainder of the year is the growing season, when conditions favor plant growth. It runs from midspring to midfall in most places—the frost-free months of mild weather and long days.

The key guideposts to the horticultural year are **FROST DATES**. Spring's **AVERAGE LAST FROST** and fall's **AVERAGE FIRST FROST** are estimated by the National Oceanic and Atmospheric Administration based on 1991–2020 climate data. You can find frost dates by city and state or zip code online at almanac.com/gardening/frostdates.

Frost dates matter because vegetables are not sown according to a universal schedule. The reason is obvious. Spring arrives earlier in one place than another, moving northward as winter retreats. Planting instructions that make sense in Mississippi, such as "sow cabbage at the first of March," would not work in Maine, where the ground would likely still be snow-covered in early March.

Sowing dates are instead expressed in relation to frost dates. For example, we are told to put out pole beans one to two weeks after average last frost, when the ground is warm. In Mississippi, that might be mid-April, while in parts of Maine it might not be until early June. By the same token, fall spinach can be planted 4 weeks before average first frost, or in late summer, whenever that happens to be in your area.

COOL-SEASON CROPS, such as lettuce, peas, onions, and cabbage, can be planted before last frost, some as soon as the ground can be worked in early spring. They are **COLD HARDY** and will survive a spring freeze; they will also live through the onset of chilly weather in fall. **WARM-SEASON CROPS**, such as tomatoes, squash, and peppers, are planted after all threat of cold has passed and will be knocked down by the first frost.

An older and subtler way to time planting is with **INDICATOR SPECIES**, which bloom in dependable sequence based on seasonal factors such as daylight hours and soil temperature. Showy spring indicators include crocuses, daffodils, redbud, shadbush, dandelions, forsythia, peonies, and iris. Every year, my mother and I compared flowering dates and determined that spring reliably reached Tennessee a month before coming to Massachusetts. By the same token, the demise of fall gardens can be tracked regionally by the most obvious of all indicators, fall foliage.

Don't rush it in the spring. "It's better to plant too late than too early," wise gardener John Coykendall told me as we sat on a swing at Blackberry Farm in Tennessee. Seeds and seedlings put out prematurely won't necessarily get a jump on the season. Late frost has killed many high hopes, and some seeds rot in cold ground. Seeds planted late will catch up during May's universal growth spurt.

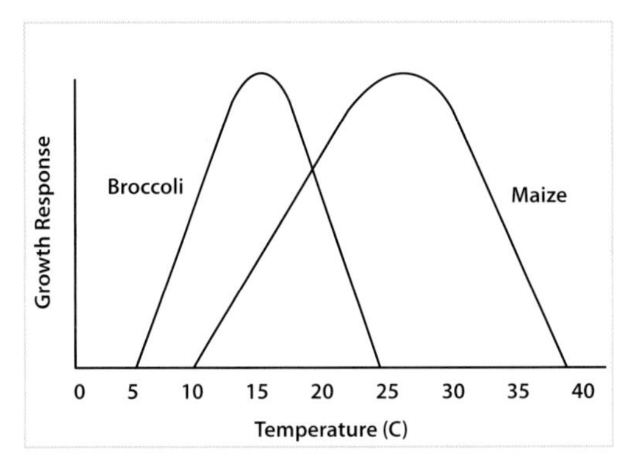

Optimum growing temperatures for cool-season broccoli compared to warm-season corn, or maize

MAY 20: More indicators: Local peaches bloom around average last frost.

JUNE 5: Lupines—all threat of frost has passed.

JULY 23: Queen Anne's Lace—8 weeks before average first frost, time for fall planting

SEPTEMBER 28: Sugar maples blaze around first frost.

JUNE 11: Most vegetables need 1 inch of rain per week. The weather sometimes cooperates. The alternative is watering.

A light freeze can occur even when the official temperature remains in the midthirties because cold air pools at ground level. Cool-season crops will ride it out. A hard freeze, around 25°F, will flatten most vegetables other than hearty greens, brassicas, some lettuces, and some herbs.

Every garden exists in a microclimate influenced by such local factors as elevation, topography, sun exposure, and even surrounding vegetation. I mentally adjust the published climate data for my area to reflect the hyperlocal reality that my garden is about 300 feet higher than the closest official weather station. On the other hand, its sloping southern exposure captures extra sun, especially beneficial in spring.

PRECIPITATION

Natural rainfall is not usually a limiting factor in planning a garden. Farmers and gardeners have used artificial irrigation in times of drought for many thousands of years. A very ancient technique called flood irrigation entails channeling water—whether from a natural waterway or ditch, drawn from a well, or released from storage—into a field, bed, or basin surrounded by a low berm. You flood the bermed area and allow it to soak in. In northwest Saudi Arabia, I visited the archeological site of Hegra, which was the southern capital of a biblical-era people, the Nabatean kingdom, who also built the ancient rock-cut city of Petra in southern Jordan. Nabatean farmers harvested rainwater in cisterns hewn from solid rock and distributed it through a system of communal irrigation ditches. Similar irrigation systems in medieval Spain, called acequias, were communally owned and collectively managed by local communities. The acequia system survives today in mountain villages in northern New Mexico, where many residents descend directly from early Spanish settlers. The dry Middle East also inspired modern drip-irrigation, a highly efficient system perfected in Israel by Simcha Blass in the 1960s and used widely throughout the world today.

Most vegetables need the equivalent of 1 inch of rain per week during the growing season. Some ancestral landrace varieties of corn and beans, which are heirlooms from arid regions such as the American Southwest, are drought-tolerant. A full discussion of irrigation practices—how to water a garden—will follow in chapter 7, Upkeep.

Incidentally, too much rainfall can also lead to trouble. Torrential downpours can cause runoff and topsoil erosion. Consistently damp weather encourages mildew, blights, and rot. Extended gloomy weather can even slow ripening. All these issues will be addressed in the chapters ahead.

I track rainfall with a rain gauge planted among the vegetables and keep notes in a garden diary. An unofficial tally of average rainfall by region can be accessed at currentresults.com/Weather/US/average-annual-state-precipitation.php.

CLIMATE CHAOS

Over long generations, farmers and gardeners have grown food using their knowledge of dependable weather patterns, including typical frost dates and annual rainfall. The mounting pace of climate change is creating climate chaos. Weather everywhere is less reliable, and not necessarily in obvious ways. A warming climate is less stable with an increased likelihood of volatility and extreme weather events. Rogue late frosts, premature freezes, erratic rainfall, droughts, floods, and fiercer winds stress the garden—and the gardener.

In the short term, we can protect our gardens by putting into practice climate-resilient strategies such as no-till, composting, mulching, cover cropping, planting for crop biodiversity, and adopting integrated pest management. Pursued at the commercial scale, climate-resilient practices take on the character of **REGENERATIVE AGRICULTURE**, which not only maintains ecological health—the organic baseline—but strives to repair damage to topsoil, watersheds, and even the atmosphere, the last accomplished by sequestering carbon in the soil.

In the long-term, individual action is dwarfed by the enormity of climate chaos. Repair at the ecosystem scale will require vast concerted changes by global agriculture and industry. Concerned gardeners, as anyone might be, can commit to political action aimed at national policy changes.

DAYLIGHT HOURS

Outside of the equatorial zone, the number of daylight hours varies over the year—longer days in summer, shorter in winter—and the greater the distance from the equator, the greater the annual shift. Some effects of seasonal daylight hours may not be self-evident. For example, spring planting should be delayed until there are ten hours of sunlight. In the southernmost tier of the United States—roughly south of the thirty-third parallel, or a line between Atlanta, Dallas, Phoenix, and Los Angeles—that's always. Which helps explain why California, Arizona, and Florida produce three-quarters of America's vegetable harvest by value. A perhaps more surprising fact is that onions are photoperiodic, meaning their growth cycle, and especially the onset of bulbing, is influenced by daylight hours. Select varieties are adapted to grow in southern regions or northern regions, as explained more fully in chapter 19, Onions and Garlic Family. The website timeanddate.com gives sunrise and sunset times by zip code, along with a full-year graph of daylight hours.

SUN EXPOSURE

Most vegetables do best in **FULL SUN**, where they receive eight hours a day of direct sunlight. Remember that the sun moves throughout the day and shifts across the growing season. The seasonal effect is more noticeable farther north. A level plot of lawn beside my house looks like an ideal garden site in high summer, when the sun is directly overhead. But for the rest of the year mature trees cast too much shade.

Before deciding where to put in a garden, inventory your sunshine. Note how much direct sun falls in the morning, at noon, and at four o'clock. Compare observations from April, July, and September. Afternoon sun is hotter. Morning sun dries dew more quickly, especially when helped by breezes, reducing the risk of mildew and waterborne diseases.

One caveat: In very hot and dry regions, such as the desert of the Southwest, full sun can be too much, causing sunburn. Afternoon shade cast by a tree or building could be a blessing.

PARTIAL SUN, four to six hours per day, will do for many vegetables. Squash turns spindly with too much shade, and beefsteak tomatoes will struggle to ripen. But hearty greens, brassicas, and root vegetables will be fine. As summer temperature rise, lettuce and other heat-sensitive crops will appreciate the dappled shadow of corn or bean trellises.

FULL SHADE, less than four hours per day—on the north side of a house or under mature trees—will doom vegetables to a slow, pitiful decline. Find another spot.

Raised beds and containers have the same sun requirements. Small containers can be moved to follow the sun across the growing season, but I personally wouldn't want to move pots daily, like a cat chasing a sunbeam. Place window boxes in a south-facing window—as much sun as you can find.

SUN REQUIREMENTS

If you have less than full sun, try growing leafy greens, root vegetables, and brassicas, which tolerate some shade. Cherry tomatoes need less solar energy to ripen than big slicers.

GROW ONLY IN FULL SUN	SUITABLE TO PART SUN/PART SHADE
Tomatoes	Cherry tomatoes
Squash	Broccoli and cabbage
Cucumbers	Salad greens
Corn	Root vegetables
Peppers and eggplant	Chard and kale
Basil and rosemary	Parsley, thyme, chervil, chives

MAY 19: Dappled morning shade gives way to midday and afternoon sun for these containers.

AUGUST 28: An ideal garden spot is sunny and level to slightly sloped.

DRAINAGE

Vegetables won't tolerate wet feet. On the other hand, they need soil that will act like a sponge to hold water between rains, and not just let it wash through like waves on a pebbly beach. **DRAINAGE** describes how quickly water passes through the soil, and it is closely related to **TILTH**, the soil's physical condition. When siting an in-ground garden, you want to avoid muddy spots and sun-baked clay, indications of poor drainage. Also avoid sandy, gravelly, and rocky ground, which may drain too quickly and require constant watering. The next chapter will explain ideal soil characteristics in detail and discuss how to improve drainage. The best cure, however, is to avoid problems at the outset.

Soggy, boggy ground is out of the question for a vegetable garden, except for watercress and any other crop that asks for it by name. Mint will thrive in a damp spot, but then mint will grow nearly anywhere.

Raised beds and container gardens circumvent underlying drainage issues. Containers must have drainage holes in the bottom. Prevent washouts with a layer of gravel, a flat rock, or a broken pot shard over the holes.

SLOPE

A level garden bed will be easiest to work. A gentle slope, not enough to prevent pushing a wheelbarrow up it, will improve drainage. A steeper slope will bc susceptible to erosion. Avoid gully-cut hillsides, which should be sown with grass or planted with fruit trees and berry bushes.

Low, flat areas will collect water. Avoid depressions that puddle after rain or hold standing water in spring.

CH.

03

SOIL

The nation that destroys its soil destroys itself.

—FRANKLIN DELANO ROOSEVELT, 1937

JUNE 28: Pete Salinetti with well-structured soil at Woven Roots Farm

The role of soil in the garden is complex. Its weight and solidity give plants a physical substance to grip—the ground. Its porous structure allows water to flow through on its way to underground aquifers, even as it holds moisture between rainfalls. Soil is also a growing medium charged with nutrients, minerals, and trace elements necessary for plant metabolism.

Where does soil come from? It is the weathered debris of the earth's mineral crust enriched with chemical elements (nitrogen comes from the atmosphere) and leavened with organic matter that has passed through the bodies of earthworms, bacteria, and other small creatures. The mineral component, if it could somehow be separated, would be inert, or nearly so. Decomposed organic matter is anything but. **HUMUS** is transitory, quickly made in comparison to mineral decay, and quickly taken up again. Dark and crumbly, it represents a brief pause in the movement of carbon between organisms, a sort of metabolic condensation, as it passes between two states of being. Both residue and inception, humus bridges death and life. See Living Soil (page 32) for more on the subject.

"The relationship with the soil is intricate and intimate," said Pete Salinetti as he and his partner, Jenn Salinetti, showed me around the beautiful garden beds at Woven Roots Farm near my house. "It's an exchange."

SOIL LAYERS

Soil develops in layers, or strata. **TOPSOIL**, as the name suggests, lies closest to the surface. The depth and quality of topsoil will vary from place to place, but it is the richest, most biologically active layer of the ground. Topsoil is where the sun and the earth meet to make life possible, as a clutter of organic detritus breaks down and covers the ground surface with fine natural mulch. Seeds sprout and plants grow in topsoil, spreading their roots for water and nutrients. Small wildlife works its way through the zone

GOD'S LOAM

Ælfric of Eynsham, an English abbot who lived from circa 955 to circa 1010, left behind a body of writings, including *The Homilies of the Anglo-Saxon Church.* In this snippet of his lengthy "Sermon on the Beginning of Creation," he uses the word *loam* in its early sense of "clay" to describe the material God used to shape man. Ælfric's original Old English text, given here to honor both the strangeness of that distant time and our linguistic ties, is followed by a modern translation.

And God þa geworhte ænne mannan of láme, and him on ableow gast,
and hine gelíffæste, and he wearð þa mann gesceapen on sawle and on lichaman;
and God him sette naman Adám, and he wæs þa sume hwile ánstandende.

And God then wrought a man of loam, and blew spirit into him,
and animated him, and he became a man formed with soul and body;
and God bestowed on him the name of Adam, and he was for some time standing alone.

LIVING SOIL

Time spent in a vegetable garden, wondering where flavor comes from, has fundamentally changed how I look at the ground beneath my feet. I used to view soil as inert, a substrate for roots to hold. I now understand it to be a subterranean ecosystem, as dynamic and biodiverse as any aboveground habitat. Specialists study soil as they do rainforests and ocean reefs, but the science is comparatively new and the complex relationships among species are imperfectly understood. Still, fertility is the condition of living soil.

Soil differs from dirt, writes Barry Lopez, in precisely this: its vitality. Dirt is dead soil ruined by overuse, drought, or pollution. Talk about living soil isn't a metaphor. The ground literally teems, although its life-forms are mostly too small for the naked eye to see. Fertile soil can be described in terms of physical, chemical, and biological markers. To organic gardeners, however, biology is paramount. Bugs and worms and microbes and fungi are not just in the soil, on the soil, and of the soil. They *are* the soil.

Animals pass organic matter through their bodies and excrete natural fertilizer. Some people refer to soil invertebrates as the micro-herd, which accurately equates the value of their micro-manure to that of barnyard livestock. Of all the soil builders, earthworms are most vigorous. They till as well as fertilize the ground. "There are few animals that have played so important a part in the history of the world as these lowly organized creatures," wrote Charles Darwin in *The Formation of Vegetable Mould Through the Action of Worms,* his last scientific work, from 1881. Earthworms are also an indicator of soil health—a robust population shows that everything underground is in good order.

At a still subtler level, healthy soil supports a vast fungal network of mycorrhizae, fungal roots. Plants benefit from so-called mycorrhizal associations because they transmit soil nutrients needed to convert sunlight into carbohydrates, which we perceive as the sweetness of newly picked peas and carrots—the flavor of a garden.

from 2 inches above ground to 2 inches below, the realm of worms and decomposers. Topsoil is rich in organic matter, fertile and dark, although its color can vary from near-black to brown, ochre, and even startling red.

The **SUBSOIL** lies beneath the topsoil. Because it is less rich in organic matter, it is lighter in color and heavier. As the legacy of decomposed rock, subsoil is rich in minerals, not yet having been exhausted by plant growth. Burrowing earthworms churn it. Some deep-rooted plants, including garden weeds, draw up minerals from the subsoil, enriching the topsoil when they die. For this reason, gathered weeds are a valuable addition to compost.

SOIL PARTICLES

The mineral components of soil sort into three sizes. **SAND**, the largest, is coarse and sharp-edged. The finest, **CLAY**, is so small you can hardly see or feel the particles—wet clay feels greasy between the fingers. **SILT** is in between.

Each has merits: Sand improves drainage and clay holds water. In excess, however, sand dries out and clay becomes sodden when wet and impermeable when dry, nature's adobe bricks. Heavy, or clay-rich, soil puddles in wet weather and warms up slowly in spring.

LOAM is one of those sturdy old Anglo-Saxon words. It designates high-quality soil displaying a favorable proportion of sand, silt, clay, and organic matter. Loam can be described more precisely by the predominate soil particle, as in sandy loam or clay loam. The USDA Soil Texture Chart at right explains more.

HUMUS

Beyond the three soil particles, the additional material of soil is organic. As already mentioned, fully decomposed organic matter is called humus, pronounced HYOO-mis, like humid, rather than HUM-is, like chickpea dip.

Soil scientists call humus amorphous because it has no characteristic form. It is more finely textured than compost, in which source material, such as leaves, remains discernible.

Sticklers will note that the terms organic matter and humus are not interchangeable. Humus is the final form of organic matter, which includes plant detritus, animal droppings, and small carcasses of voles, crickets, ants, and countless species that die beyond the limit of our concern. Humus results after these rougher materials have passed through the bodies of earthworms, bugs, and soil microbes. Humus is the completion of a metabolic cycle; it is equally the beginning of a new cycle, a latent vitality to be reincorporated into plant bodies, which will in time decay into humus, and so on.

Humus adds nutrients to the soil, chiefly nitrogen, but also carbon-rich bulk and absorbency. Sticky gums associated with humus aggregate soil particles. Humus is what makes a garden.

Adding organic matter to the soil to create humus is the single most important investment you can make in a garden's long-term health and productivity, as described in

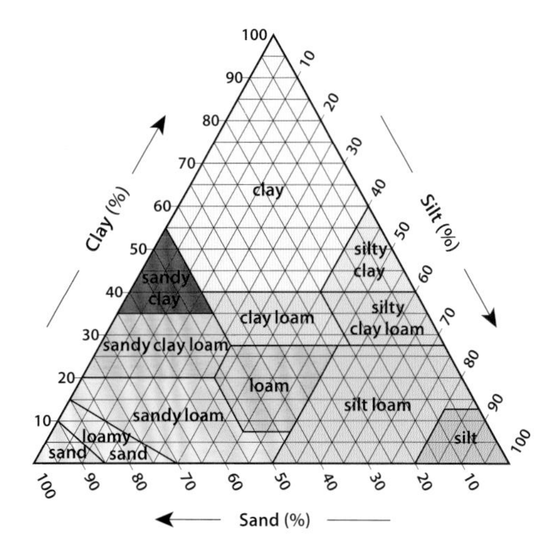

USDA SOIL TEXTURE CHART

the section that follows on compost. Prudently safeguarded as the soil's capital asset, organic matter will accrue value over time and repay an inexhaustible annual return, like interest from a trust fund. Spent willy-nilly, humus will be depleted, bankrupting the soil. Slash-and-burn agriculture as it is practiced in parts of the tropics is often presented as a blatant example of wasteful farming practices. But today's industrial agriculture exemplifies "deficit spending," frittering away soils that sustained the American experiment. Future consequences include soil exhaustion, erosion, and desertification—the death of agriculture. Humus is the antidote.

SOIL STRUCTURE

Ideal garden soil drains quickly thanks to an open porous structure that combines tensile strength (it holds together) and friability (it breaks apart). That sounds oxymoronic. But see the photo on pages 30 and 31 for an example of **STRUCTURED SOIL**, also called well-drained soil. It shows enough tensile strength that farmer Pete Salinetti of Woven Roots Farm in Tyringham, Massachusetts, can lift a clump in his hands. But it is friable enough that he can crumble it between his fingers.

Well-drained structured soil absorbs rainfall quickly, eliminating runoff and its evil side effect, topsoil erosion. Downward-trickling water coats every pore and cavity with a film of moisture—hole-y but saturated, like a kitchen sponge. Plants tap into the column of soil moisture at every level. While the upper ground will dry between rains, deeper soil layers hold what amounts to an underground reservoir all the way down to permanent groundwater. Paradoxically, well-drained soil is a gardener's best protection again drought.

When you see a gardener take a handful of soil and squeeze it, that's to assess texture and, by extension, drainage. The analysis is all in the feel, and you learn through practice. If the moist soil compacts and holds an impression of your fingers, it has a lot of clay, which indicates potential drainage problems. If it won't hold together, there's a lot of **SAND**—the sign of fast drainage.

A soil percolation test is less subjective. Dig a straight-sided hole the size of a gallon paint can—roughly 8 inches across and 12 inches deep. Fill it with water and let it drain overnight. The next day, fill it with water again. Lay a rake handle or pipe across the top and measure the distance to the water's surface. Set a timer and remeasure at 2-hour intervals as the water percolates into the surrounding soil. Ideal drainage is 2 inches per hour. One inch per hour is slow. Four inches per hour is fast.

IMPROVING DRAINAGE

The mechanical fix for poorly drained soil is tilling, digging, or cracking open the soil layers with a garden fork. Longer-term soil structure can be improved by adding organic matter to the soil. Humus contains sticky compounds called humic acids that glue **SOIL PARTICLES** together. Expressed in the technical jabberwocky of soil science, such compounds will *flocculate the colloids,* in which colloids are fine soil particles and flocculation is the process by which small particles adhere into larger aggregates called flocks. (This sense of *flock* relates to a tuft of wool rather than to a flight of birds.) Aggregated soil particles create a more porous structure, improving water filtration. Garden lime made from pulverized limestone can have the same effect.

Organic matter is the cure, as well, for fast-draining sandy soil. Because it is absorbent, it holds water that otherwise would flow through coarse mineral sand.

SOIL FERTILITY

The biological potential of soil, its fertility, exists as dynamic flux. Easier to quantify are the three elements essential to plant metabolism:

NITROGEN, **PHOSPHOROUS**, and **POTASSIUM**, represented by the symbols N, P, and K and always given in that order. Phosphorous gets P, so the mnemonic goes, because it has two of them.

The numbers printed on a package of commercial fertilizer indicate percentages of N, P, and K. For example, a hundred-pound bag of 10–10–10 has ten pounds of each. The remainder is inert filler for easier application. The blue-colored Miracle-Gro granules my mother used to dissolve in water for her potted flowers was 24–8–16, a reminder of how powerful artificial fertilizers can be. On the other side of the coin, a bag of worm castings, a rich organic supplement, has an **NPK** content of roughly 2–1–1. Fresh compost, while variable depending on the source material, is not far off, in the range of 1.5–0.5–1. These invaluable soil-builders show the limitations of NPK as a measure of fertility.

WHAT IS A FARMER?

E. B. White's classic *One Man's Meat* recounts the author's move from New York to Maine, where he tested his mettle against a small saltwater farm. His stepson Roger Angell, another talent from the *New Yorker* stable, later described the setting as inspiration for what is surely White's best-known work, *Charlotte's Web*. White found the humor in his new calling as a farmer.

> *Mr. Highstone [a Maine neighbor], being himself a practicing farmer, knows one important truth about country life: he knows that farming is about twenty per cent agriculture and eighty per cent mending something that has got busted. Farming is a sort of glorified repair job. This is a truth that takes some people years to discover, and many farmers go their whole lives without ever really grasping the idea. A good farmer is nothing more nor less than a handy man with a sense of humus.*

Nitrogen

Plants use nitrogen to build stems and leaves. Its effects show in the vibrant green of new growth and the deeper hues of mature leaves. Plants take up nitrogen quickest during their spring growth spurt. Heavy feeders—tomatoes and corn—will also appreciate a midseason application, or **SIDE DRESSING**, of compost or a nitrogen-rich supplement. Because nitrogen is water soluble, it leaches from the soil, and in excess becomes polluting agricultural runoff.

Too much nitrogen makes plants weedy, "rank" in the Shakespearean sense, as they become overexcited, like kids after birthday cake and soda. A tomato growing in "hot" soil, ground overcharged with nitrogen, will sprawl unproductively—all vine, no fruit. Pests and disease thrive on the soft and babyish leaves.

A gardener has multiple natural sources for nitrogen. Compost adds it in balance with other soil nutrients—a full meal for plants rather than a sugary snack. Worms steadily deposit nitrogen as they metabolize organic matter. And the legumes—beans, peas, and clover—take nitrogen from the atmosphere and mineralize it, making it available to plants. Even industrial agriculture harnesses legume's superpower by planting soybeans in annual rotation with nitrogen-hungry corn.

In time, every plant returns its unused share of nitrogen to the soil as it decays. Animals drop their excess nitrogen frequently while still alive: fresh manure is hot.

Nitrogen deficiency often first appears in plants as yellow leaves. Side-dress puny plants with compost or scratch in a nitrogen supplement. Blood meal, a dried by-product of slaughterhouses, is 12–1.5–0.5. Alfalfa meal and seaweed extract are vegan alternatives. Fresh grass clippings are nitrogen-rich and free (4–0.5–2). Sprinkle a light layer around heavy feeders, provided the lawn hasn't been treated with herbicides.

Phosphorous

Phosphorous builds strength through root growth. If nitrogen boosts youthful vigor, phosphorous supports balanced maturity. It derives from both mineral and organic sources. Phosphorous deficiency can cause reddish or purplish leaves, but its symptoms are less apparent than with nitrogen deficiency. A **SOIL TEST** (see page 37) is the only sure way to determine phosphorous deficiency. Remedy with bone meal, which is approximately 3–15–0.

Potassium

Potassium makes plants vigorous, better able to resist pests and disease. Jean-Martin Fortier, a noted farmer and author in Quebec, believes potassium also improves the keeping qualities of root vegetables as well as the size, color, and even flavor of fruiting vegetables. Potassium is the freebie of the major soil elements—already abundant in most soils.

SECONDARY NUTRIENTS AND TRACE ELEMENTS

Other elements used by plants include calcium, magnesium, and sulfur. The longer list of trace elements extends to iron, boron, zinc, copper, and other metals. In broad strokes, if one is missing, plants can suffer. But human health also requires minuscule amounts of zinc, copper, iodine, and other micronutrients. They're all part of a balanced diet. Same with the garden. If you're taking care of N, P, and K organically, then you're likely covered.

A note on calcium: Tomatoes and peppers suffer from a disease called blossom-end rot, which blackens the fruit opposite the stem. It is common in some areas, including New England, and the given cause is calcium deficiency. The subtle wrinkle is that the deficiency is in the plant rather than in the soil. In other words, the disease occurs when plant roots can't absorb adequate calcium, usually because of inconsistent irrigation during hot weather. Regular watering and a layer of moisture-conserving mulch will help. Scattering eggshells or oyster shells around the plants, two folk remedies, don't deliver calcium fast enough to counter the disease once established. Lime and wood ashes, two other calcium sources, can help longer term but might open a can of worms called pH, see opposite.

ARTIFICIAL FERTILIZERS

You don't need them! Any amateur gardener, no matter how novice, can grow backyard vegetables without artificial fertilizers. There's no zero-sum tradeoff between natural practices and garden success. Organic math is additive. The core techniques—composting, mulching, and biodiverse plantings—yield healthy soil *and* healthy plants *and* delicious meals. I came to this conclusion through direct experience. Because I didn't like handling nose-tingling, skin-burning chemicals, I decided I would accept a lesser harvest in return for more pleasurable garden work. The outcome was opposite of what I expected: The vigor, beauty, and flavor of my organic vegetables showed conventional produce to be of lesser quality—oversized and underflavored.

During the so-called Green Revolution of the mid-twentieth century, petroleum-derived artificial fertilizer was paired with hybrid seed to invent industrial agriculture. The unreckoned costs of soil-based capitalism included soil erosion, watershed pollution, and the degradation of food nutrition and flavor. Applied fertilizers increased short-term yields, but assembly-line thinking led to assembly-line quality and failed to account for long-term soil degradation.

Artificial fertilizers decimate soil microorganisms. A perverse result is that artificially boosted soil declines after the initial jolt. A crash follows the sugar high. A negative feedback loop sets in, as depleted soils require heavier doses in a vicious cycle akin to addiction. Nor does artificial fertilizer fix underlying soil problems related to drainage and tilth.

For the home gardener, artificial fertilizer isn't a fix-all but a trap to avoid.

pH

The peculiar acronym pH is short for *potenz*, "power" in German, and the symbol for hydrogen. It indicates acidity on an inverse scale from 1 to 14: A low number means high acid. Low acidity is also referred to as an alkaline or base condition. Remember that pH is a description rather than a measure, as with soil fertility. It relates to fertility, however, because pH affects a plant's mineral uptake. Neutral to somewhat acid soil, between pH 6.0 and 7.0, suits most vegetables and is also ideal for microbial activity.

Soil pH varies by region, based on the underlying continental crust. The Northeast generally has acidic soils, and those of the Great Plains and Southwest tend toward the alkaline. Environmental factors matter at the local and hyperlocal level. Tree leaves and conifer needles acidify the soil as they break down, for example.

Check pH with a soil test (see below). Attempts to adjust pH should be made cautiously and by degrees. To raise the pH of acidic soils, sparingly apply lime (powdered limestone) or hardwood ashes. Consult package recommendations for lime. For hardwood ashes, the University of Wisconsin Horticultural Extension recommends no more than a single 5-gallon bucket (15 to 20 pounds) per 1,000 square feet per year. Err on the side of caution and reapply the following year, if needed.

Organic matter will acidify alkaline soils. Peat moss is often prescribed, but peat bogs are ancient carbon sinks developed over vast timescales. (*Gathering Moss, a Natural and Cultural History of Mosses,* by Robin Wall Kimmerer, quoted in another context on page 45, changed the way I think about peat harvesting.) Pine needles and leaves are renewable options. Both are mildly acidic, but earthworms buffer the effect—their castings are nearly neutral.

SOIL TESTS

A soil test is a laboratory analysis of chemical and physical properties including pH, NPK levels, organic content, and soil-particle size. Sending away a soil sample for testing is like getting your vitals checked at the doctor's office. It won't tell you everything, but if the essentials are out of whack, you know to treat symptoms and seek an underlying cause. Testing is done by local agricultural extensions, state universities, and private labs. Results typically come with suggestions for amendments.

The testing sample should represent an average profile of your garden soil. For accuracy, take soil from several spots. Scrape back mulch and other organic matter. Using a nonreactive plastic or aluminum trowel, dig a small hole 6 or 8 inches deep and take a cross section from the side, removing it whole, like a slice of cake. Put it in a nonreactive bucket and repeat at each sampling spot. Mix well. The lab will specify a quantity of the blended sample to send for testing—usually 2 cups.

DIY test kits, which cost around $20, can give a quick read of soil pH and NPK levels.

CH.

04

COMPOST

MAY 25: Building a compost pile on the Nearing model (see page 42)

Plants eat the soil. A vegetable garden is extractive, like logging or grazing. You take something to the kitchen—a cucumber—and that reduces the amount of whatever grew the cucumber. Everything we eat represents a debt to the soil. In a home garden, the currency of repayment is compost.

WHAT IS COMPOST?

Compost is thoroughly rotted organic matter. It mimics the natural decay that happens in a forest, for example, where everything that once lived eventually hits the ground. Autumn leaves, dry fern fronds, wisps of woodland sedges and the twigs of shade-tolerant shrubs, fallen fruit, nuts, acorns, and the scattered wastes of animals and birds—it all mixes where it lies and breaks down into a crumbly layer of coarse organic material, which, over time, further resolves into humus. The middle stage of that process, the midpoint between a hodgepodge of coarse detritus and the subtle humus to come, is wild compost.

Everyone talks about *making* compost, but in fact the gardener merely assembles the material and then steps aside for two natural processes, one bacterial, another fungal.

Source materials include tree leaves, grass clippings, kitchen scraps, pulled weeds, spent pea vines, unwanted zucchini, tomatoes that split after a heavy rain and began to sour in the heat, a footlong cucumber you somehow overlooked when it was eating size, and the numerous other plant residues that exist in and around the garden and kitchen.

What you can't use are lawn clippings treated with weed killer or other chemical agents, nonvegetarian kitchen scraps (such as bones), dog droppings and cat litter, colored newspaper and glossy printed material, and leaves from eucalyptus, black walnut, and cedar trees, all of

BARREN SOIL

The eighteenth-century painter Joshua Reynolds, founder and first director of the Royal Academy of Arts in London, developed an elaborate metaphor gardeners can appreciate.

> *The mind is but a barren soil; a soil which is soon exhausted, and will produce no crop, or only one, unless it be continually fertilized and enriched with foreign matter.*

which suppress plant growth. Obviously avoid anything nonbiodegradable, such as plastic. Also be wary of "compostable" plastics or plastic-alternatives. Leave them for large-scale municipal composting facilities.

See Browns vs. Greens (page 44) for a list of composting ingredients.

MAKING COMPOST

A compost pile can be assembled by formula or by feel. The elements required are organic materials, oxygen, and water. Hot composting uses frequent turning to encourage frantic microbial activity, stoking internal temperatures to above 130°F, hot enough to kill weed seeds and many disease pathogens. Hot compost also extracts the maximum fertility from its ingredients. Cold composting, ruled by fungal decay, runs slower, and it takes less effort. Either way, compost is the gardener's most valuable resource. The flavor of a garden starts in the compost pile.

Making compost resembles cooking in many ways. You put together a bunch of ingredients that are transformed by heat into a nutrient-rich final product. And, as with cooking, there is no one definitive method. Gardeners will find their own composting style, as cooks do in the kitchen.

In either case, a few general guidelines apply. A compost pile needs to be layered with a suitable mix of materials. It should be large enough to hold its moisture and sustain heat. And it should be looked after, as you would look after a sourdough starter or pickling crock.

PLACEMENT AND CONTAINER

Put a compost pile in or near the garden, or any place you are likely to cross it in your daily routine. A remote compost pile, like a gym membership across town, will be neglected and finally abandoned. A compost pile doesn't stink unless you abuse it, but it will crawl with little creatures, so avoid putting it against a house or garage.

Make compost on the ground—a pile. Direct ground contact invites worms, bugs, and microbes. A protected or partially shaded spot will keep the pile from drying out. Avoid a soggy bottom.

A bitty pile will dry out quickly and lose heat. Three feet wide by three feet deep by three feet tall is large enough to maintain steady internal conditions. A walled bin or enclosure will hold it together and keep the area tidy. Latticed or gapped siding (skipping every other plank) improves ventilation.

If constructing a permanent bin with wooden slats or wire stretched between wood, leave one side open or make it removable for access. Twinned bins make for easy turning, back and forth across the divider. A three-bay bin allows you to keep two independent piles going simultaneously, one nearing maturity, another freshly built.

The simplest enclosure is made from a length of woven-wire fence looped back on itself. Sixteen feet of wire will make a circular bin five feet across. Ten feet of wire will give you a three-foot pile, the minimum.

I operate two different compost systems. My home pile is a sturdy, permanent three-bay bin

SEPTEMBER 9: Black gold at a Vermont dairy's honor-system farm stand

COMPOSTING TESTAMENTS, OLD AND NEW

The aptly named *An Agricultural Testament* by Sir Albert Howard is a foundational text of the organic movement—an old testament, perhaps *the* Old Testament.

Born in 1873, Howard was an agricultural researcher in service to the Empire who spent his career in India trying to improve crop yields and boost profits on tea, sugar, cotton, and sugarcane. Compost, he realized, was the supreme achievement of India's traditional agriculture. His contribution was to systematize a technique called the Indore Method, in which organic materials were layered, watered, and turned to speed decomposition. *An Agricultural Testament,* published in 1941, lays it all out, step by step.

The composter's New Testament is *The Rodale Book of Composting*, released in 1992 by the publishers of *Organic Gardening* magazine. J. I. Rodale (1898–1971) was a New York City–born businessman who, inspired by his reading of Howard, decided with his wife to "get a farm at once and raise as much of our family's food by the organic method as possible." They later founded Rodale Press. Before internet searches answered all questions, Rodale's printed output advised America's organic-food movement.

Rodale was also a wingnut. An anti-vaxxer before the word existed, he ranted against sugar, gorged on vitamins and dietary supplements, and published quack ideas, such as the one expressed in the title of his 1970 book *Happy People Rarely Get Cancer.* He also committed to print appallingly racist notions.

Setting all that aside, if one can, Rodale's compost book, edited by Deborah L. Martin and Grace Gershuny, offers sound practical advice and lots of scientific detail for those who want to go deeper.

ALL IS COMPOST, COMPOST IS ALL

Contemplative gardeners probe the mysteries of vegetative life during solitary hours among the plants. A few of them, destined to teach, become garden gurus.

The first type of guru proselytizes a One True Way. Charismatic teachers, they inspire passionate followers with their mystical glimpses of Paradise and issue commandments on how to follow in the master's footsteps. Rudolf Steiner and Alan Chadwick were this type: inscrutable yet doctrinaire, shamans whose rulemaking rivaled that of French bureaucrats.

Other sages, more clear-headed if perhaps a shade less luminous, talk plainly about their own trial and error. Direct and practical, they provide manageable to-do lists. Yet over time they draw initiates into subtleties not imagined at the outset. Simple precepts, in their case, derive from complex knowledge. Sir Albert Howard was a guru of this variety, as are the organic movement's current power couple, Eliot Coleman and Barbara Damrosch.

Radical homesteaders Scott and Helen Nearing, Coleman's mentors, embodied aspects of each lineage. Their 1954 book *Living the Good Life* inspired disciples with its utopian vision of Back to the Land, and yet it also gave direct advice for how to make vegan compost.

Gardeners are denominationalists, and each sect has its hymnal. I've found only one constant among them. The path to garden enlightenment always runs through the compost pile.

inspired by Donald Judd's furniture and built of stout four-by-fours by Deeva Gupta, a student at my alma mater, Deep Springs College, who interned with me one summer. The other is an impermanent Nearing-style structure built up like Lincoln Logs of six-foot elm poles Del and I harvested from the thicket behind my house. As suggested by the Nearings, I drive two metal T-posts into the ground at the center of the first layer. As the pile rises, I scissor apart the T-posts to open a ventilation shaft.

Occasionally I'll throw a tarp over mature compost if torrential rain is predicted, but it's not essential to cover a pile as it cooks. That said, a black plastic wrapping will hold in moisture and absorb heat, jump-starting fermentation. Chicken-wire will keep out animals.

LAYERED BROWNS AND GREENS

A compost pile works best when thick layers of dry, coarse, high-carbon browns are interspersed with thin layers of dense, juicy, nitrogen-rich greens. Browns alone lack the nutrients and moisture to fuel rapid bacterial growth. Too many greens will suffocate the pile as the heavy wet stuff compresses into a slimy mass, depriving bacteria of oxygen. An overly rich pile will fester and stink.

I aim for roughly three times as much brown as green, measured by eyeball. Composting allows for a wide margin of error. The order is browns on bottom, followed by green, with additions described on pages 48 and 49.

Browns include fallen leaves, straw, spoiled hay, dead and dried vegetation, shredded paper, corrugated cardboard, peanut shells, and fully dried grass clippings. Coarser material breaks down more slowly. The first brown layer—the ground contact layer—can included stalky or twiggy material for drainage. (Tree branches and brush decay too slowly.) Tough wood chips will "lock up" nitrogen as they decompose. Use them sparingly; they are best suited to pathways and mulching under ornamental shrubs.

Greens include fresh garden debris such as the ragged outer leaves of lettuce and the coarse ruff from a head of cabbage, vegetable scraps from the kitchen, fresh weeds, spent or bolted vegetables, and green grass clipping.

If you're unsure of which type of material you have, consider how recently it was alive. Anything fresh enough to eat (kitchen scraps) or bright green in color (grass clippings, pulled weeds) is full of nitrogen. Faded, dried leaves are mere husks, the empty carbon architecture.

CARBON AND NITROGEN IN COMMON COMPOSTABLE MATERIAL

MATERIALS HIGH IN CARBON	C:N
autumn leaves	30–80:1
straw	40–100:1
wood chips or sawdust	100–500:1
bark	100–130:1
mixed paper	150–200:1
newspaper or corrugated cardboard	560:1
MATERIALS HIGH IN NITROGEN	**C:N**
vegetable scraps	15–20:1
coffee grounds	20:1
grass clippings	15–25:1
manure	5–25:1

The recommended ratio of carbon to nitrogen in a compost pile is 30:1. The above chart, Carbon and Nitrogen in Common Compostable Material, is interesting for the sake of comparisons. Did you know coffee grounds can have more nitrogen than manure? But in practice, I don't fret about the ideal 30:1 ratio. Instead, I rely on the more approximate guideline of three parts browns to one part green.

Anxious first timers might be tempted to add a store-bought compost "starter." Don't bother. A well-built compost pile starts itself, thanks to natural microbes found and the nutrition supplied by an animal activator (or vegan alternative) as described below.

BROWNS VS. GREENS

BROWNS	GREENS
High carbon	High nitrogen
Dry	Juicy
Fluffy	Dense
Light	Heavy
Fallen leaves	Kitchen scraps
Dry crop residues, like cornstalks and tomatoes after frost	Green crop residues, like pea vines and bolted greens
Brushy coarse weeds	Juicy green weeds and fresh sod
Sawdust (use sparingly)	Herbicide-free grass clippings (use sparingly)

ANIMAL ACTIVATOR (OR VEGAN ALTERNATIVE)

A nitrogen-rich amendment kicks off the bacterial furnace—food for the billions. I think of this launch energy as the animal activator because a common source is manure. Chicken droppings contain about 1 percent nitrogen by weight. Sheep manure has slightly more, goats somewhat more, and rabbit pellets over twice as much, at 2.4 percent. Fresh cow manure has 0.6 percent and inoffensive horse manure less than 0.1 percent. All are golden. Be aware that pig manure can potentially carry intestinal worms and must be composted hot. Never use dog, cat, or human solid waste, all of which can transmit pathogens. Urine-soaked straw or sawdust bedding from stables is effective. And for the uninhibited, one's own fresh urine is safe, pathogen-free, and NPK-balanced.

Other potential activators include animal by-products such as blood meal, bone meal, and crustacean meal (ground crab and shrimp).

There are animal-free activators as well. Scott and Helen Nearing were famous among their followers for their compost piles and their strict vegetarianism. The latter applied to the former. They activated their many piles—built up within frames of stacked birch logs—with wheelbarrow-loads of fresh seaweed gathered along the shore of their saltwater farm in Maine. I sometimes use nitrogen-rich comfrey leaves as an activator.

I think of the activator as the red layer, because it is "hot." It follows browns and greens.

LIST OF POSSIBLE ACTIVATORS

ACTIVATOR	PERCENTAGE NITROGEN
Blood meal	15%
Guano	10%
Stinging nettles	5.5%
Crustacean meal	4%
Bone meal	3.5%
Alfalfa	2.5%
Dried poultry manure	>2.5%
Rabbit manure	2%
Livestock urine	Up to 2%
Wilted comfrey	1.8%
Dried seaweed, kelp meal	1%
Fresh cow, horse, or sheep manure*	<1%
Human urine	0.6%

* Dried manures are up to five times higher in nitrogen.

WATERING COMPOST

Think of composting as large-scale fermentation—analogous to what happens in a pickling crock or, for that matter, in the human gut. Bacterial activity requires water, like any other living metabolism. A dry pile will not decompose.

Add moisture as you build the pile for consistency—a "blue layer" on top of brown, green, and red.

Once the pile is cooking, test the interior moisture regularly. Break open the top layer. It should release a muggy but not unpleasant tropical sigh. The interior should feel hot and damp, with no dry or soggy spots. Squeeze a handful: I like to see a little seepage around my fingers, but no drips. The target is moist but not wet, like a wrung-out kitchen sponge. In dry weather, you'll need to water the pile, although not as often as you water the garden.

TURNING

Mixing the contents of a compost pile breaks up coarse organic material and stimulates oxygen-greedy decomposers.

The usual practice is to turn, or aerate, based on interior temperature. A newly activated pile will reach 130°F or hotter in 10 days. Then the temperature will fall off as microbial activity declines, until it cools to below 100°F. At that point turn the pile to stir aerobic bacteria back to life. Fungal activity progressively takes over as the temperature falls off.

Use a long-probed soil thermometer to track the temperature or plunge in your hand to gauge—warm, hot, hotter, ouch! Or just turn the pile every 2 to 4 weeks.

REAL-WORLD PRACTICES

A compost pile, like a casserole for the oven, should be assembled all at once, bottom to top, then left alone as it cooks. When you go back to turn the pile, you don't add new material, any more than you would add raw ingredients to a half-baked casserole.

That's the ideal. The reality is that you don't necessarily have enough of everything at once. You can still make compost.

Start base layers of brown and green with whatever material you have. Keep surplus browns, such as raked leaves, on hand for whenever you have more kitchen scraps or garden residues to add. Follow the normal progression. The layers will settle. Add more layers whenever you have fresh material. Once the pile has topped out, let it go for several weeks before turning. From this point on, don't add fresh material.

FINISHED COMPOST

Finished compost is friable, or crumbly, with a fine, homogeneous texture. The original organic materials won't be immediately discernible. The mixture will be pleasant: dark in color, moist, and odorless apart from a woodsy smell. If you find chunky bits—a tangle of tomato stems or a tough broccoli stalk—chuck them into your next compost pile for another go.

GATHERING MOSS

by ROBIN WALL KIMMERER

Compost is a group effort. Various life-forms chew through organic refuse until all that remains is crumbly, dark, all-natural plant fertilizer. In this anecdote from *Gathering Moss: A Natural and Cultural History of Mosses* by Robin Wall Kimmerer, a professor of environmental biology at SUNY, it seems that one prospective composter misses the point.

> *A soil ecology colleague tells the story of a panicked phone call from a woman who had followed the direction in a pamphlet he'd written to start a compost pile in her backyard. Several weeks had elapsed and when she checked the pile of leaves and salad scraps, she was horrified to find it was full of bugs and worms. She wanted to know how to kill them.*

The texture of a pile's source material, the interior temperature achieved, and the rate of turning all determine how quickly compost will mature. Thinly layered fine material, such as fresh grass clippings and shredded leaves, will heat up quickly in warm weather and stay hot with frequent turning. If you turn the pile every 2 weeks during the growing season, you will have finished compost within 6 to 8 weeks.

If portions of the pile remain uncomposted, it's likely too dry. Telltale signs are brittle, papery leaves and compacted khaki-green grass clippings that send up clouds of dust and mold spores. Break up the dry material and water thoroughly, or else add to a new compost pile.

BANNED FROM COMPOST

Don't add meaty or oily kitchen scraps or starchy leftovers.

Don't use dog droppings or cat litter.

Don't compost any plant with *poison* in its name.

Don't compost yard or garden trimmings that might contain herbicides, such as clippings from a lawn treated with broadleaf-weed killer.

Don't overdo wood chips or sawdust.

Don't use colored printed paper or glossy paper.

Don't use "compostable" cutlery or other biodegradable plastics.

Don't compost mixed garbage, plastics, or aluminum foil.

Don't use charcoal or coal ashes, although you can sprinkle in a scant handful of hardwood ashes from a fireplace or woodstove.

OTHER SOIL AMENDMENTS

In addition to compost, organic gardeners might apply natural amendments to fortify garden soil. Evan Thaler-Null of Abode Farm in New Lebanon, New York, compared compost to a healthy, balanced diet and soil amendments to vitamins.

Amendments include worm castings, guano, blood meal, bone meal, alfalfa meal, fish meal, kelp meal, and others. Liquid seaweed extract is strong medicine. A side dressing of organic fertilizer applied at a plant's base early in the season and again at midseason supports heavy feeders.

Agricultural lime is applied to raise pH (or "sweeten" acidic soil), thereby boosting nutrient uptake and improving tilth, as described earlier. Hardwood ash has similar effects and supplies calcium.

Other mineral supplements include rock phosphate for phosphorus and granite dust for potash (potassium). Greensand, another source of potash mined from ancient undersea deposits, additionally improves nutrient absorption, soil texture, and moisture retention. Rock powders are typically applied in fall and winter. They break down slowly, releasing nutrients over time, and can be applied with a heavy hand, perhaps

JULY 26: Two months and two turnings after the image on pages 38 and 39, the compost pile at rear is ready. A new pile, foreground, is layered with comfrey leaves.

10 pounds per 100 square feet. Apply every third or fourth year.

GREEN MANURES

A **COVER CROP** is a thick stand of beneficial plants sown on fallow ground, or a portion of the garden taken out of normal use to rest and regenerate. Through the power of photosynthesis, cover crops leverage nutrients banked in the soil to produce copious growth. Turned under before flowering (or chopped for mulch), they create so much enriching organic matter that cover crops are also called **GREEN MANURES**.

Common cover crops include **NITROGEN-FIXING** vetch, clover, and field peas. Deep-rooted buckwheat and vigorous Sudan grass grow fast to outcompete weeds. Tillage radish drills compacted soil with its girthy taproot, hence the name. Yellow-flowered mustard has added value as a "biofumigant" that suppresses soilborne fungi and nematodes if incorporated into the soil at peak bloom.

"THIS COMPOST"

by WALT WHITMAN

Walt Whitman's ecstatic democratic vision extended even to "This Compost," a poem from 1856, about the regenerative potential of the American continent after the Civil War. Whitman marvels that the nation's land, sown with diseased carrion and the "foul meat" of soldiers, somehow "grows such sweet things out of such corruptions."

He is, as ever, tireless in his singing. Below is just one of the poem's six stanzas.

Behold this compost! behold it well!
Perhaps every mite has once form'd part of a sick person—Yet behold!
The grass of spring covers the prairies,
The bean bursts noiselessly through the mould in the garden,
The delicate spear of the onion pierces upward,
The apple-buds cluster together on the apple-branches,
The resurrection of the wheat appears with pale visage out of its graves,
The tinge awakes over the willow-tree and the mulberry-tree,
The he-birds carol mornings and evenings, while the she-birds sit on their nests,
The young of poultry break through the hatch'd eggs,
The new-born of animals appear—the calf is dropt from the cow, the colt from the mare,
Out of its little hill faithfully rise the potato's dark green leaves,
Out of its hill rises the yellow maize-stalk, the lilacs bloom in the dooryards,
The summer growth is innocent and disdainful above all those strata of sour dead.

COMPOST RECIPE

There are many recipes in the chapters ahead. This is mother to them all. The key is to balance a generous amount of dry, bulky, carbon-rich browns with just enough juicy, dense, nitrogen-rich greens. An all-brown pile will sit there: It's too lean, not nutritious enough to feed the hungry microbes. An all-green pile would be too rich and collapse into a slimy mass. My analogy is making coleslaw in the kitchen. A bowl of shredded cabbage is not very palatable. Those are your browns. Your greens are like mayonnaise, too rich to be eaten alone. You combine the two in proper proportion and "season" the pile with sprinklings of animal activator.

Like sourdough starter or a bubbling crock of dill pickles, compost is a living microbial ecosystem. It actively ferments. Your goal is to balance the ferment—to not upset the digestion.

Compost is best made in a big batch. It scales up easily but there's a limit to scaling down. Composting in a shoebox hardly makes sense. A pile 3 feet square by 3 feet high is large enough to "cook" and small enough to fit into most yards.

A final note: Locate the pile where it is easy to get to, right out in the middle of things. Composting is a dignified and useful act, and there's nothing immodest about the process. A pile won't stink if properly built. A pile should be shaded and/or covered in arid regions.

BROWNS
BLUE
RED — repeat for 3 feet
GREENS
BROWNS

GROUND

3 parts browns, carbon-rich organic materials (see Browns vs. Greens, page 44), including some extra-coarse stemmy or twiggy material
1 part greens, nitrogen-rich organic materials (see Browns vs. Greens, page 44)
Handfuls of animal activator (see chart, page 44) or a vegan substitute
Water
Special equipment:
A container, roughly 3 feet square × 3 feet tall or larger (see Placement and Container, page 40)
A tarp (optional)
A long-probed soil thermometer (optional)

1. Begin with browns. Put down a base layer of extra-coarse twiggy material to provide drainage and airflow. Top it with a 3-inch layer of browns. If your browns are newly raked leaves and airy enough that kids and dogs will want to romp in them, make the layer 6 inches. Shredded browns will break down more quickly. Cardboard should always be shredded. Newsprint and plain white paper can be separated into sheets and crumpled.
2. Add a 1-inch layer of greens. Bulky kitchen scraps, such as kale stems, can be added more generously. The same holds for lightweight greens such as carrot tops. Be sparing with fresh grass clippings and wet scraps, such as potato peels—anything that will compact into a soggy mess. Multiple thin layers are better than a single overly thick one.
3. The animal activator, your "red layer," is thin. If fresh, such as wet manure or seaweed, lightly cover the previous layers with it. If using a dry activator, such as blood meal or bone meal, apply a few handfuls, a heavy dusting, or follow package instructions.
4. Water is the "blue layer." Sprinkle or spray enough water to wet the pile, but not so much that it runs out of the bottom.
5. Repeat until the pile is 3 feet high, or as high as your container is wide. Top the pile with a final layer of browns. Water well. In very dry climates, optionally cover with a tarp to conserve moisture.
6. After 3 days, stick your hand in the pile to check moisture. Water from the top if necessary.
7. After a week, check the pile's interior temperature with a long-probed soil thermometer, or stick in your hand. It should be warm to hot. Water the top only if the interior feels dry and cold. Ten days in, the temperature will rise to 130°F or more.
8. After three weeks, when the temperature has dropped below 100°F, turn the pile. For example, lay a tarp alongside, fork everything onto it, then return it to the bin. By now, the organic matter should be partially broken down and more homogeneous than when fresh. If nothing has happened to begin transforming the raw material, the pile was too dry. If you come across slimy or stinky clumps—the pile should smell damp and a bit rich, but not sour, putrid, or mildewed—break them up and stir well.
9. Continue to monitor the pile for moisture and temperature. After turning, it will heat up, and again cool off. Turn the pile again after 2 to 4 weeks, if you want. Otherwise leave it to finish under the slower powers of fungal breakdown. With no further turning, in 2 months you'll have somewhat coarse compost, ideal for mulch. The pile will continue to cure for a year or more. The texture becomes finer with time.

CH.

05

BREAKING GROUND

MARCH 21: Aerating no-till beds with a garden fork. The near bed, spread with compost, is ready for planting.

In the cold months you think about a garden and daydream among the pages of a seed catalogue. At last there will come a spell of fine weather and you know it's spring, or close enough to it. Time to go outside and get dirty—the start of the growing season.

CONTAINERS AND RAISED BEDS

Containers and raised beds are compact, suited to small spaces, easy to plant, and easy to care for. Plus you don't have to dig up a corner of the yard, which for any gardener is a daunting chore. Small terra-cotta pots, the right size for herbs, can thrive on a porch step or even on a windowsill, if there is enough sun. Larger pots and wooden planters, the kind you place on a patio or deck for the whole summer, will have room for cherry tomatoes or peppers. Containers can be fitted with trellises for pole beans and cucumbers.

Whatever container you choose, be sure it is drilled for drainage. Cover the hole or holes with a layer of gravel to prevent washouts. Fill the container to within 2 inches of the rim with a potting mixture composed of equal parts compost and topsoil. (Or use bagged potting soil.) Level the surface. You're ready to sow seeds or put out transplants.

Size the container to the plant's mature size. Chives or thyme will make it through a season in a 6-inch pot. Lettuces have shallow roots and spreading leaves, so give them a wide pot. Slow growing vegetables will need to reach their roots into a large, deep container. A 5-gallon bucket or equivalent will hold one tomato plant, one pepper, or a summer squash.

Containers dry out quickly, especially wooden window boxes and unglazed terra-cotta pots. Water frequently, even daily in hot weather. (Save the shards of broken terra-cotta pots to cover drainage holes next time.)

If you have yard space for a raised bed, build a four-sided, bottomless 4 × 8-foot frame of a weather-resistant material, such as 2 × 6 cedar or locust planks. (The sides of the bed can be any height; taller beds require less bending over, making them accessible for gardeners with mobility issues.) Place the frame directly on the ground, in a level spot that receives full sun. If it covers the lawn, don't worry about the grass. If needed, mow or chop weedy ground cover and line the bottom of the bed with corrugated cardboard before filling.

When siting a raised bed, leave a 2-foot perimeter for durable mulch, such as wood chips or gravel. With two beds, leave 4 feet between, wide enough to pass with a wheelbarrow. Four or more evenly spaced raised beds makes a handsome grid. Expand as needed, without limit.

JUNE 6: A raised bed planted with lettuces, peas, and walking onions

IN-GROUND GARDENS

To establish an in-ground vegetable garden will likely require converting a patch of yard from its former use, probably lawn. The second year will be easier.

There are two strategies: **TILL** and **NO-TILL**.

A tilled garden turns the soil—think of a home-sized version of a farmer's plowed fields. The goal is to blend stratified layers and disperse amendments such as compost throughout the root zone, the top 12 inches.

No-till gardening never turns the soil, but aerates with a garden fork instead. Organic matter is left on the surface in the form of compost and mulch. Earthworms and other small helpers do the mixing. Over time, the soil in a no-till garden shows clear stratification, with an obvious upper layer of humus-rich topsoil. No-till practitioners believe their approach approximates the soil-building process observed in forests and prairies.

Till and no-till represent two paradigms, two sets of practices nested within larger ideologies. There is, as you might imagine, a cultural component. To paint the stereotypes, tilling represents the ag establishment of John Deere men who graduated from land grant universities and manage thousand-acre farms. No-till is countercultural: It walks barefoot, trailing patchouli and New Age eco-philosophy. The Ag Men might concede that no-dig has benefits, such as carbon sequestration, but dismiss it at the world-feeding scale of commercial farming. No-till true believers sniff at tillers as fossil-fuel addicts who strip-mine the soil. The reality is more complex. I've seen no-till practiced at a commercial scale, and my West grandfather, a John Deere man through and through, viewed his farming as stewardship of God's creation.

For a small first garden, up to maybe 100 square feet, many people will want to buy a shovel and dig in—turning the soil holds an intuitive logic. At a somewhat scale larger, maybe 20 × 20 feet, it becomes easier to create and maintain a no-till garden, and I believe the soil in my garden has improved using no-till practices. All the same, at some still-larger size, perhaps 100 feet square (10,000 square feet), a garden will require far less labor if manpower is aided by a gas-powered mechanical tiller.

My approach combines both paradigms. I find tilling useful to establish a garden and to manage large plots. I've also been impressed by the improvements in soil quality while using no-till practices. In the next section, I'll expand on both paradigms and describe how to establish and maintain a garden using each.

THE TILLING PARADIGM

Tilling, the physical turning of the soil, can be accomplished with hand tools, a motorized tiller, a tractor, or—more common in the past than today—a horse-drawn plow. Tilling physically loosens and aerates the soil, breaking clods to prepare a smooth **SEEDBED**, or a finely textured surface for sowing. Sod and cover crops are turned under to enrich the soil with organic matter.

The etymology is useful. The earliest sense of *till,* traceable to documents written in Old English around the year 900, was to exert oneself or strive through one's labor. By the thirteenth century, it came to mean the specialized work of cultivating the land. And a couple of centuries after that, the word settled into its modern meaning, to plow the ground. A plow or plough was the area of land a man with an ox could till in one day.

Tilling has fed empires. An excavation in India uncovered a plowed field from circa 2800 BCE, while still older evidence suggest that plowing followed the domestication of cattle, thousands of years earlier. Practiced with care, as on the small-scale farms observed by Franklin Hiram King during his Asian tour in the early twentieth century, tilled gardened plots can support intensive vegetable production without a falloff in soil fertility—as long as 4,000 years in King's observation. Eliot Coleman, who has helped popularize no-till techniques, describes the market gardens outside nineteenth-century Paris, origin of an intensive form of tilling called double digging, as the finest vegetable gardens ever grown.

MAKING A TILLED GARDEN

The tools required for a tilled garden do not necessarily include a tiller. Start small with hand tools: a D-handle garden fork, a shovel, a steel rake, and long-handled hoe. One person could install a 10 × 10-foot garden on a pleasant afternoon and tend it with a few hours of work per week. With experience and an extra set of hands, a couple could comfortably manage 400 square feet as a weekend hobby. Scott and Helen Nearing, in their prime, worked their 10,000-square-foot garden with hand tools and kept up an old-age garden of 50 × 50 feet until he was in his nineties and she in her eighties. Most couples are not the Nearings.

For a start, break ground in early spring, or as soon as the soil has thawed. Wait a half-week after rain for the ground to dry. Mark out the plot with a garden hose or twine stretched between stakes. Include a 4-foot perimeter buffer, enough room to maneuver a wheelbarrow or stand back to admire the garden at its midsummer peak.

Clear away any brush or coarse weeds. Using a shovel, turn the sod under. Work in an orderly fashion, one shovelful at a time, from one side of the plot to the other. Try to dig to a depth of 12 inches. Chuck stones to the side. Don't worry too much about breaking up clods. Leave the newly opened ground a week or, better, two. Come back to turn the plot a second time, preferably with a garden fork, to refine the texture. Work the soil as deeply as you can. Plunge the fork forcefully into the subsoil. Rock the handle back and forth to aerate the deeper levels.

Now work in soil amendments. Scatter 2 to 3 inches of compost—you can hardly use too much—along with organic amendments and lime, if indicated by a soil test. Fork everything in. Do not incorporate fresh manure or other uncomposted organic matter, such as leaves, fresh lawn clippings, or wood chips. To ask the soil to grow vegetables and simultaneously decompose masses of organic matter is akin to asking someone to eat a steak while running a marathon.

Finally, rake out the soil surface. Level it. Use the back of the rake to smooth the seedbed. Mark your rows or beds with twine, or "draw" on the soft soil with a stick. You're ready to plant.

Jump ahead to page 56 for sowing and transplanting instructions.

THE NO-TILL PARADIGM

Again, no-till leaves soil layers unturned. The stratified soil structure is meant to be porous and latticed, like the crumb of homemade sourdough bread. To aerate and create vertical passageways for air, water, and roots, the soil is "cracked" with a garden fork or an extra-wide broad fork. Compost and soil supplements are left on the surface; earthworms carry fertility deeper. Often a no-till garden consists of permanent beds separated by paths. The no-till approach has come in from the fringe, but still has the air of progressive innovation about it. One commercial grower I spoke to called it "faddish." It's a whole different gestalt.

In practice, no-till seeks to keep the ground covered as much as possible with an ongoing succession of crops, cover crops, and mulch. No bare soil! No-till also seeks to minimize inputs, including human energy. A no-till garden strives to harmonize with nature, rather than master it, in theory meaning more vegetables for less work.

MAKING A NO-TILL GARDEN

To establish a new garden, lay down a tarp or other artificial mulch to smother what grows there—the awkward term of art is occultation.

Four to six weeks will kill off most grasses and annual weeds. A few persistent perennials such as bindweed, quackgrass, and crabgrass might have to be dug out by hand, but in their feeble state they will more readily succumb to a hand trowel.

The smothering layer can be almost anything opaque. Cardboard shipping boxes are cheap and ubiquitous. Plus, they don't need to be removed because they quickly decompose. Instead, leave them in place. First, aerate the soil (see immediately below), then lay down a single layer of corrugated cardboard and cover with 6 inches of compost or topsoil. Sow or plant into the seedbed without further ado. Earthworms will convert the refuse of a mail-order economy into fertility.

Plastic sheeting, plastic tarps, and canvas drop cloths are alternatives. They must be removed after smothering.

AERATING

Aerate *before* you lay down cardboard or *after* smothering with tarps. In coming seasons, aerate beds again before replanting.

To aerate the soil, plunge the tines of a garden fork or broad fork vertically into the ground. Lever back the handle. You might raise a slight mound, but there's no need to overdo it and lift the soil. Move your fork 4 to 6 inches backward and repeat. You progress backward to avoid standing on the soil you've just aerated. It doesn't take much practice to find a smooth three-beat rhythm: push, crack, step back.

To finish, spread a 2-inch layer of compost and rake it smooth for your seedbed.

MARCH 21: Stratified soil layers after several years of no-till: darker topsoil over mineral-rich subsoil

CH. 06

SOWING AND TRANSPLANTING

JUNE 5: Starting cucumbers and squash in divided trays or "cells" to thwart cutworms

You can put *in* a garden or put *out* a garden. In either case, the raw material—apart from the soil itself—will be seeds or seedlings. The latter are commonly known as starts or transplants, because they are started indoors (in a greenhouse, for instance) by sowing seeds in small pots or "cells," the pocketed six-packs seen at garden centers.

The advantages to starting a garden from seed are low cost and maximum variety. Commercial transplants are not necessarily the tastiest varieties nor best suited to your garden conditions. To discover the full potential of varietal cooking requires seed catalogues. A catalogue is a living archive of garden biodiversity, cultural knowledge, and culinary potential—a library of flavors. (In chapter 9, I'll discuss saving seed from vegetables you've trialed and approved.)

The advantage to transplants is a head start on the season. You begin the garden indoors long before outdoor planting is viable and put out robust little seedlings as soon as the weather permits, when otherwise you'd first be pushing seeds into the bare soil.

I mix it up. I'll start some seeds inside on a windowsill or small table set up in a bright front room. As soon as commercial transplants are available, I'll buy a flat of baby lettuces to put out, and, at the same time, I'll sow a row of lettuce seed. The starts will be ready to pick as the seedlings fill in. Then I'll sow another round of seeds to grow up along the first. This method of sowing and harvesting in staggered generations is called **SUCCESSION PLANTING**.

SEED BASICS

A seed is the reproductive structure of a flowering plant. (It is also something more. See What Is a Seed? on page 59.)

Sowing a seed where you intend it to grow until maturity is called **DIRECT SOWING**. An alternative approach is to **START SEEDS** or **START INDOORS** in pots, cells, or trays placed in a bright, warm spot. Another version of "indoors" is the **COLD FRAME**, a glass-topped wooden box covering an outdoor seedbed in the manner of a mini-greenhouse. Seedlings started indoors are moved to their permanent location when four to six inches tall, once they have established a root system sturdy enough to survive the shock of transplanting.

To sow a seed, you put it in contact with the soil, where it germinates in the presence of water and warmth. (While seeds will sprout on a damp paper towel in the dark, the emergent seedling soon needs light.) The embryonic root stretches to reach water and nutrients, and the tiny leaf-like **COTYLEDON** unfurls to initiate the plant's lifelong work of photosynthesis.

Warmth controls the pace of sprouting. Germination and emergence speed up with rising temperatures, at least to a point. (The seeds of some cool-season crops, such as lettuce, enter **THERMAL DORMANCY** in hot weather to avoid sprouting in unfavorable conditions.) In the spring garden, peas will germinate in soil as cold at 40°F, but they come up quicker at 50°F and will pop up at 60°F. Bear in mind that daytime soil temperature is warmer than the air because dark ground soaks up solar energy. Often, a plant's ideal germination temperature is slightly higher than its ideal growth temperature. When starting seeds indoors, an old trick is

GARDENING WITH THE GALATIANS

"Whatsoever a man soweth, that shall he also reap," wrote the apostle Paul in his letter to the Galatians. Gardeners beware: Between sowing and harvest comes a spell of hard work. Paul continued, "And let us not be weary in well doing, for in due season we shall reap, if we faint not." As with the everlasting soul, so it goes with peas and tomatoes. A chief cause of weariness in the garden is unwise ambition. Let us not soweth too much.

The temptations will be many. There are dozens of vegetables and countless varietals in the world's seed catalogues. But even a single seed pack typically supplies more than you had planned for, more than you need, and more than you could possibly use. Two hills of butternut squash will satisfy even ardent squash eaters, but a seed pack might have a dozen or thirty seeds. Sown at once, they amount to a ruination of work or a shame of avoidance that will weigh on the conscience like a stack of unread *New Yorker*s.

Moderation in all things, especially seeds. There's no need to plant a whole packet at once. Save some for later, to mitigate against future disaster, like when crows steal your first planting of corn, or cutworms fell your cucumbers' seedlings. Leftover seed will remain viable at least until the next planting season if stored in a cool, dry, dark place. (The minor exception is perishable onion seed.)

To avoid overplanting, make a garden plan (see page 88) and stick with it. If you try to wing it outside on a fine spring day with a handful of seed packets, you will lose yourself in the willy-nilly romance of sowing. Discipline yourself. Overplanting leads to weary work, which in turn will cause the heart to grow faint. Heed Saint Paul: sow wisely.

to place a newly sown seed tray on top of the refrigerator—a warm spot in a chilly house. Once the seedlings emerge, move the tray to a bright window. Specially designed seedling heat mats—low-temp waterproof heating pads—can be adjusted to stimulate germination or growth.

Some seeds with extra-thick coats benefit from an overnight soak in tepid water. I always soak okra and peas to improve germination; beans, corn, and squash can be soaked. The ornamental morning glory vine, a sweet potato relative, has such tough seeds that they should be nicked with a file or emery board.

Spacing

How much space a plant needs to grow depends on its mature size and feeding habits. Small leafy arugula takes less space than bushy, deep-rooted potatoes and much less than galumphing pumpkins. Crop-specific spacing recommendations are always printed on the seed packet and will be included for reference in the chapters ahead.

At maturity, plants should be close enough that their leaves touch without crowding each other. Each individual needs room to gather sunlight and spread its roots for water and nutrients. At the same time, an unbroken canopy shades out weeds and holds in ground moisture, a living mulch. Properly spaced plants create a microclimate favorable to their own growth. Even in the heat of summer, the soil beneath a row of beans will feel cooler and damper than the exposed soil between rows. Dense planting

WHAT IS A SEED?

A seed contains an embryonic plant within a hard casing, the seed coat, plus a store of nutrients for the emergent seedling.

A seed is the mechanism of sexual reproduction. It develops after the female ovule is fertilized within the flower by male pollen, an intricate and delicate process that is nonetheless as dependable as May. The fertilized ovule, a soft capsule, dries down at maturity to become a hard seed. Many of the world's staple foods—corn, rice, wheat, barley, and beans among them—are edible seeds.

Some seeds are carried within a fruit, which is the mature ovary. Many vegetables we eat are **FRUITS**, including tomatoes, squash, peppers, and eggplants. A single fruit can contain dozens, perhaps hundreds, of seeds.

A seed persists through a long night of dormancy—winter—safeguarding the plant's life force from one generation to the next.

A seed begets stories, one of which is a genetic code spelled out in its DNA molecules. Its other stories are transmitted by gardeners: where the seed came from and what you do with it. A seed's horticultural history—its cultural DNA—is expressed in the name, lore, recipes, yarns, folk wisdom, tall tales, and myths that travel with it.

Seeds expire. If a crop is not renewed periodically through replanting, it will go extinct. Century-old seed catalogues record vegetables celebrated in their day for supreme culinary quality. It is dismaying to learn how many have been lost. For whatever reason—a change in taste or limited attention in a marketplace flooded with novelties—countless varieties fell out of favor and then disappeared entirely. A lapse of one generation breaks the chain.

Global crop biodiversity remains at risk due to pressures from industrial seed conglomerates. A backyard gardener is akin to a museum curator who preserves and protects a cultural legacy—and in this case tasty vegetables—for future generations.

JUNE 28: Cabbage sown in beds with offset rows, at Woven Roots Farm, for maximum planting density

also makes the most of limited garden space, a practical necessity for raised beds.

SEED SPACING describes the distance between seeds when you sow them. It is always closer than **MATURE SPACING**. Why? You make allowances for loss. Some seeds won't germinate. Others will be duds and runts. Drought and heat take a toll. Cutworms, voles, and clumsy-footed gardeners interfere. It is common practice to overseed by 100 percent or more. Get used to it. Corn is sown every 3 to 4 inches and weaker sprouts are progressively thinned to leave a mature spacing of 10 to 12 inches. Some small crops that don't mind crowding can be thickly sown in **BANDS** 2 to 3 inches wide. Cocktail onions, radishes, and small greens are all candidates.

ROW SPACING, the distance between **ROWS**, is wider than spacing within rows to accommodate spread. I think of a line of military cadets. In single file, they are close together, but each file is widely spaced to allow them room to do jumping jacks. Widely spaced garden rows allow the gardener room to walk between, water, and weed.

An alternative to planting in rows is planting in **BEDS**. For convenience, beds should be 30 to 36 inches wide and separated by paths that are 12 to 24 inches wide. (A foot-wide path is quite narrow in practice.) Plant on a grid within the bed, with each plant equidistant from the others, or, to maximize vegetables per square foot, devise an offset grid—staggered ranks. Sometimes you'll **BROADCAST**, or thinly scatter, the seeds of small crops. This works well for looseleaf lettuce and cooking greens.

Sprawling vegetables such as pumpkins are sown in widely spaced **HILLS**. Sometimes there is an actual mound, to encourage drainage, but a "hill" might also be flat, or ringed with a low berm to catch water. Overseed each hill, just as you would rows. I sow 5 pumpkin seeds per hill and thin to the strongest 2 or 3. A singsong maxim advises planting 4 zucchini per hill: one for the cutworm, one for the crow, one to rot, and one to grow. I add a fifth for good luck, then select the strongest 2 seedlings.

Planting Depth

Every seed has its ideal **PLANTING DEPTH**. Seed size gives a rough indication. The largest seeds—beans, peas, corn, and squash—can be planted an inch or more deep. Somewhat smaller, BB-sized seeds—beets, chard, and okra—go ½ inch deep. Disc-shaped tomatoes, eggplant, and other medium seeds should be covered with ¼ inch of soil. The brassicas have tiny, bead-like seeds that should also be covered by ¼ to ½ inch. Lettuces and small greens are barely covered with ⅛ inch of soil or simply pressed into the soil with your palm. Confirm planting depths in the chapters ahead or use the information printed on a seed pack.

A useful tip on sowing comes from John Coykendall at Blackberry Farm, whose country epigrams have stayed with me. He says to "overplant and undercover." In other words, sow more seeds than you think you need and cover them less deeply than you think you should.

Days to Maturity

The planting information on a seed packet includes **DAYS TO MATURITY** or **DAYS TO HARVEST**, the amount of time it takes for a vegetable to grow to eating size. Radishes are ready in as little as 21 days. Silver Queen corn needs 92 days.

Many factors determine a vegetable's real-world performance—weather, planting date, watering schedule, and so on. Days to harvest usually does not include the time it takes for seeds to germinate, which also varies by species. A cucumber emerges in as little as 3 days. Early spring parsley can take nearly a month in cool weather. Most vegetable seeds sprout within 1 to 2 weeks.

Vegetables grown for their leaves (lettuce, arugula, chard) will be ready for harvest soonest for the obvious reason: Plants put out leaves quickly because photosynthesis powers further growth. Roots, no less vital, also grow quickly, in rough proportion with above-ground parts. Small-rooted vegetables such as radishes and Hakurei turnips will be ready within a few weeks, while larger-rooted beets, parsnips, and carrots take more time to bulk up.

Other edible structures mature later. Stems, such as celery and fennel, emerge soon enough but thicken by slow degrees. Flowers come later in the life cycle, including broccoli and cauliflower, which are dense clusters of tiny buds. Fruit, the reproductive structure, also requires maturity. Those fruits eaten green, in the sense of unripe—including green beans, green chiles, green peppers, green peas, cucumbers, summer squash, and zucchini—will be ready sooner than fruits left to ripen. Red bell peppers need 2 weeks more than green, and butternut squash matures long after zucchini has run its course. Size, too, makes a difference. Cherry tomatoes come in before beefsteaks.

Finally, some plant families are innately sluggish. The alliums, so nimble in the kitchen, dawdle in the field. Garlic is planted in late fall (or very early spring in mild climates) and reaches fat-bodied maturity after midsummer. Tightly wrapped cabbage can't be hurried. Collards, by comparison, are essentially nonheading cabbage and the leaves can be picked immature, weeks earlier.

Seedbed Moisture

Sprinkle a newly sown seedbed with water, or "water in," to establish good contact between seed and ground. When sowing in very dry, dusty conditions, you might preemptively sprinkle in advance to make the seedbed receptive.

From this point on, the seedbed *must* be kept consistently moist during germination and emergence. Once watered, a seed is "activated," or summoned to life. The days that follow represent peril, a crisis of germination, a now-or-never emergency. The subterranean embryo is vulnerable and the emergent seedling hardly less so.

Keep the top inch of seedbed consistently watered; poke your finger in the soil to check. Light but frequent sprinklings are best. In hot or windy weather, it's not too much to water twice a day. Think of a gentle rain and try to duplicate its effects with a watering can or a spray nozzle. Gushing or splashing water washes out weightless seeds. Inundations and wet-dry cycles cause crusting.

Emergence

During emergence, the seedling develops symmetrically, with the aboveground cotyledon increasing in proportion to the belowground radical, or embryonic root. Monocots, such as corn and other grasses, have one cotyledon—they look like the proverbial green shoots of economic recovery. Dicots such as beans have

AUGUST 4: Chard thinnings for the kitchen

two—the cartoon version of a newly emerged seedling. If you split open a roasted peanut, you can see all the parts in miniature: the cotyledons, stem, and radical.

A so-called true leaf emerges next; in shape it will resemble the mature plant more than the cotyledon. The seedling now races to establish itself before depleting the picnic lunch of energy carried within the seed.

Established plants give the gardener a wide margin for error, but helpless seedlings need care. Water regularly. Protect from pests as much as you can. (More on this to follow.) Drought in the top inch of soil, the infantile root zone, can kill without warning. You will find yourself on all fours looking for the wispy remains.

The crisis passes with the second or third set of leaves. By then, the root system is established, and a growth spurt will begin. Surging young plants remain somewhat unsteady and, like teenaged humans, will need continued oversight and care, but they can be trusted on their own for days at a time.

Thinning

A week or so after germination, selectively **THIN** less vigorous seedlings until you reach the recommended mature spacing. I usually hedge my bets by thinning lightly at first, taking more as the young plants fill in. I will let salad greens, cooking greens, radishes, and root vegetables go longer, until they reach "baby" size, large enough for kitchen use. Chard and other hearty greens can be thinned progressively: first at 2 inches for the salad bowl, and again at 6 inches to sauté.

DIRECT SOWING

This is the gardener's iconic gesture—stooping to put a seed in the ground, an act of hope, of foresight, of nurture. "To sow a seed" is a symbol of beginnings.

Most vegetables can be sown directly at the proper time. A handful of crops, including carrots and parsnips, are always sown directly because they don't like being transplanted. Likewise, tubers and bulbs—potatoes, garlic, shallots, onion sets—always go straight into the ground.

Cool-season crops, such as peas, lettuce, broccoli, cabbage, and kale, potentially have two sowings per year. For a **SPRING GARDEN**, seeds go out before last frost. A **FALL GARDEN** is sown in July and early August for late pickings around and sometimes after first frost.

Warm-season crops, which include the suite of high-summer vegetables, are not sown directly until the ground warms. I put out tomatoes, corn, peppers, beans, okra, and squash 1 to 2 weeks after last frost. Tomatoes, peppers, and eggplants are often started indoors for transplanting. Corn, beans, and okra rarely are.

Prepare a seedbed for sowing by raking smooth a 2-inch layer of fine-grained, fully matured compost. Its dark color will soak up heat to encourage germination, as well as mulching and fertilizing the sprouts during their rapid initial growth. Mark out rows or beds with twine, or just draw furrows with your finger.

AUGUST 13: Mustard greens dusted with diatomaceous earth (to discourage slugs) and ready for thinning

MAY 19: Transplants give quick results.

Drop seeds in place at the recommended spacing by rubbing them between your thumb and forefinger. Cover to the recommended depth. For larger or more widely spaced seeds, such as squash, I'll individually push them into the soil. Planting a gridded bed is easiest if you dibble all the holes first, then place and cover the seeds. When broadcasting seed over a bed, I get more consistent coverage if I mark quadrants and sow each in turn.

Keep a seedbed moist during windy or hot weather by covering with damp burlap. (Sprinkle the burlap with water daily and remove as soon as the seedlings come up.) Slow-to-germinate carrots especially benefit from this trick.

If a late frost threatens tender new seedlings, protect them with a lightweight fabric called **FLOATING ROW COVER**. It lies directly on the seedbed, "floating" over the plants, and it can be left in place long-term for protection against pests as well. Its open weave allows water and sunlight to pass through. Weight the edges with stones or boards. In a pinch, old bedsheets or even thin flour-sack kitchen towels can be used to cover plants overnight during a cold snap, but remove them promptly in the morning.

STARTING SEEDS INDOORS

Several benefits justify the chore of starting seeds indoors. The most obvious is an earlier harvest from plants germinated indoors weeks before direct sowing could begin in the garden. In addition, some seeds, such as onion, celery, and parsley, germinate very slowly or require specific temperature conditions that are easier to achieve indoors. A few vegetables, in particular Brussels sprouts, have such a long growing season that northern gardeners might not have enough frost-free days otherwise. Finally, transplants gain an advantage over weeds and pests.

You start seeds several weeks before the target planting date, enough time for them to establish roots before handling. But not too early: Larger, older transplants are not necessarily better. The goal is to have compact, vigorous toddlers that are just ready to take off running, rather than leggy tweens that have outgrown their home, leaving them rootbound and stunted.

Four-pack and six-pack trays (peat or plastic cells) hold just enough soil to get seeds going. Any other little pot with a drainage hole will work, too. I have a collection of 2-inch terra-cotta pots for starting seeds. Biodegradable pots made from peat, manure, or plant fiber can go into the ground with the plant, reducing the risk of transplant shock. Flat, shallow plastic propagating trays—some are fitted with a clear domed lid—are useful for sprouting seeds en masse.

The most reliable growth medium for starting seeds indoors is bagged **SEED STARTER MIX**, which is sterilized soil rich in organic matter and salted with bits of porous perlite or vermiculite to hold moisture. I use sifted compost, although it can carry disease spores.

Unfortunately, pathogens are a greater risk with seedlings started indoors—crowded as they are in a consistently warm and damp environment—than in the garden, which is continuously cleansed by hygienic sunlight and fresh air.

Fill each container with seed starter mix and mist with water. Sow 2 or 3 seeds per cell/container, and thin to the strongest. Beets, turnips, and onions can be sown 3 to 5 seeds per cell. Don't thin. Once in the garden, they will nudge each other out of the way as they grow. Scallions can be sown 10 or 12 seeds per cell for a ready-made bunch.

To propagate in a tray, fill with at least 2 inches of growing medium and mist with water. Make a shallow furrow at the recommended sowing depth, drop in the seeds, and cover. Leave 2 inches between rows. Label everything at once. "A short pencil is more reliable than a long memory," my mother always said.

The soil surface should remain consistently moist, but not waterlogged. Soggy conditions encourage damping off, a fungal disease that cuts seedlings dead at ground level. To reduce the spread, irrigate containers from the bottom, placing them in a shallow pan of water for 10 minutes. Moisture will wick up through the drainage holes. Frequent light misting is second-best.

Cover trays and containers with a clear dome or plastic wrap stretched over toothpick tentpoles. Keep in a bright spot, but out of direct sun. Fluorescent grow lights and heat pads complete the Cadillac version of a seed-starting setup, but recycled yogurt cups in a sunny windowsill will get the job done. (Don't forget to punch drain holes.) The transplant will be ready for the garden when 3 to 6 inches tall and rooted but not rootbound. See the photo on page 67.

Pricking Out

A tray of sprouts will soon need to be separated and transplanted into cells or growing pots. **PRICKING OUT**, as the delicate process is called, happens as soon as you can see the first set of true leaves. For each seedling, fill a cell or container with seed starter mix and dibble in a hole. One by one, gently pry each seedling from the soil using a popsicle stick or butter knife while lightly grasping its leaf, holding it "by the ear" rather than pinching its vascular stem, or "throat." The root at this stage is a branching thread and will release from the soil. Ease it into the dibbled hole and press soil around it. Water well. Transplant to the garden when 3 to 6 inches tall, as for other starts.

Hardening Off

Plants started indoors, like children with overprotective parents, are tender and easily bruised by the rough-and-tumble of the outside world. They need to be gradually acclimated to direct sun, wind, and outdoor temperatures, or **HARDENED OFF**.

A week before transplanting, place seedlings, uncovered, in a sunny spot outside for 1 hour, then bring back indoors. The next day, increase the outside time by an hour, and so on every day until the plants can spend 8 hours outside. Reduce

watering to every other day. For seeds started in a cold frame, prop open the lid progressively longer every day and cut back on watering.

OUTDOOR NURSERY BEDS

Even in warm weather, some gardeners start transplants in an outdoor nursery bed. Ira Wallace, the heroic garden godmother behind Southern Exposure Seed Exchange, explained that it's much easier for her to water a nursery bed than to drag a hose to every corner of her garden. My uncle David likewise uses a nursery bed to sprout "late" tomatoes, his second succession planting, since a bed is easier to protect against a late cold snap.

TRANSPLANTING

As mentioned earlier, starts are ready to transplant to their permanent location when 3 to 6 inches high, well rooted, and sturdy enough to survive the disturbance.

Buying transplants is akin to bringing home a puppy that's already housebroken—you skip the bother and go straight to the fun part. Select small, robust plants with the fresh look of new growth. Avoid plants with yellow or wilted leaves. Once home, water everything and plant promptly.

To establish a transplant (or any container-grown plant) in the garden, dig a hole twice as wide as the container. Invert the container and ease out the plant while holding the stem, shaking gently to release. Keep the root ball intact. Most vegetables except tomatoes prefer to be planted at the same ground level as in the container. Don't bury too deeply. Refill the hole and firm the soil around the roots. Water in.

Tomatoes *should* be planted deeply. Remove the vestigial cotyledons and the first set of leaves. Bury the bare stem up to the leaf crown—6 or 8 inches deep is not too much. The stem will root out. Peppers and eggplants, related nightshades, will also root from the stem.

Squash, cucumbers, and pumpkins should be transplanted as young as possible and with extra care to protect the roots.

If a transplant is rootbound from being held too long, with knotted or densely woven roots, tease the roots apart or, worst case, vertically slice into the root ball in several places before planting.

It will be easier to water tomatoes, squash, and cucumbers—the widely spaced large plants—if you create a shallow basin around each by hilling up a ring of soil, perhaps 1 inch high and a foot in diameter. (See the photo on page 65.) Lightly mulch the basin with compost or leaf mold to prevent soil crusting. Closely spaced starts such as onions and leeks can be planted in a shallow trench, a so-called sunken row.

The next chapter will discuss watering and tending a garden.

JULY 3: Transplants should be rooted but not rootbound.

CH.

07

UPKEEP

In the sweat of thy face shalt thou eat bread, till thou return unto the ground; for out of it wast thou taken: for dust thou art, and unto dust shalt thou return.

—**GENESIS 3:19**

JULY 2: Beautiful greens at Soul Fire Farm

There will be weeds. Pests will come. Your plants will suffer drought and break under torrential rain. You will work against adversity to cultivate the harvest. To eat by the sweat of your brow is the cost of expulsion from the Garden.

"Fertility, water, weeds, pests—that's universal," said Max Morningstar as we walked the fields at M|X Morningstar Farm in the Hudson Valley. "We all have the same problems."

On a brighter note, the trick to gardening, if there is one, is knowing when to do light tasks with ease to prevent future heavy work. Vigilance and upkeep stave off drastic remedies. You keep an eye on things. You check in. Inspect your containers and beds. Walk the garden every day and size up your vegetables, individually and in troops. You'll see what needs to be done, or at least you'll see that *something* needs to be done. "We must cultivate our garden," says Candide. "Sow. Hoe. Grow," answers Louis Van Deven in his charming self-published growing guide *Onions and Garlic Forever*.

This section covers the basics of how to tend a garden: watering, **WEEDING**, fertilizing, and dealing with pests and disease. Tending a garden is firstly about learning to see and, secondly, about learning to interpret the signs. Among the many reasons a garden should be made beautiful is that beauty holds the eye. Beauty makes us look. In looking, we see. And in seeing, we learn.

WATERING

Plants evolved to live in changeable conditions, the back-and-forth between sunshine and rainy days. Most vegetables will hold steady for a week without water though they will flag during longer spells of what my grandfather called "drouthy" weather. Mulching conserves moisture by reducing ground-surface

THE ONE-STRAW REVOLUTION
by MASANOBU FUKUOKA (1913–2008)

"You could not step twice into the same river," said Heraclitus, a philosopher of flux who saw in the world the constant *becoming* of perpetual change rather than the settled state of fully realized *being*. His kindred spirit across time and geography was Japanese farmer-philosopher Masanobu Fukuoka, for whom man is inseparable from nature.

Fukuoka's landmark book, *The One-Straw Revolution*, became an inspiration for organic growers and an intellectual cornerstone of the permaculture movement. In it he wrote:

> *Even though it is the same quarter acre, the farmer must grow his crops differently each year in accordance with variations in weather, insect populations, the conditions of the soil, and many other factors. Nature is everywhere in perpetual motion; conditions are never exactly the same in any two years.*

For Fukuoka, as for Heraclitus, nature is flux. He stood against the false promise of the Green Revolution: that growing vegetables could be standardized, as if farming could imitate the assembly line. A gardener, Fukuoka countered, does not manipulate uniform materials and industrial processes but instead responds to natural conditions that vary from year to year, even on the same narrow plot of land. It's never the same garden twice.

evaporation. Still, plants continuously lift water through their roots to their leaves, where it escapes as vapor, a process called transpiration. To irrigate—from the Latin *to introduce* + *moisture*—is to replenish the supply.

Plants will let you know they're thirsty. They droop, then wilt. Leaves might turn yellow and drop. With extended drought, plants will become stunted.

As mentioned in an earlier chapter, most vegetables need the equivalent of 1 inch of rain per week. Thirsty celery likes 2 inches per week. Water your garden as needed in dry weather. I aim for every third or fourth day unless there's been a good soaking rain—the kind that keeps you inside for several hours. A gardener learns to pay attention to the weather, and perhaps even jot down a record of when and how much it rains. (The summer I began this book, my rainmaking efforts mostly failed, and only 4 of the 19 weeks between the start of May and the end of September received a full inch.)

Develop the habit of regularly checking soil moisture in the garden and, crucially, in containers. Poke your finger into the soil to test. If the top inch feels dry, it's time to water. The subsurface should feel cool and moist. Raised beds will require more watering than an in-ground garden. Containers dry out faster still and might need daily watering.

Regular watering encourages steady healthy growth. Lettuce, arugula, chard, kale, and other quick crops (radishes, Hakurei turnips) develop the finest flavor when grown quickly in rich soil with plenty of water.

Consistency is the key. Erratic watering causes misshapen vegetables and even affects flavor. Zucchini and cucumbers won't fill out. Onions turn sharp and garlic can become too hot. Brassicas take on a bitter edge.

Be advised that squash and cucumbers sometimes play possum. They flop in the midday heat as if dying. Don't worry. Overnight they will likely rebound as cooler temperatures slow the rate of transpiration. Check on them the following morning, and if they still look limp, give them water.

Dry-climate gardeners should consider planting for drought tolerance. Deeply rooted tomatoes and pumpkins, which can access groundwater over their long harvest cycle, will be less reliant on irrigation. Pole beans tolerate drought better than bush beans. The ancient blue corns of the American Southwest have evolved to thrive in arid conditions.

How to Water

On a hot summer afternoon, my inner child wants to aim a hose at the garden and let water fly. The instinct is wrong. It's better to keep water at the ground level. Disease spores are spread by splashing, and wet foliage encourages mildew and fungus.

Hold a hose or watering spout close to the ground and aim the stream in circles around a hill of zucchini, for example, or use a figure-eight pattern among a row of beans. A useful visual clue for how long to stay in one place comes from *Alan Chadwick's Enchanted Garden* by Tom Cuthbertson. Water until the ground shines, he writes. Wait a few moments for the liquid film to absorb, then water a second time before moving on. Aim for the roots, not the empty ground between plants. For containers, flood the pot, let it sink in, then flood a second time. Stop when water trickles from the drainage hole.

Watering is most effective at the ends of the day. In the gentle hours between evening and morning, plant leaves open microscopic pores called stomata to draw in humid night air and bathe in the morning dew. The stomata close in defense against midday heat and sun.

I like to water in the slant afternoon light, between four o'clock and cocktail hour, giving moisture the night to soak in. However, if mildew is afoot in your garden, water instead in the morning, when the sun will dry splashes and reduce the spread of contagion.

WEEDING

Every square foot of arable earth comes preplanted with weed seeds. Unchecked weeds will choke a garden. The solution to weeds is weeding.

Always be weeding is my best advice. Never walk past a weed if you have a moment to take it out. Never believe that weeding will be easier tomorrow or next week. Learn at the outset to distinguish vegetable seedlings from weed seedlings—weedlings—so that you don't pull up that which you endeavor to grow. Learn to find the baby in the bathwater.

Weed seeds germinate at or very near the surface. The goal is to eliminate tiny weedlings when they are most sensitive to disturbance and avoid turning up more seeds to sprout behind them.

JUNE 6: These onions should have been weeded a week ago, but I was waiting for the goosefoot, an edible weed, to be skillet-sized.

APRIL 8: Edible weeds for the first garden meal of the year: stinging nettle tips, garlic mustard greens, cress, stray garlic shoots, and self-sown chervil

For the smallest weedlings, ones you can see only at kneeling height, break the contact between soil and thread-like root by scratching the ground with loosely curled fingers, the way you'd scratch a person's back. A few days later, when the weeds are a bit taller and thick as hair, scrunch up a handful at a time using the same gesture as scrunching the folds on the back of a dog's neck. At this stage, the Korean hand hoe, or ho-mi, is wonderfully effective. Its thin, curved blade skims beneath the soil's surface to sever roots.

Once weeds are big enough to see from standing height, take up a long-handled hoe. Chopping out mature weeds, hacking at the dense roots, illuminates the meaning of the phrase "a tough row to hoe." Hoeing works best in dry weather.

Hoes come in many forms. I like the maneuverability of the batwing hoe's narrow profile. A scuffle hoe looks like a stirrup. The bottom "tread" is a two-edged blade that scuffles along just beneath the surface, cutting on both the push and the pull.

Pulling up weeds by hand goes better after rain, when the soft ground releases. To protect a vegetable seedling closely surrounded by weeds, lay your palm flat on the ground with your fingers snug around the vegetable. Press down to hold the soil in place as you tug the weed with your other hand. You'll feel the mingled roots pull apart.

For all their nuisance, weeds are also remarkable plants. They are thrifty and adapted to life at the raw edge of human activity—around footpaths, roadways, building sites, gardens, and anywhere else the ground has been disturbed. Opportunistic to a fault, they leap up with extraordinary life force. Many weeds are edible. Those that are not contribute their vigor to the compost pile. Weeds enrich the garden ecosystem and add to the diversity of life in and on the soil. Know thy enemy.

MULCHING

A lazy-man's approach to garden maintenance comes down to Water, Weed, Mulch. But the greatest of these is mulch because it traps moisture and blocks weeds. Organic mulch also improves soil over time as earthworms and other decomposers convert it to humus. And still more: Mulch cools the ground in high summer and disperses heavy rainfall to stop soil erosion. As with compost—itself a supreme mulch—you can hardly overdo it.

I use **LEAF MOLD**, partially decayed tree leaves that have been raked, heaped, and overwintered. Four inches piled around vegetables in the spring, as soon as they're large enough not to be smothered by it, isn't too much. Alternatively, cover every inch of your garden plot with up to a foot of newly raked leaves. Fluffier when fresh, they will settle over winter and smother weeds until the start of planting season.

See The Mulch Method (page 74) to read about mulch advocate Ruth Stout's method of growing backyard vegetables.

JUNE 6: Onion beds mulched with leaf mold

THE MULCH METHOD

When I first horned in on neighbor Del's garden, I followed his lead as he plowed under leaves to enrich the soil with organic matter. By degrees, the garden became "mine" as I spent more time on its care. My practices evolved. I tilled less, letting leaves lie aboveground as mulch. The new method revolutionized my garden life. I had reinvented the wheel.

Garden writer Ruth Stout, now largely forgotten, was once an authority celebrated for her successes with "no work" gardening. Her practices were expounded with tart, funny stories in the 1955 bestseller *How to Have a Green Thumb without an Aching Back*. Stout's grand theory came down to 8 inches of mulch. "Mulch gardening," she called it.

A second mulch mentor, Masanobu Fukuoka, called his methods "natural farming" or "do-nothing farming." His annual garden to-do list was more like a don't-do list. Do not till. Do not use fertilizer or prepared compost. Do not weed. Fukuoka got away with such contrarian principals because he laid on the mulch.

Additional organic mulches include straw and spoiled hay, although the latter can be seedy. Newspaper, corrugated cardboard, and even old canvas tarps would work. Straw and pine needles are extra tidy. Spread them around cucumbers, summer squash, and greens to prevent muddy splatters. Organic mulch "disappears" over time, rotting away. That's a good thing. Reapply as needed.

Tree leaves and pine needles acidify the soil as they break down. That said, earthworm activity seems to buffer the effect. Steve Cunningham at Berkshire Bounty Farm told me he has used leaf mulch liberally for over a decade. When I asked about long-term effects on soil pH, he showed me a picture on his phone of a bed of freshly raked of leaves. The ground was covered with pH-balanced castings produced by a well-fed population of leaf-eating worms.

Slow-to-decay sawdust and wood chips are not good mulch for vegetables; save them instead for garden paths where their longevity is a virtue. Use only herbicide-free grass clippings, and scatter thinly. Applied too thickly, clippings become slimy and gross.

Don't mulch seedbeds before emergence, other than perhaps a very light scattering of straw to discourage birds. Put down mulch once the seedlings have sent up two or three pairs of true leaves.

One disadvantage: Mulch insulates against spring warmup, which can slow down your spring planting. I rake it off my garden beds a week or so before planting to let the ground soak up the sun.

A final note: Black plastic sheeting or plastic "mulch" is much used on commercial farms to block weeds, trap moisture, and minimize soil contact for cleaner vegetables. It also absorbs the sun like a solar panel, turning the ground into a heat sump. Garden beds covered with plastic in early spring can be worked sooner than bare ground. And in northern climates, planting through black plastic sheeting extends the growing season and delivers "solar power" to heat-loving summer vegetables. Melons and sweet potatoes will benefit from black plastic mulch in cooler climates, and even tomatoes, eggplants, and peppers will appreciate the solar gain.

First prepare the bed, then stretch the sheeting and weight the edges. Punch or burn holes through the plastic to put out transplants.

PESTS AND DISEASE

Creatures great and small will pester your garden, and disease is an aspect of garden life, by which I mean that plant-damaging bacteria, fungi, and viruses are living microbial participants in the garden ecology. I don't welcome pests and pathogens. But raising vegetables using organic methods requires a proper mindset, one that is both passionate and sane. A gardener must cultivate a sense of proportion. Most insects, including airborne pollinators and subterranean decomposers, are beneficial. We learn to deal with the detrimental others. Good garden hygiene reduces the risk of disease. And still, despite best efforts, sometimes a garden catches cold, so to speak. A gardener learns to carry on.

Pests

The list is long. Japanese beetles, cucumber beetles, Colorado potato beetles, flea beetles, white flies, and cabbage moths come back to my garden every year. Voles, deer, groundhogs, and raccoons make an occasional appearance.

Some pests are specialists, such as the Colorado potato beetle. Others cause general havoc: The destructive capacity of deer is shocking. If you live in an area with deer, I recommend fencing all vegetables except garlic, cucurbits (squash, cucumbers, pumpkin), and nightshades (tomatoes, potatoes, eggplants, peppers). I used to think shallots were distasteful to deer. Wrong.

As for insects, I encourage you to remember that a garden grows outdoors. Bugs are part of nature. Biologist E. O. Wilson established that biological life is a complex web of relationships among diverse species. The more species in an ecosystem, the more productive and resilient it will be. In a pesticide-free garden, the defense against insects is more insects. Nations of spiders, ladybugs, praying mantises, hunting wasps, and other predatory insects hunt sap-sucking bugs, destructive larvae, and soft-bodied caterpillars.

Plants also guard themselves with biodefense systems, which we can perceive in the bitter taste of brassicas or the heat of capsaicin in chiles. Stout and compact organically grown vegetables show stronger defenses. Overall plant health is a measure of their capacity to withstand pests. I've

THE MAPLE SUGAR BOOK

by SCOTT NEARING (1883–1983) and HELEN NEARING (1904–1995)

Scott and Helen Nearing, both born to privilege, left New York City at the depths of the Great Depression to test themselves against the Vermont countryside. They homesteaded and, living without electricity or running water, labored mightily at self-sufficiency. The chronicles of their life on the radical fringe filled books. Most famously, 1954's *Living the Good Life* explained "how to live simply and sanely in a troubled world." The Nearings wrote brisk, almost curt, prose. Self-assured by birth and education, they purged themselves of drippy sentiment about "country living" and fashioned their good life from the Yankee granite of hard work and the sturdy oak of empirical learning. Rarely did they wax lyrical.

Yet in 1950's *The Maple Sugar Book*, their first joint publication, the Nearings indulged in something approaching poetry. Their inspiration was tree leaves—for many gardeners a cheap and abundant source of soil-building mulch.

> *The leaves are the fluttering flags that attract the sunlight and through the action of sun and air transform those qualities into nourishment for the parent tree. They breathe in sunlight and breath out moisture. They also, when dead, play a part in forest culture and human economy by adding yearly to the carpet of the forest's ferny floor and thus maintaining a necessary precondition of life—a rich and healthy soil.*

repeatedly noticed in my potato patch that runty, puny plants draw more pests than healthier neighbors on either side. Weak plants are more vulnerable, so the best defense is a healthy garden. I know it sounds tautological, but the claim makes sense within an ecological mindset.

When garden pests do arrive, as they will, safe remedies include hand-picking, companion planting, physical barriers, organic pesticides, and crop rotation.

Hand-Picking

Inspect plants often. In June, the first thing I do when I go to the garden is to walk the rows of potatoes. The key to controlling insect damage is to stop the spread early. Bugs have incredible reproductive powers, so a minor problem can increase exponentially. Get to know the common pests in your region. Take pictures of your garden bugs and research their identities online. Also get to know common beneficial insects, including larval forms. Everyone knows the ladybug. At the immature phase, this key beneficial predator wears an all-black coat brightened by orange spots, like an art collector adorned with amber jewelry. Check the website of your state university's agricultural extension for information on pests in your region.

When you find pests, immediately remove them by hand and kill them. Steel yourself. There is no such thing as a vegan garden. Every meal of vegetables comes at the sacrifice of small lives. An alternative to crushing bugs with your fingers or underfoot is to fill a small bucket with warm soapy water and knock in the bugs. (The soap breaks surface tension so they sink.)

Inspect plants closely. Look underneath chewed bean leaves for Japanese beetles. Pull back a cabbage's skirt and peek inside with frank immodesty to find caterpillars. Rifle the mulch around brassicas to uncover slugs.

Also search for egg clusters, which are often yellow, orange, or shiny brown. Inspect each plant as thoroughly as checking a dog for ticks. Crush the eggs between your fingers.

Companion Planting

An old tradition channels the witchy knowledge of garden sages, cloistered friars, and curanderas to create vegetable gardens layered with flowers and herbs that repel pests. Marigolds, garlic, nasturtium, thyme, petunias, summer savory, rosemary, mint, and tansy are only a few of the plants said to protect vegetables against insect attack. In my own experience, I noticed far fewer Colorado potato beetles after I interplanted tansy, a fragrant herb with an ornamental flower that is also useful in compost for its high potassium content. (Note that tansy spreads aggressively and requires containment.)

Louise Riotte's charming book *Carrots Love Tomatoes* expands on the topic of companion planting, including herb-based pest controls. The author was a lifelong gardener in Oklahoma, and her writing combines the authority of long experience with a teacherly love of correctness.

Companion planting can also benefit the gardener. I plant summer savory among green beans to remind me to pick them together for the kitchen. (Savory is also said to repel Mexican bean beetles.) Marigolds, nasturtium, and tansy may be repellent to insects, but they beautify the garden, and stopping to smell fragrant herbs is a tireless pleasure.

Physical Barriers

Floating row cover, which is used to protect seedbeds and seedlings against frost, also protects against many pests, including deer. Covering arugula is the only way to keep off flea beetles. Spread the cover over newly sown beds and leave it in place as the plants grow or else use flexible rods to create a tunnel.

Chewing and sucking bugs, including slugs, snails, and grasshoppers, are discouraged by a mineral product called diatomaceous earth, a powder of fossilized plankton. It looks and feels like cornstarch but coats plants with microscopic grit. To apply, put a cup of the powder in the toe of an old sock and tie it off. Shake the pouch over leaves to coat them like a powdered donut. (See the photo on page 63.) Reapply after rain. Durable buckwheat-shell mulch also repels slugs and snails thanks to its sharp edges.

Cutworms destroy young tomatoes, peppers, eggplants, squash, and cucumbers by severing them at ground level. Neighbor Del showed me how to make his mother's cutworm collar. Tear strips of newspaper 2 inches wide by 4 inches long. Wrap a strip around each transplant's stem. Dig in the transplant such that the ground-level collar is held in place by surrounding soil.

Organic Pesticides

Bacillus thuringiensis (Bt) is a species of soil bacteria identified by Japanese scientist Shigetane Ishiwatari in 1901. Its spores disrupt the guts of soft-bodied larvae, causing them to starve. Liquid Bt—the spores suspended in solution—is sold as biological insecticide. It is supposed to be species-specific, although there is some cross-species collateral damage. The *aizawai* strain, targeted to moth caterpillars such as the cabbage worm, is deadly to honeybees. Bt is widely used by both organic and conventional farmers, although for the backyard gardener it is likely to be a last resort.

Crop Rotation

The practice of changing where you grow a given vegetable from one year to the next is **CROP ROTATION**. Among multiple benefits, rotating controls pests and disease. (See Succession, page 92, for more benefits.)

Many pests are host-specific, meaning they depend on a single species or genus to complete their life cycles. Pest larvae overwinter in the soil and seek the host when they emerge. Crop rotation breaks the cycle at this vulnerable stage when pests are least able to disperse. Rotating crops also reduces disease pressure, for the same reason that pathogens overwinter in the soil.

Rotation is based on crop families rather than individual species. Refer back to the chart on pages 10–12 for a refresher on garden taxonomy.

An ideal rotation allows three or four years before a plant family returns to ground where it previously grew. Some organic gardeners develop elaborate rotational practices. In a small backyard garden, it's enough to simply be aware of needing to move vegetables annually. Mix things up from one year to the next.

Look ahead to chapter 9, Planning the Next Garden (page 88), for practical tips and a simple rotational plan.

DISEASE PREVENTION

Sunlight and fresh air keep a garden healthy. Frequent tidying limits the spread of disease, especially at the end of the season. Avoid clutter. Rotating vegetables to prevent the buildup of pathogens in the soil is the best cure of all.

In humid climates, or when mildew becomes a problem, plant vegetables widely, giving them more than the recommended spacing to maximize airflow and sun exposure.

FERTILIZING CONTAINERS

A final note on upkeep is specific to container gardens. Vegetables grown in containers have a small reservoir of soil fertility to draw on, and frequent watering leaches nutrients. It's good practice to fertilize every 2 weeks with a diluted commercial liquid fertilizer, such as kelp concentrate or fish emulsion, or with homemade compost tea. To make in a 5-gallon bucket, cover one part compost with four parts unchlorinated water, such as rainwater. Stir the mixture and steep for several days, then strain out the solids. Dilute the infusion by half with fresh water before applying.

CH.

08

HARVEST

SEPTEMBER 14: Near the end of the tomato season

"Corn is coming in!" my West grandmother would say to my mother over the phone. My garden diary captures that same exclamation-point excitement. "July 2: new potatoes!"

Knowing when and how to harvest backyard vegetables is partially intuitive, or at least honed by experience. "Vegetables have their time," said Ria Ibrahim Taylor as she showed me around Soul Fire Farm in upstate New York. If you've shopped at farmers' markets, you've seen what good vegetables look like. In your own garden, you'll just know when the chard is ready and the wax beans look right. The time arrives when you can't wait any longer. My day-by-day assessment goes something like this: not ready, not ready, irresistible. Some vegetables have a built-in ripeness indicator. Tomatoes glow with color and butternut squash turns from green to bronze.

But there is an essential difference between buying vegetables at a farmers' market and picking them from your garden. At the market, you select whatever looks best that day. In a garden, you choose the day on which the vegetable looks best. You harvest at a moment within a longer life cycle, dipping into a stream of continuous growth. Harvest is a culmination and an end. Do you want little veggies for a crudités platter—see the inch-long okra and tiny haricots verts served raw with charred eggplant puree on page 445—or do you want great big, mineral-rich beets for hearty soup? Some vegetables can be picked once—cabbage has but a single head to give—while others grouped together under the cringey name "cut and come again" provide multiple harvests.

Homegrown produce straight from the garden will be shaggier and dirtier than what you buy. It will need trimming and washing. Beets and turnips and potatoes will need scrubbing. A head of cabbage is protected as it grows by a ruff fit for a Renaissance queen. The adornment will need to be removed and the head inspected for stowaway caterpillars and roly-poly sow bugs. The same holds for lettuce, greens, broccoli, cauliflower, and so on. There's a knack to choosing vegetables in their prime, revealing their beauty, and keeping them fresh until ready to use. Read on.

HARVEST CYCLE

Early-season vegetables have a wondrous quality, a kind of I-can't-believe-it-ness. Full of life, not yet toughened by hardship, they are mild and sweet, not bland but delicate. Young salads go to the table dressed in a vinaigrette as light as a linen shirt. The first corn hardly needs cooking, just a warming through to release its sweetness. I'll season early spring pickings with a few grains of salt, a dusting of ground black pepper, and flecks of chervil with its fugitive scents of chlorophyll and anise.

Late-season vegetables, by contrast, swagger in with big flavors and blazing colors—eggplants, hot chiles, and tomatoes weighing a full pound. Some vegetables change personality as they mature. I'll pull tiny turnips to braise in butter, cooking their tops and tails together, almost like a pan of sauteed greens. Later, I'll peel fist-sized turnips and slice them to bake in a bubbling gratin with cream and whole thyme branches.

At the outset of the book, I talked about growing what you like to eat. Now the idea is harvesting as you like to eat it. There is no single right time to pick a vegetable. New potatoes are delicious and, left in the ground, they will grow to baking size and be no less delicious later. Baby chard can be tossed into a mixed green salad or melted in warm butter to roll inside an omelet. A month later, big piles of mature chard can be sauteed in a wide skillet and pelted with the robust flavors of toasted pine nuts, raisins, and lemon juice. Your choice.

Small vegetables can be charming to look at and fun to eat—bite-sized radishes and miniature corn—but most are better left to grow up a bit before picking. Baby zucchini lacks flavor. Also, there is an economic rationale to harvesting vegetables at maturity. The first peas of spring can be so small there's no way to fill a pot with them, and April's tiny green garlic would have formed a head of fat cloves by August. Anyone concerned with yield will want to reckon the opportunity cost of picking early. (Of course sometimes the reckoning works the other way. A tender small cucumber not picked today will turn into a tough-skinned lug by next week.) The recipe chapters in the second half of the book will pick up the theme of harvesting for flavor.

Just as some vegetables announce their time has come—the ripe tomato—some also let you know the moment is about to pass. A head of broccoli, which is a tight cluster of flower buds, will open two or three yellow blossoms as warning sparks before the whole thing explodes in bloom.

SINGLE HARVEST

With some single-harvest vegetables, the part we pick is irreplaceable—a turnip's root, the tightly clustered flower bud that is cauliflower, the decapitated cabbage.

Corn is single-harvest for a different reason. Pollination in corn happens as a brief spasm. The entire crop matures at once, give or take a few days, with no cadet generation coming up behind. Winter squashes and pumpkins differ again in that their fruit ripens very slowly. By the time you pick orange pumpkins in September, the weary vine is too wracked to produce a second crop and the season is too late.

REPEAT HARVESTS

Many vegetables encourage you to come back again and again. Cucumber fruits are immature at prime eating size. The vine's response to picking is to furiously rebloom. Green peas and green beans are likewise stimulated by frequent picking. (If you want dried beans, let the plant run its course until the end of the season.) It is the same with other fruiting vegetables including tomatoes, peppers, eggplants, okra, and summer squash.

Salad greens and cooking greens can also be managed for repeat harvests. Lettuce, arugula, chard, kale, and collards leaf out from a central basal rosette. When taking the outside leaves, avoid damage to the central bud, and the plant will continue to produce until the end of the season or exhausted by its efforts.

AUGUST 4: Single-harvest Ruby Perfection cabbage

JULY 1: Selectively remove the largest outer leaves of Silverado chard and you can go back for repeat harvests through late fall.

AUGUST 24: Larger-than-ideal yellow crookneck squash are still good for soup.

Some vegetable species can go either way. Big-headed broccoli is one and done, but sprouting or blooming broccoli, which invests less heavily in a central head and thus has the wherewithal to regrow smaller side shoots, is cut and come again.

HARVEST INDICATORS

When a usually staid vegetable such as lettuce **BOLTS**, or goes to flower, its calm domestic training flies out the window. It jumps up, flings it arms in the air, and creates a florid ruckus. This is the frenzy of sexual reproduction, plant style, and it happens to leafy greens and some brassicas if they are not harvested in time. Spinach is first to go in my garden, as soon as temperatures rise; Asian greens also grow fast and die young in a spectacular bolt.

Vegetables will often give notice before they go berserk. Round-head lettuce turns pointy, and romaine will pull itself taller, visibly restless. (The photograph on page 104 shows radicchio harvested on the verge of bolting.) A day or two later, a flower stalk escapes its gripped leaves. Bolted lettuce becomes very bitter, but stemmy spinach is still good. Sprouting broccoli and broccoli rabe lose nothing in flavor and gain in looks when their blooms first winkle open.

WHOLE VEGETABLE, WHOLE LIFE CYCLE

Media savvy chefs have done much to popularize the concept of root-to-stem vegetable cooking, a spin-off of tail-to-snout butchery. What it means is you use the carrot's green top in addition to its orange root. There are practical limits—how much carrot-top pesto can anyone use?—but by championing the cause, chefs have drawn attention to food waste. The unconscionable fact is that one-quarter of all food produced in the world gets thrown out.

Traditional cooking is no less thrifty due to its allegiance to the garden, where a vegetable's value is clearest. The catchall stockpot condenses flavor from peelings and scraps. A chard-rib bone pile can be trimmed and baked into a rich gratin. In the end, a kitchen discards no vegetable wastes at all, because a compost pile absorbs them. Still, thrift extracts as much flavor as possible before surrendering the odd bits to compost. The chapters ahead will give specific recommendations for how to use all parts of each plant.

In a similar vein, consumers have been trained to expect vegetables harvested at a consistent, unchanging moment in the life cycle—the perfect 7-inch zucchini. Earlier in this section, I argued for letting "baby" vegetables mature to realize their full adult flavor. Here I'd like to suggest also finding ways to use past-prime vegetables. Taste what grows in the garden—always be tasting—and use with an open mind whatever has potential. Half-bolted bok choy or romaine can be grilled, sauced with umami-rich anchovy dressing, and carved at table, like a haunch of venison. Garlic's flower stalk, or scape, which is removed to increase bulb size, can be prepared like green beans, or else processed as pesto.

Bolted cilantro is not a lost cause, but a second chance. The flowers are delicious. (The same applies to chives, thyme, rosemary, borage, marjoram, etc.) More surprising, perhaps, are the tender green seeds that follow cilantro blossoms. They shine like lacquer beads and pop with remarkable flavor, half cilantro and half coriander, which is the spice-rack name for dried cilantro seeds.

Fennel is commercially grown for its thickened stem, like its cousin celery. But I always leave one or several to grow for the herbal fronds, a wonderful herb, and the fragrant stalks, which make an aromatic bed for roasted fish. Yellow fennel blossoms draw pollinators and shed fennel pollen, culinary fairy dust gathered by shaking the flower over a sheet of paper. Left until fall, the flowers produce fennel seeds. I gather a handful to get me through winter.

HARVEST TIPS

- Harvest frequently. At the peak of summer, vegetables will pass their prime in a flash. Zucchini are notorious on this count, and cucumbers hardly better. Green beans will not stay tender from one weekend to the next, and 5 days of hot weather will ruin radishes and arugula. Corn will come in and go entirely if you're at the beach for 2 weeks in August.
- Harvest thoroughly. Search the plant all over—under leaves, along each branch, viewed from different angles. You don't want to miss anything. If you see one glossy serrano chile lifted above the leaves on proud display, there will surely be others hidden from view.
- Harvest everything that's ready. "Ripeness is all," says Edgar in *King Lear.* If a vegetable is ready to pick, don't leave it there. Ripe quickly turns overripe, and for many vegetables picking increases yield. If you can't eat all the corn that's ready, you still need to bring it in. Freeze the surplus or give it to friends rather than let it stand on the stalk until tough. Ugly tomatoes, once trimmed up with a sharp knife, will taste the same as the beauties.
- Harvest early in the day, when vegetables are cool and turgid after a night of drawing up moisture. In droughty weather, water the day before. This matters most for leafy vegetables.
- Always take a cutting implement or two with you into the garden—scissors for greens, secateurs for squash, a short-bladed knife for broccoli—as well as a lightweight container, such as a wicker basket or cardboard tray. Wide-mouthed buckets are fine for potatoes and squash but too deep for squishable tomatoes.
- Harvest neatly. Move from one end of a row to the other. Mentally divide a bed into quadrants and work through one at a time as you select choice leaves or roots.
- Don't be rude to the mother plant. Cut stems and stalks instead of yanking or twisting. With greens, trim the leaf as close to the base as possible without nicking it. Cut summer squash, cucumbers, melons, and winter squash with 1 inch of stem. Snip off tomatoes or peppers with clippers to avoid snapping the brittle plants or accidentally damaging unripe fruit or flowers. Picking beans is hard on the plants; try to minimize disturbance. Pole beans and trellised English peas dangle in clear view, but snatching at them can break the vines.
- Use what you harvest as soon as possible. Peas and corn are perhaps the most perishable because their sugars convert to starch soon after picking. Plan to cook them (or freeze) right away. Most fruiting vegetables—eggplants, peppers, cucumbers—will hold for several days on the counter. Green beans, squash, and scallions will keep in the refrigerator for nearly a week, cabbage longer. That said, hardly any vegetable improves with keeping, other than the half-ripe tomato placed on a windowsill. Leafy vegetables always hold better in the garden, attached to their roots. Vegetables brought in for long-term storage, such as potatoes, carrots, turnips, and late green tomatoes are all in a different category. Consult the vegetable chapters ahead.
- Harvest with kitchen eyes. The fun of gathering in dinner is hard to quit—the enchantment of a garden. I'll come back inside with what I thought was a bunch of chard, only to realize I have an armful, enough for dinner three times over. My advice is to carry the same container to help learn visual quantities. I use a stainless steel bowl that holds just enough salad greens for dinner with friends. A larger bowl is my measure for bulky braising greens. It doubles as a wash tub for rinsing them outside with a hose.

- As you harvest for the kitchen, remember to collect waste, discards, and debris for the compost pile. Contribute spent pea vines and bare cauliflower stalks. Clip tangled or tired growth. If you turn up a rotten pepper or bloated cucumber, compost them. While you're at it, always be weeding. The compost pile lives on residues and scraps; it eats untidiness.

THE END

"A wise farmer knows when to stop," said Elizabeth Keen, co-owner of Indian Line Farm and a cornerstone vendor at the Great Barrington Farmers Market. To keep putting resources—not the least of which is a gardener's time—into vegetables that are no longer needed is a kind of waste. Her advice flashed in mind one day in late August when I felt afflicted by the sight of another 10 pounds of yellow squash from my overenthusiastic planting. I stopped picking right then. The vines stopped fruiting and rode out the summer as stately as idled parade floats. Their spreading leaves served as ground cover until frost, when I pulled the blackened vines to compost.

SAVING SEED

If you've ever carved a pumpkin, you know first to scrape out the seeds. Air-dried, they will keep for years. Likewise, a cucumber matured on the vine will fill its length with ivory seeds, just as bolted lettuce will launch parasols from its puffball seedheads, a dispersal tactic shared by others in the aster family, also known as the composites (*Asteraceae* or *Compositae*). Most other garden vegetables also produce seed that can be saved, which is to say gathered from the mother plant and replanted from one year to the next. Heirlooms, by definition, are grown from seed that is handed down through generations. Seed savers rely on so-called **OPEN POLLINATED (OP)** varieties that produce viable seed through the natural agency of wind, insects, or self-pollination rather than human intervention. For an OP variety to persist, it must **SOW TRUE**, or perpetuate the hoped-for traits into the next generation. Sometimes a robust strain of exceptional merit will be singled out and marketed as **IMPROVED**, such as Long Green Improved cucumber.

Simple as it sounds, seed saving has many complications. Seeds are, after all, the result of sexual reproduction, and genetic mischief can result. For example, some *Cucurbita* species (summer squash, winter squash, and pumpkins) will readily crossbreed, as I learned one year when I saved seed from cushaw, a green-striped crookneck winter squash. The next year, less than half of squash grown from the seed looked like cushaw. The rest, having been pollinated by other squash varieties, were rogues: round or torpedo-shaped and ranging in color from white to celadon to pumpkin-orange. To perpetuate a desirable squash takes skill. The mother plant must be widely separated from other varieties, or else a chosen flower is hand-pollinated and covered with a net or bag to prevent contamination. Corn's windborne genetic dust travels even farther than squash's insect-borne pollen. A variety grown for seed must be isolated by as much as a half-mile from other corn.

Beans, by contrast, are self-fertile, meaning the flowers have the necessary male and female parts to reproduce internally, so to speak. They are the gateway for an aspiring seed saver. Varieties can be separated by as little as 15 feet. When the vines begin to bear, designate a half-dozen pods from among multiple plants—to ensure genetic diversity—and leave them to mature. Harvest when the dry beans rattle within their faded husks. Shell them out and store in a cool, dark place.

HYBRIDS are the offspring of a controlled cross. They are often endowed with desirable traits, a consequence of so-called hybrid vigor. The second generation generally isn't viable or won't sow true, however. Hybrid seed can't be saved, so farmers have to buy fresh seed annually.

Incidentally, genetically modified crops (GMO) do not come from crossbreeding, as hybrids do. They are the result of laboratory alterations to the plant's genetic code, bits of DNA inserted to introduce traits such as disease resistance, drought tolerance, or, notoriously, resistance to the chemical herbicide Roundup.

The commercialization of hybrid and GMO seed under the ownership of a shrinking number of global corporations has created fierce backlash among small-farm activists, none more outspoken that Vandana Shiva. Advocates for OP crops argue that seeds are a common good, a shared cultural legacy representing vital crop biodiversity, and shouldn't be owned by any corporation. Open-pollinated seeds are akin to open-source technology, developed through collective creativity and available for universal use. Politics aside, there is no question about the value of saved seed. Shared among growers and preserved through generations, shared seeds were the bedrock of traditional agriculture and remain a reservoir of genetic diversity among the world's major crops.

Seed saving is a fascinating topic and beyond the scope of these pages. To learn more, visit websites for Seed Savers Exchange and Southern Exposure Seed Exchange.

CLEANUP

After the party comes the cleanup. The last chore of the growing season is to put away the garden. In cooler climates, frost runs down most crops even before a freeze lays them flat. Mild-climate gardens will become sluggish with shorter days. Collards and other hardy vegetables might overwinter, lingering but dormant under weak winter sun; they won't put on fresh growth until spring.

As with any kind of cleanup, closing out the garden goes quicker if you've tidied along the way. Throughout the season, put away each vegetable in its own time. Cut it to the ground when production stops and carry spent greenery to the compost pile—first zucchini and beans, then tomatoes and corn, then squash and pumpkins, and so on until nothing of summer remains. You don't need to rake out a garden, but you'd best dispose of clutter that shelters pest and disease spores. Some cool-season vegetables will hold steady until late fall—cabbage, Brussels sprouts—but at least pick up their shed leaves. Cover bare ground with a thick layer of mulch or sow a fall cover crop suited to your region. Spent container gardens can be emptied into the compost pile and the pots or boxes stored out of the weather.

NOVEMBER 5: Saving Turkey Craw seed

CH.
09

PLANNING THE NEXT GARDEN

AUGUST 25: Green Zebra and Sungold tomatoes

By the end of the growing season, I'm ready to be done. The summer runs down, and each month exhausts my zest for it. Then fall arrives, and, unexpectedly, the new season brings new enthusiasm and new cravings unmet: I'm ready to eat stew and bitter chicory salads garnished with freshly picked apples and walnuts. The change of weather—a coolish morning in August when you notice the light's slant or an October night when you want a blanket—releases hope. It is a cozy feeling to know the garden will overwinter throughout the house: potatoes in the cellar, garlic hanging in the back stairwell, onions and shallots in a cold extra bedroom, canned tomatoes in the pantry, corn and a great deal more in the freezer, and a cupboard of dried herbs, chiles, fennel seed, coriander, and dill. A few hardy plants survive the first snow, but mostly the garden is dormant now—thank goodness. The gardener is also ready to overwinter. What's ahead is a season of long-cooked meals and holidays, recordkeeping and research, books and seed catalogues. The end of one gardening year begins the next. As chef Erin French put it to me, "Menu planning starts in January."

Which is why this chapter on garden planning comes now and not earlier. A first garden in particular often happens haphazardly, an unplanned project. You get a wild hair and buy a pack of seeds. On a whim you pick up a tomato plant or a potted herb or a six-pack of mixed lettuces from a sidewalk display in front of the grocery store. You take them home and put them out hurriedly, with hope, as a trial. The upkeep is freewheeling. Worries are hazy and minor. And things work out. The tomato gives you tomatoes. The herbs grow like weeds. At some point during the summer you find excuses to go check on the plants. You look forward to picking something for dinner. You start to think how you'd do things differently the next time. You begin planning your next garden.

Here, then, are thoughts on how to prepare with foresight.

WHAT TO GROW NEXT YEAR

Success with gardening lies in specifics—the specifics of your taste, your climate, the narrow spot of earth where you garden. A variety of Romano bean that thrives for me might not do well for you. San Marzano tomatoes, bred under the hot Mediterranean sun, can be superb in California but might be less suited to your Minnesota garden than Amish Paste, a variety developed in Pennsylvania. The variability of growing conditions from one place to the next justifies the number of varietals in the world. There exist hundreds and probably thousands of unique beans. At least one of them will be ideal for your climate and your taste. In the opening chapter, I suggested keeping a garden diary to know what to replant. Now is the time to consult it. Even if you didn't take notes, think back on your successes and failures. Which vegetables proved to be carefree and prolific? Which varieties? Which ones tasted best in the kitchen? Which underperformed on any count? Next year's garden can and should include both sure bets and new gambles, balancing reliable producers with experimental trials.

Because a garden is an invitation to grow your knowledge and expand your palate, try adding a new variety or an unfamiliar vegetable each year. Shop multiple seed sources for wider inspiration. Hit up gardening friends who might be willing to share saved seed—their favorites could become yours. You'll discover your preferences by growing out the options.

And never let published advice discourage your own homegrown experimentation. "Don't trust the seed pack," Jen Salinetti told me as we talked on her porch during a summer downpour. A gardeners' joke attributed to Eliot Coleman holds that plants don't read books, meaning that all published advice is subject to real-world revision because plants don't know what has been written about them. There is precedent in the garden, but little is ironclad. A garden is a perpetual experiment. If you take a notion about something, try.

SEED CATALOGUES

A cook-gardener interested in the varietal approach to growing and cooking tasty vegetables has no greater resource than seed catalogues. Individually and collectively, they contain worlds of flavor. Each has its own character and slant. The following list includes a few favorites.

Baker Creek Heirloom Seeds, based in Missouri, specializes in rare and forgotten vegetables, including nineteenth-century survivors, global varietals, and oddities. Its annual printed *Whole Seed Catalog* runs to over five hundred pages. *Best for the tireless treasure seeker, the ambitious completist.*

Botanical Interests offers a tight selection of kitchen-garden favorites. The illustrated seed packets are printed inside and out with detailed growing instructions and erudite varietal information. *For the varietal-curious in search of interesting options, but not too many.*

Fedco Seeds and Supplies, despite its agri-business-sounding name, is a Maine-based cooperative specializing in crops and varietals selected for northern-tier growers. The company's back-to-the-land ethos is rigorous but friendly. The black-and-white newsprint catalogue and no-nonsense website are filled with practical information and historical references. *For serious growers, especially in the north, looking for the real deal.*

Filaree Farm in Washington State specializes in garlic (one hundred–plus varieties) along with other alliums, potatoes, sweet potatoes, and asparagus. *For garlic enthusiasts.*

Johnny's Selected Seeds, based in Maine, is a catalogue-based superstore for hardcore organic gardeners and small-to-medium-sized professional farmers. The employee-owned company, founded in 1973, rigorously trials its seeds (both proven heirlooms and improved hybrids) and equipment, including no-till tools. *For professional-minded growers.*

Maine Potato Lady sends out a simple catalogue, as plain as a church circular, rich with culinary treasures. The website adds photographs of every variety—and there are dozens. Indispensable. *For potato lovers.*

Row 7, cofounded by influential chef Dan Barber, helps develop and bring to market delicious new varieties. The offerings are few and expensive but extraordinary—future heirlooms. Honeypatch winter squash from plant breeder Michael Mazourek is a masterpiece. *For the flavor hound.*

Seed Saver's Exchange has a high-minded mission: to steward America's food-crop legacy. Founded in 1975, SSE deserves a chapter in the great annals of our national agricultural history, having rescued or reintroduced essential heirlooms, including Jimmy Nardello peppers and Brandywine tomatoes. The user-friendly catalogue offers superb selection without overwhelming. *For the heirloom enthusiast and history buff.*

Southern Exposure Seed Exchange, founded in 1982 in Charlottesville, Virginia, explores the rich biodiversity of the farming South. Its introductions include Cherokee Purple and Charlie's Mortgage Lifter tomatoes, and, under the direction of Ira Wallace, a collection of rare collards of distinguished culinary quality. *For a deep dive into regional flavor.*

Sow True Seed is another Southern seed specialist, more boutique in scale, of interest to gardeners around the country because of the exceptional flavors found in its tightly curated catalogue. Some delicious varieties, such as Seven Top turnip greens and Country Gentleman corn, were once widely esteemed but are today rarities. *For a glimpse of the tripartite soul of our national cuisine—Native American, African, and Anglo-European crops.*

Territorial Seed Company in Oregon offers seeds for vegetables, herbs, flowers, and cover crops. Its selection covers heirlooms, newer OP varieties, and hybrids, as well as basic equipment such as seed-starting materials. The catalogue is exceptionally informative and user-friendly. *For an all-in-one garden catalogue with a West Coast perspective.*

PLANNING FOR YIELDS

Deciding how much of something to grow requires two bits of information. The first is entirely subjective. How much do you like, say, English peas? The second is estimated yield, which is calculated by seed companies, government agencies, and other resources. (The chapters ahead will provide guidance.) In the case of peas, a 10-foot row should yield 3 to 5 pounds of pods, which will shell out to a scant 1 cup per pound. Not a lot. If you love peas, plant double.

So how much seed is required for a 10-foot row? No rule of thumb covers all vegetables because seeds vary widely in size, shape, and spacing requirements. For peas, it would take a half-ounce of roly-poly seeds (about 60 total) to fill the 10-foot row. As a point of comparison, a half-ounce of zucchini (about 100 seeds) properly spaced would stretch nearly 70 feet—enough to feed a neighborhood.

Seed companies size their seed packets to meet the typical needs of a home gardener. It usually works out. A bulging packet of pea seeds will come with the right amount to fill a 10-foot row, whereas a slim packet of zucchini seeds will sow several hills with extra to resow in case of cutworms. If you love a vegetable and worry one seed packet won't be enough, make the second pack a different variety. Perhaps choose one early variety and one late. Or find two colors—green snow peas and purple snow peas.

As mentioned elsewhere, surplus seed will keep for later. Stored in dark, cool conditions, most seeds remain viable for several years or longer, except for perishable onion seed.

SEED VIABILITY

SEED TYPE	LONGEVITY UNDER PROPER SEED STORAGE CONDITIONS
Arugula	3 years
Beans	3 years
Beets	4 years
Broccoli	3 years
Brussels Sprouts	4 years
Cabbage	4 years
Carrots	3 years
Cauliflower	4 years
Celery/Celeriac	5 years
Chard	4 years
Collards	5 years
Corn	2 years
Cress	5 years
Cucumbers	5 years
Eggplant	4 years
Endive/Escarole	5 years
Fennel	4 years
Kale	4 years
Kohlrabi	4 years
Leeks	1 year
Lettuce	5 years
Melons	5 years
Mustard Greens	4 years
Okra	2 years
Onions	1 year
Peas	3 years
Peppers	2 years
Pumpkins	4 years
Radish	5 years
Rutabagas	4 years
Spinach	2–3 years
Summer Squash	4 years
Tomatoes	4 years
Turnips	5 years
Watermelon	4 years
Winter Squash	4 years

GARDEN LAYOUT

A garden is a puzzle composed of rows, beds, hills, trellises, and walkways. A square or rectangular layout enjoys obvious advantages, although a creative mind could devise an expressive and free-flowing garden design that still respects the fundamentals of spacing and row spacing. A few thoughts to consider.

Before planting the first seed, it is very helpful to commit a garden design to paper, regardless of your artistic ability. I have no talent for drawing, but at least I can use graph paper to outline boxes for garden beds and draw lines for pathways. Make your drawing approximately to scale for it to be useful outdoors in the real world. Remember to draw pathways at realistic scale as well.

Accept recommended spacing guidelines as givens and don't try to overcrowd a raised bed or small garden. When you put out young seedlings and transplants, they will look absurdly small at first, as if lost in a lonely wilderness. Trust that summer squash really will grow to fill the recommended spacing of 3 feet apart. Design for mature plants.

Be attentive to the orientation of sun and shadows. Trellises, pole-bean teepees, and tall crops such as corn can shade out shorter crops. Place them on the north edge of the garden. In general, grow smaller vegetables in the foreground of taller plants to maximize sun exposure.

That said, some vegetables tolerate light shade and cool-weather crops such as lettuce can benefit from dappled shadow during the heat of summer. Refer to chapter 2 for more information.

SUCCESSION

Succession planting extends the harvest window and maximizes yield within a limited space. There are multiple ways to apply the concept. For example, you could sow a packet of peas in staggered phases—putting out half in March for a

May harvest and the rest in April for a June harvest. In the same spirit, Thomas Jefferson advised in his one horticultural work, a "General Gardening Calendar" published in the May 21, 1824, issue of *American Farmer* magazine, that gardeners sow a thimbleful of lettuce seed every Monday from February 1 to September 1. Another type of succession is to replace an early crop immediately after harvest with something else, for example following peas with pole beans on the same trellis.

A related concept, especially useful for a small garden, is **INTERPLANTING** or intercropping. It means to grow something small and fleeting between larger, slower crops. Fill the gaps between young broccoli with bok choy and it will be ready to harvest long before broccoli leaves touch. Turnips can go into the bare spots between kale. My approach to interplanting tends to be spontaneous and opportunistic. When I find a gap somewhere, I'll stick in a fast-growing something I need in the kitchen, such as cilantro or chervil. Interplanting tomatoes and basil makes sense because I use them together, and I'll stick a parsley plant or two among lettuces as a reminder to take some in—you always need parsley. A nurse crop is a quick-to-sprout species interplanted with a slow grower. Radishes, for instance, pop up seemingly overnight and provide cover until dilatory carrots finally emerge.

A fun part of garden layout is maximizing yield through clever succession and interplanting. Once you designate a block for radishes, for instance, keep in mind they will be gone by early summer. The game is to figure out what should follow. I pencil in approximate planting dates for each succession, such as "4/15: Cherry Belle radish, 5/21: Fordhook Giant chard."

Likewise, a garden plan helps visualize the opportunities for interplanting. When I see the sheer square footage required for Brussels sprouts or cabbage, it reminds me to interplant with lettuce, cilantro, bok choy, or another jump-up crop.

Save garden layouts from year to year to compare and to ensure proper rotation.

ROTATION PLANS

Crop rotation was discussed in chapter 7 as a means of pest control and disease suppression. Moving vegetables in an annual sequence also improves soil quality and prevents soil exhaustion. Some crops feed more heavily than others. You don't put corn in the same spot two years running because it is a nitrogen hog. Instead, greedy corn follows generous beans, which add nitrogen to the soil. Even industrial agriculture rotates corn and soybeans on a vast scale. A plant's physical structure further affects how it mines soil nutrients. Lettuce's network of hairlike roots spread shallowly, while a pumpkin's strong, ropey roots probe the depths for minerals and water. As mentioned earlier, crops are rotated by family rather than individual species.

A rectangular garden divided into quadrants simplifies rotational practices. There are many, many plans out there. One comes as a jingle: roots, fruits, beans, greens. In other words, root vegetables (including potatoes and alliums) followed by fruiting vegetables (tomatoes, squash) followed by the nitrogen-fixing legumes followed by green vegetables, including brassicas.

YEAR 1

Roots	Fruits
Greens	Beans

YEAR 2

Greens	Roots
Beans	Fruits

Another option is to sort crop families based on nutrient requirements: soil builders, heavy feeders, and light feeders, which I also think of as efficient scavengers. See the box Nutrient Use by Family (page 94).

Rotation is a puzzle with no one perfect solution. I grow lots of potatoes, so an entire quadrant of my garden goes to the nightshade kin. That was my starting point for a rotation plan, and I fitted in the other families based on what else I like to grow. See My Basic Rotation, at right. The plan isn't textbook-perfect. Cucurbits, corn, and okra share a quadrant because they are on the same planting schedule and all need space. Heavy-feeding brassicas and the scavenging root vegetables split a quadrant in my plan because I don't need to grow more of either.

If all this starts to sound complicated and confusing, go back to the fundamental principal, which is simple and easy to follow. Don't grow any one vegetable in the same place year after year. The rest will sort itself out with experience.

MY BASIC ROTATION (moves clockwise)

Nightshades: potatoes, eggplants, peppers, tomatoes	Cucumbers, squashes, corn, okra
Brassicas, hearty greens, roots, onions	Legumes, cover crop

NUTRIENT USE BY FAMILY

Large, long-season vegetables take more from the soil than small, fast-growing vegetables. Among soil builders, nitrogen-fixing legumes include both edible crops and green manures. Rotate annually between the categories below.

HEAVY FEEDERS	LIGHT FEEDERS	SOIL BUILDERS
Brassicas (broccoli-type, cabbage, collards, kale)	Alliums	Legume crops (beans and peas)
Celery	Chard	Legume cover crops (vetch, clover, alfalfa, field peas, fava beans)
Corn	Herbs	Other cover crops or green manures
Cucurbits (cucumber, squash)	Lettuce	
Nightshades (tomatoes, eggplants, less so peppers)	Potatoes	
Okra	Root crops	

AUGUST 21: Turkey Craw pole beans climbing Country Gentleman corn in a Three Sisters planting

CH.

10

GLOSSARY

JUNE 28: Fast-growing Asian cabbage interplanted with slower Brussels sprouts

ANNUAL A fast-growing plant that completes its life cycle in a single growing season. Beans, peas, corn, and many other crops are annuals and must be resown from year to year.

AVERAGE FIRST FROST, AVERAGE LAST FROST See **FROST DATES**.

BAND A thickly sown strip; a wide rather than single-file row; practical for small crops such as baby greens, radishes, and pearl onions.

BED A planting plot within the garden, often a permanent feature in no-till gardens. Seeds within a bed can be planted in rows or on a grid. If **BROADCAST** over the bed, the result is a thick stand of vegetables that can be picked as needed, such as a Southerner's greens patch of mixed mustard, collards, and turnips. See also **RAISED BED**.

BIENNIAL A plant that lives 2 years. The first goes into vegetative growth, the second to flowering. Some biennial vegetables hold well at maturity, such as heading cabbage. Others including root vegetables and alliums provision themselves with an underground storehouse of energy—the part we eat.

BOLT To flower and go to seed; a late stage in the life cycle of some vegetables. Bolted lettuce turns bitter, but other bolted greens, including spinach, remain edible.

BROADCAST To sow seed by scattering. See also **BED**.

CLAY The smallest **SOIL PARTICLE**.

COLD FRAME A glass-topped wooden box covering an outdoor **SEEDBED** in the manner of a mini-greenhouse.

COLD HARDY Describes a plant that is resistant to low temperatures. The degree of cold hardiness varies by species and even variety.

COMPANION PLANTING The practice of growing vegetables, herbs, and flowers together for the benefits they mutually confer. Marigolds are said to repel pests, for example. The knowledge of companion planting lives in folk wisdom and ancestral lore.

COMPOST Decomposed organic material used in the garden as a soil amendment, fertilizer, mulch, seedbed, and cure-all; the organic gardener's irreplaceable asset.

COOL-SEASON CROPS Vegetables that thrive in cool temperatures, including peas, carrots, and broccoli. They can be sown early for a spring garden or in midsummer for fall harvest.

COTYLEDON The first leaf-like structure to emerge from a seed. It contains a storehouse of nutrition to establish the infant plant until its first true leaves appear. Dicots such as beans produce paired cotyledons. The monocots, including corn, send up a single cotyledon that emerges as a grasslike shoot.

COVER CROP Thick vegetation grown to protect fallow ground and enrich it when turned under or chopped down for mulch. Plants in the legume family (clover, vetch, cowpeas) also fix nitrogen as they grow. The soil-building value of cover crops is such that they are sometimes called **GREEN MANURES**.

CROP ROTATION The practice of planting vegetables in a different location from one year to the next in order to break the pest cycle, reduce the spread of plant disease, and prevent soil exhaustion.

CULTIVAR A named cultivated variety.

DAYS TO HARVEST, DAYS TO MATURITY How long after emergence until a vegetable matures; often placed in parenthesis, such as Sungold (57 days).

DIRECT SOW To place seeds in the ground where they will grow to harvest. An alternative is to start seeds indoors or in a nursery bed before transplanting seedlings to their permanent location.

DRAINAGE How quickly water moves through the soil. Coarse sandy soil exhibits fast or sharp drainage. Heavy clay soil drains slowly. See also **STRUCTURED SOIL**.

FALL GARDEN Refers to cool-season crops sown in mid- to late summer for a fall harvest; also called a late garden.

FERTILITY The biological potential of soil. Fertility can be described by its physical, chemical, and biological conditions. Fertile soil is rich in organic matter (physical). It supplies plants the essential nutrients nitrogen, phosphorous, and potassium (chemical). And it teems with subterranean life and microbial activity (biological).

FLOATING ROW COVER A lightweight, translucent fabric that forms a physical barrier against pests and frost. Because it allows water and sunlight to pass through, the fabric can be left in place to "float" over the plants as they grow.

FROST DATES The average dates for last frost in spring and first frost in fall. The bookends of the growing season. Frost dates are determined by the National Oceanographic and Atmospheric Administration using historical weather data. Sowing and transplanting are set by frost dates. For example, you might sow corn 1 to 2 weeks after the average last frost date in your area.

FRUIT The reproductive structure of seed-bearing plants. Many vegetables are fruits, botanically speaking. Zucchini and cucumbers are immature fruits. Red chiles and winter squash are ripe ones. Tomatoes and peppers are a specific type of fruit, the berry.

FULL SHADE Less than four hours of sun a day; insufficient for vegetables.

FULL SUN Eight hours a day of direct, uninterrupted sunlight.

GREEN MANURE. See **COVER CROP**.

GROWING CONDITIONS The set of environmental factors, such as temperature, sunlight, water, and soil pH, that favor a plant's growth.

GROWING SEASON That part of the year suited to a plant's flourishing; more or less the gardening season. In many regions, it is bookended by **FROST DATES**.

HARDENED OFF Seedlings started indoors must be acclimated gradually to variable outdoor conditions. A week before transplanting, they are placed outside for 1 hour. The next day, outdoor time increases to 2 hours, and so on until the seedlings are hardy enough to survive the garden full time.

HEIRLOOM A time-honored vegetable variety passed down through generations to preserve a noteworthy trait such as deliciousness, early ripening, or drought tolerance.

HILL Widely spaced vegetables such as cucumbers and squash are often said to be planted in hills, whether or not the soil is mounded.

HUMUS The final stage of organic matter in the soil after it has passed through the bodies of creatures and microorganisms. Humus represents a transient phase between decomposition and new growth; a pause between death and regeneration; the wealth of soil and nations.

HYBRID A novel variety created through a controlled cross of dissimilar parents, sometimes even different species, as crossbreeding a horse and a donkey produces a mule. Hybrids can synthesize the best of each parent while resembling neither, a sort of multiplicative effect known as hybrid vigor. Hybrid seed cannot be saved: It will be sterile or unstable in the second generation. Instead, it must be bought annually from the breeder or seed company, whose legal ownership can be patent-protected.

IMPROVED A new strain of vegetable selected from an existing variety. For example, by saving and propagating seeds from the most disease-resistant bean plants, the grower could market disease-resistance as an improvement.

INDICATOR SPECIES Showy plants and animals whose life cycles mark the seasons. For example, blooming forsythia and frog song announce spring, while southbound geese and blazing maples herald fall. Indicators can guide garden work. In New England, old timers say to plant potatoes when dandelions bloom.

INTERPLANTING A strategy of sowing quick-to-mature crops in the gaps between slower ones. In the picture on pages 96 and 97, bok choy (50 days to maturity) is ready to harvest by the time Brussels sprouts (100-plus days) start filling in.

LEAF MOLD Partially rotted tree leaves; very desirable for compost and mulch.

LOAM An ancient word now used to describe a soil type that exhibits the three **SOIL PARTICLES** plus ample organic matter in favorable balance.

MATURE SPACING The recommended distance between vegetables at maturity, which is typically greater than the spacing between seeds at sowing. See also **SPACING**.

NITROGEN An essential nutrient for plant health, responsible for new green growth. Natural sources of nitrogen include organic matter in the soil and the action of **NITROGEN-FIXING** bacteria that live on the roots of legumes (beans, peas, clover). The bacteria mineralize nitrogen from the atmosphere, or convert it to a form that plants can use, "fixing" it in the soil.

NO-TILL A set of gardening practices that avoids turning the soil or disturbing soil layers. Its hallmarks include aerating the soil with a garden fork, layering compost and mulch on the soil surface, covering weeds to smother (occultation), and cover cropping.

NPK The shorthand notation for nitrogen, phosphorus, and potassium, the three major plant nutrients. The **NPK RATIO** is the percentage by weight of each nutrient in fertilizer. A ten-pound bag of 10–10–10, for instance, contains one pound each of nitrogen, phosphorus, and potassium. Inert ingredients make up the balance.

OPEN POLLINATED A crop variety grown from seed produced through natural agency, such as wind pollination. See also **HYBRID**.

PARTIAL SUN Four to six hours per day of direct sun exposure.

PERENNIAL A plant with a multiyear life cycle. A perennial will likely delay reproduction for a year or several years while it invests energy in its permanent structures, such as roots and stems. Flower gardeners love perennials—the peonies, irises, and delphiniums you plant once and enjoy for years. The vegetable garden has relatively few, asparagus, artichokes, and sunchokes among them. Permaculturalists prefer perennials.

PHOSPHOROUS An essential plant nutrient for healthy growth. In the NPK ratio, phosphorus gets P, so the mnemonic goes, because it has two of them.

PLANTING DEPTH How deeply to cover seeds with soil when sowing. Large seeds such as corn and peas might be pushed into the soil by an inch or more, while lettuce and cress seeds are barely covered.

POTASSIUM The third essential plant nutrient, with the symbol K. Readily abundant in most soils, it contributes to overall vigor.

PRICKING OUT To transfer days-old seedlings from the tray in which they germinated to the container in which they will grow; an eye-straining task.

RAISED BED A bottomless planter or four-sided box filled with soil; a quick way to establish a small garden. Because the sides can be made any height, typically between 8 inches and 3 feet, raised beds are more accessible than in-ground plantings for gardeners with limited mobility.

REGENERATIVE AGRICULTURE As used conscientiously by well-intended practitioners, an ecological approach to food production that aims to improve environmental conditions and repair ecosystem damage caused by chemical-based farming practices. Its goals are to rebuild topsoil, sequester carbon, bolster climate resilience, improve watershed quality, and foster biodiversity. In the home garden, regenerative practices begin with the organic baseline (composting, crop rotation, green manures) and extends to no-till, mulching, pollinator plantings, and so on.

The term is controversial among organic advocates and food activists because it is squishy. The word *organic,* by contrast, enjoys legal protection under USDA organic certification guidelines. *Regenerative* is undefined in law, and its meaning is subject to abuse and deception in corporate marketing and advertising lingo, as with the similarly debased *artisanal.*

ROW A single-file planting. See also **BAND**.

ROW SPACING The distance between rows. See also **SPACING**.

SAND The largest **SOIL PARTICLE**.

SEED A structure for sexual reproduction in plants; the fertilized ovule. A seed contains within its protective coat the embryonic plant and a storehouse of energy. More mysteriously, a seed contains the germ of life transmitted across time.

SEED SPACING. See **SPACING**.

SEED STARTER MIX A fine-grained, moisture-retaining growth medium used to germinate seeds in containers or trays.

SEEDBED Groomed soil ready for sowing. See also **BED**.

SIDE DRESSING Compost, fertilizer, or other amendment scratched into the soil around a plant's roots, often applied at midseason. Synonymous with top dressing.

SILT The second-largest **SOIL PARTICLE**.

SOIL The uppermost layer of the earth; the biologically active zone conducive to plant growth. Soil accumulates over vast expanses of time, eroded from the underlying bedrock and enriched with decayed organic matter. Plants root into it for support, nutrients, and moisture. Soil ruined by overuse or pollution is dirt.

SOIL PARTICLES The mineral components of soil, sorted by size into clay (the smallest), silt, and sand. See also **DRAINAGE**.

SOIL TEST The analysis done by a professional laboratory or using an at-home kit to discern such chemical and physical conditions as NPK, pH, and organic content.

SOW TRUE Used to describe seeds that produce offspring resembling the parent; equally applies to a variety that perpetuates desirable traits from one generation to the next.

SPACING The distance between seeds in a row or bed; also called **SEED SPACING**. By contrast, **MATURE SPACING** is the recommended distance between fully grown plants. Seeds are typically sown more closely than mature spacing (to account for losses) and **THINNED** as they grow in. **ROW SPACING** is the distance between rows and is greater than the spacing within a row to allow plants room to spread and gardeners room to pass through.

SPRING GARDEN Sometimes called an "early garden," it comprises cool-season crops sown early, perhaps as soon as the ground thaws, and harvested before the onset of hot weather. See also **FALL GARDEN**.

START SEEDS/START INDOORS To germinate seeds in a container or tray before transplanting the seedlings to a permanent growing location.

STARTS Synonymous with **TRANSPLANTS**.

STRUCTURE, as in **STRUCTURED SOIL** and soil structure. A term to describe how soil particles fit together and hold together; i.e., the three-dimensional physical characteristics of the soil matrix, including its porosity, tensile strength, and friability. Well-structured soil is porous, coherent (you can pick up a clump), and friable (the clump crumbles easily). It drains well thanks to its open pores (somewhat analogous to the airy crumb of sourdough bread). Paradoxically, well-structured soil also absorbs and holds water well, like a sponge, buffering against drought.

SUBSOIL The mineral-rich layer beneath the **TOPSOIL**; formed from underlying bedrock.

SUCCESSION PLANTING Any of several sowing strategies to extend the harvest cycle and maximize yield. For example, a single crop might be planted in multiple stages or successions: one-quarter row of lettuce this week, one-quarter the next week, and so on. Or an early crop might be followed by a late crop on the same spot: spring lettuce then summer zucchini. **INTERPLANTING** could also be considered a type of succession—two crops sown simultaneously, harvested in succession. Finally, you could sow two varieties of the same crop with different maturity dates: early Latte corn (68 days) and late Silver Queen (92 days).

THERMAL DORMANCY A tendency of lettuce and some other seeds to resist germinating when soil temperatures rise above 75°F.

THIN Due to inevitable losses, seeds are sown more thickly than **MATURE SPACING** requires, then selectively removed as they grow. Many **THINNINGS**, such as greens, root vegetables, and herbs, make for delicate eating.

JULY 9: A vegetable can be any part of the plant. Zucchini is an immature fruit; broccoli's head is clustered flower buds.

TILL To turn the soil. **TILLING** is an ancient set of farming practices and the basis of human civilization. Its purpose is to open the earth with a shovel, plow, or other instrument and regularly invert the soil layers in order to distribute fertility, control weeds, and incorporate organic matter and other amendments. See also **NO-TILL**.

TILTH The physical condition of soil; its structure.

TOPSOIL The fertile, biologically active uppermost layer of the ground; formed over time through the breakdown of organic matter; above the **SUBSOIL**.

TRANSPLANTS Seedlings sprouted in containers and moved to a permanent growing spot after established enough to survive the disturbance. Cucumbers and squash are susceptible to **TRANSPLANT SHOCK**.

USDA PLANT HARDINESS ZONE A map compiled by the Department of Agriculture that sorts the country into climatic zones based on average winter minimum temperatures; a key reference for gardeners.

VARIETAL A cultivated variety; the named selection of a plant species, typically chosen for a distinctive trait, such as color, size, taste, drought tolerance, or disease resistance.

VEGETABLE The part you eat. There is no botanical definition. A vegetable could be a shoot, leaf, stem, root, flower, or **FRUIT**.

VOLUNTEER A self-sown vegetable or flower.

WARM-SEASON CROPS Cold-intolerant vegetables sown or planted outdoors only after the ground has warmed and all threat of frost has passed, often 1 to 2 weeks after **AVERAGE LAST FROST**. Many were domesticated in the tropics, including the high-summer suite of tomatoes, summer squash, eggplant, and peppers.

WEED A plant growing where you hadn't planned for it; a self-sown wild edible; a vigorous deep-rooted grass or forb that pulls up nutrients from the **SUBSOIL** and contributes them to the **COMPOST** pile; the cause of **WEEDING**.

WEEDING To eliminate vegetative competition around favored plants, such as vegetables and flowers; after sowing and harvesting, the defining task of a gardener's life; a chore not to put off.

JULY 10: Indigo radicchio on the verge of bolting

PART

02 VEGETABLE KNOWLEDGE

When the vegetables were ready to be picked it never occurred to us to question what way to cook them. Naturally the simplest, just to steam or boil them and serve them with the excellent country butter or cream that we had from a farmer almost within calling distance. Later still, when we had guests and the vegetables had lost the aura of a new-born miracle, sauces added variety.

—ALICE B. TOKLAS
The Alice B. Toklas Cookbook

CH. 11

COOKING FROM A GARDEN

AUGUST 31: Impromptu noodles with farmers' market mushrooms and green beans, opal basil, cilantro, and green chiles from the garden

Homegrown food simplifies kitchen life. Masanobu Fukuoka writes in *The One-Straw Revolution,* "If you do not *try* to make food delicious, you will find that nature has made it so." I heard similar opinions in my West grandmother's kitchen and at Chez Panisse, a training ground for two generations of chefs. The flavors are in the garden.

Taken together, the previous chapters were a gardening primer: how to grow what you want to eat. The chapters ahead, the book's larger half, will explain how to cook what you grow.

My way of making dinner is intuitive, homey, unfussy, and efficient, but not impatient. It is plant-based, not vegetarian. I season with unsalted butter, cream, pan drippings, rendered bacon fat, fish sauce, anchovies, and other animal products—but most recipes can be adapted to dietary preferences. I grew nearly all the vegetables and herbs used to develop recipes and shoot photographs for this book, but what of it? You don't have to. There are other ways to cook from a garden. (Refer back to the Preface, page xi.) I invariably stop at farmers' markets and roadside honor-system farm stands. The recipes ahead will work with good ingredients from whatever source. That said, the simpler the recipe, the likelier it is to miss if made with shabby produce. There's no way around the quality difference—and therefore the taste—between freshly picked Green Magic broccoli from an organic garden and cellophane-wrapped heads from a nitrogen-addled field three thousand miles away.

Vegetables aren't sides on my table. They are the main event, and I plan the meal around what's growing. Depending on the number of people at table, I'll put out two or three or five varied but compatible dishes, mixing salads and cooked vegetables for balance. To me it's a very Southern way of eating, akin to the farm-based meals I grew up with. My two grandmothers never saw kombu, shiitake powder, Aleppo pepper, picholine olives, za'atar, colatura di alici, and any number of other flavor-boosters I've gathered from the global pantry, but they would have instantly recognized the tie between my garden and my kitchen.

Cooking from a garden varies throughout the year. An ideal spring meal at my house might come together with a salad of tiny greens and chervil, scallions in cream with chive blossoms, and a ragout of buttered peas with asparagus tips and tiny radishes—maybe spooned over pan-roasted monkfish. Summer: sliced cucumbers and onions, sautéed zucchini with new garlic, cream-style corn, new potatoes tossed with pesto, sliced tomatoes and a grilled hanger steak. Fall: little chicories with sliced apples, baked winter squash filled with cream, braised bitter greens, slices of roast pork with potatoes slow-cooked in the drippings. Winter: buttered Brussels sprouts, baked home-canned tomatoes with dried savory, mashed potatoes, and chicken roasted with shallots and garlic.

Protein is the side on my table and I like best those that improve the next day—pork in chile verde, braised chuck roast, a pot of beans. Lunch is leftover vegetables with an herb omelet or a fried egg.

These days I don't have the time or inclination to spend forever in the kitchen. A generous hour should be enough to make dinner. Each dish must come together quickly—hence my bent toward simple techniques elevated with pinches and dashes, herbs and edible flowers, chiles and sneaky "secret ingredients," such as garlic grated on a Microplane or a dab of tomato paste or a pour of sauerkraut juice. All in all, it's traditional home cooking—what Diana Kennedy and others copying her called "nothing fancy"—but I've tried to bring flavors up to date. I like to think of spontaneous meals from the garden as Tuesday Night Cooking, and many of my recipes fall into this category. Other recipes ahead take a bit more planning and time—Saturday night dinner. And, finally, I include a few gala dishes, like a showstopper roasted whole pumpkin stuffed with chestnut dressing.

A shopping list is not where I start making dinner. I'll go out to the garden to check what's ready. Or on Saturday mornings I'll go to the Great Barrington Farmers Market to see what Elizabeth Keen and her crew have brought. When visiting family in Tennessee, I go religiously to the Maryville Farmers' Market, and during regular trips to Los Angeles I return to the Wednesday Santa Monica Farmers Market, my home market for years. It's inspiring to learn from passionate growers about their favorite varieties. There is education in the gardens of others.

Some of the finer culinary flourishes in the book were suggested by my kitchen collaborator Dashiell Nathanson, a friend who flew in from California to help develop recipes, style photo shoots, and explore the garden's potential. Dash sharpened the California influences I picked up during two formative periods in my life. The first was when I lived in Berkeley during a gap year from college and hung around Chez Panisse; the next was a long stint in Los Angeles, from 2004 to 2016, years when the future of American cuisine was taking shape in Southern California. The wild mash-up of culinary styles that defined the city—chaos cooking—arose from the freestyle creativity of chefs who were equally fluent in European gastronomy and multiple cuisines from Latin America and Asia. I think of the style as *Blade Runner* cooking, a vision of the flat-world future beamed into the here and now. In Los Angeles, all that creativity is anchored by amazing California produce. It makes sense that Dash and I convened—creatively and literally—in the garden.

As for myself, I'm a capable home cook, not a chef. The recipes ahead are within anyone's reach. Some might seem, at first glance, almost too simple. Do you really need a recipe for cut corn barely heated through with butter? Maybe not, but every time I make it for friends, they go crazy. What is the secret, they ask? I'd like for you to know. I'll include a handful of such foundational techniques, culinary basics, for my

JANUARY 13: Baked stuffed Robin's Koginut

most-used homegrown vegetables. In doing so, I mean to remind you how satisfying and delicious simple food can be. Plain cookery is a joy of eating at home. *Brutto ma buono,* say the Italians. Ugly but good. When inspiration strikes, take my simplest recipes as prompts to personalize and make your own.

As a cook-gardener, I've become happier and more relaxed in the kitchen. There's less for me to do now that I understand my job to be a modest one. I am nature's helper, and nature is a genius.

CH. 12

BEANS AND PEAS

LEGUMES

Fabaceae* or *Legumaceae
THE PEA FAMILY

Cicer arietinum
GARBANZO BEAN

Glycine max
EDAMAME AND SOYBEAN

Phaseolus coccineus
SCARLET RUNNER BEAN

P. lunatus
LIMA BEAN

P. vulgaris
THE COMMON BEAN

Pisum sativum
THE GREEN OR ENGLISH PEA

P. sativum var. *macrocarpum*
SUGAR PEAS, INCLUDING SNOW PEAS

Vicia faba
FAVA BEAN

Vigna unguiculata subsp. *sesquipedalis*
ASIAN LONG BEANS

V. unguiculata subsp. *unguiculata*
COWPEAS OR FIELD PEAS

AUGUST 24: Emerite pole bean on a birch-sapling tripod

Productive, tasty, beautifully vined, easy to pick, and quickly prepared in the kitchen—the legumes are a prize of the backyard vegetable garden. The number of bean varieties approaches the uncountable. Some are spectacularly tasty: Turkey Craw, Rattlesnake, Lazy Wife Greasy, Northeaster. Others are plain spectacular. Scarlet runner beans climb 8 feet, froth with red flowers, and produce seeds as lovely as enameled jewelry. Green peas show less range but no less vigor: Champion of England can climb 10 feet.

The legumes' harvest season spans the growing year. Peas can be sown in cold soil for spring picking, a triumph of anticipation. Heat-loving beans are planted after all danger of frost has passed, and their maturity marks the onset of peak summer harvest. Pole beans bear until frost, and a late crop of peas, if timed right, gives you a fall picking. Dried beans provide a year-round protein-rich staple, but they do require some space, a bean patch. (Thoreau put out two acres at Walden.) The bonus crop is shellies, which are full-sized but still tender shelling beans, a special in-between stage that cooks quickly and has wonderful flavor. Field peas and chickpeas can also be picked at the in-between stage.

The legumes' superpower is to fertilize the soil. *Rhizobia* bacteria live in nodules on their roots and mineralize atmospheric nitrogen, converting it to a form plants can take up. A typical crop rotation follows beans with heavy feeders (tomatoes, corn, or brassicas). Garden centers sell powdered *Rhizobia* spores to inoculate seeds before sowing, akin to a dietary probiotic for the garden; but most gardeners can count on compost to foster a healthy soil microbiome.

Legumes require a deer-proof garden.

JULY 24: Bush beans

AUGUST 24: Pole beans

BEANS

Green is not a color, but a phase that beans pass through on the way to drying out. A dried bean, called by some old timers a fall bean, is mature. The summertime bean is not, and the colors of its youth run to yellow, deep purple, speckled mauve, mottled red, and—as everyone knows—jadeite. Most of the countless bean varieties belong to the species *Phaseolus vulgaris,* the common bean.

The common bean originated in the New World tropics, a blooming vine domesticated early, eons before Columbus. The large, smooth seeds are a pleasure to handle, and seedlings emerge within a week. The bent stem shows first, like the neck of a downcast swan, then pulls the cotyledons from their tight casing. A tentative start is followed by rampant growth. Beans bask in the summer heat. Their small, orchid-shaped blossoms convey a memory of the tropics, followed two weeks later by the first picking.

The vocabulary required to make an informed journey through the bean pages of a seed catalogue is like an overnight bag stuffed with a week's clothing, jumbled and lumpish.

To start, bush beans are those that raise themselves only to knee level before their ambition tops out. Pole beans climb and climb . . . and climb. They differ from Jack's beanstalk in that they will need staking, whether a simple pole or a permanent support, as at Monticello, where scarlet runner beans, the species *P. coccineus,* draw hummingbirds to a long arbor. My neighbor Del, who is considerably taller than 6 feet, grouses about folding himself in half to pick bush beans. I hear fewer complaints as he works his way around the tripods and teepees set up for pole beans. The less common half-runner bean is like an unruly bush bean. Generally low to the ground, it will here and there let fly stray tendrils, like so many wild hairs.

To continue with the bean lingo: Green beans are sometimes called snap beans by those who break them into manageable lengths before cooking. Several old-fashioned varieties, some of them renowned for flavor, are called string beans for the long fibers running down each seam. The string, removed before snapping, is a defect all but eliminated by modern plant breeders. In my childhood, some people believed string beans tasted better, especially men who didn't have to destring them. In recent years I've grown Appalachian heirlooms known as "greasy" beans for their slick pods, which lack the familiar grippy fuzz. Greasies are harvested after the bean has filled the pod, like some other old-fashioned varieties. Slow-cooked

THE LOST BEANS OF BLOUNT COUNTY

Shopping at Horn of Plenty Market in Maryville, Tennessee, I came across a bin labeled white half-runner beans. They cooked up with silky pods and a flavor that reminded me of my West grandfather's white half-runners, the name of which went missing from family memory. The market manager told me the variety I'd taken home was State. "It was very popular around here until twenty or thirty years ago," he said. "Now it's only grown by a few old timers." An online search led me to an article in *West Virginia Explorer* magazine. In 1933, it read, a gardener named Mrs. Euna Poling Hall gave heirloom bean seeds to the J. Chas McCullough Seed Company in Cincinnati, which likely were used later by West Virginia University to develop State and a similar variety called Mountaineer. Given local habits among Blount County farmers of my grandfather's generation, State is likely the "lost" bean I remember from his garden.

SEPTEMBER 21: Clockwise from top left: Kentucky Wonder, Northeaster, Shantyboat Lima, Kenearly Yellow Eye Bean (in pods and shelled), Scarlet Emperor, Dragon Tongue, Lazy Wife Greasy

with a ham hock (or a glug of olive oil), greasy beans have an archaic flavor, the deliciousness also found in long-stewed greens.

Still more bean names: The wax bean has a pod as yellow as tallow. Filet or French beans—the bistro's haricots verts—are skinny and often served whole in a boastful show of stringlessness. Roma or Romano beans have big, flattened pods that are wonderful even when 10 inches long. They stand up to stewing.

Shelling beans, mentioned prior, are removed from the pod—shelled—before eating. They cook much more quickly than dried beans and can be combined with snap beans for a delicious all-bean ragout.

Dried beans, harvested at the end of the season when they rattle within their desiccated pods, come colored (cranberry), speckled (Jacob's Cattle), or white as Carrara marble (borlotti). Such polished beauty seemingly owes to a sculptor's hand. The culinary potential of dried beans has been perfected by home cooks of many nations. Black beans, pinto, navy, tepary, cannellini, flageolet, Anasazi, and endless others make me wish I grew more. But gardener beware: Dried beans require long rows to grow usable quantities. In smaller gardens, opt for productive pole varieties.

To end with other bean species: Scarlet runners include some white-seeded varieties that produce elephantine beans of the sort known in Greece as gigantes. Eaten young and fresh, they are sometimes called butter beans, a point of confusion with lima beans, *Phaseolus lunatus*. Other legumes hail from Old World species. The garbanzo bean is a very ancient chickpea, *Cicer arietinum*. Asian long beans, *Vigna unguiculata,* dangle from climbing vines at impressive lengths, although not quite the "yardlong" attributed to them. They are immediate kin to African cowpeas or field peas. Edamame and its mature form, the soybean, represent an entirely different species *and* genus, *Glycine max*. Fava beans belong to still another genus, *Vicia faba*.

In the Garden

Beans are sown directly, not started indoors. Plant successions of bush beans at 3-week intervals to ensure a steady supply. A second planting of pole beans in mid-July will bear until frost. Plant 5 seeds per pole and select the 3 strongest to grow. To trellis on corn or sunflowers, give the supports a 2-week head start, then sow 2 beans per stalk.

Green beans can be harvested quite small. Pick often because the fast-growing pods toughen quickly; picking stimulates the vines to set more flowers. I go over my bush beans every 4 or 5 days, and get in three passes before the vines flag. Picking is hard on beans; the fragile plants are left rumpled and broken. Pole beans bear longer. Green beans will keep in the refrigerator, wrapped in a kitchen towel, for 3 to 5 days.

Among the estimated 130 named green bean varieties, the heirlooms Blue Lake (bush and pole) and the nineteenth-century classic Kentucky Wonder (pole) stand out for flavor. Rattlesnake (pole) are beautiful fresh or dried. Dragon's Tongue (bush), a nineteenth-century Dutch heirloom wax bean, has a flattened, red-striped pod. Northeaster (pole) is a superior Romano-type. The purple Velour (bush) is an early filet bean, and Provider (bush) is an undemanding and productive green snap bean. Typical maturity for green beans is 50 to 60 days.

Dried beans are represented by four thousand varieties in the collections of Seed Savers Exchange. Many grown today come from Native American seedstock. Among them are Jacob's Cattle (from the Passamaquoddy tribe in Maine), the True Red Cranberry, first described

JULY 3: Green Arrow and Strike peas

by the seventeenth-century French writer Marc Lescarbot as being planted with corn by indigenous agriculturalists, and speckled Anasazi beans from the Southwest. Dried beans typically mature in 90 days.

Seed catalogues offer a daunting list of beans. My advice is to select a bush bean and a pole bean, or two of each if space permits. Note their days to maturity and aim for a staggered harvest. Northeaster, for example, matures at 55 days, 2 weeks ahead of Kentucky Wonder and Rattlesnake.

For a small garden, stick with pole beans, which like city skyscrapers fill airspace above a compact footprint. You can also grow pole beans in a large planter, such as a 5-gallon bucket. Plant 6 to 8 evenly spaced seeds in the container and thin to the strongest 4 or 5. Provide a trellis for the vines.

For most bush beans, 1 ounce of seed will sow a 10-foot row, enough to yield 5 to 7 pounds. A same-sized row of dried beans will shell out to a scant cup, just enough for a small pot.

BEANS—BUSH VARIETY

DIRECT SOW	DEPTH	SPACING	ROW SPACING	DAYS TO EMERGENCE	THIN TO	DAYS TO MATURITY	MATURE SIZE
After last frost	1 inch	2–4 inches	24–36 inches	8–16	4 inches	50–65	18–24 inches

Note: Sow lima beans 2 weeks after last frost.

BEANS—POLE VARIETY

DIRECT SOW	DEPTH	SPACING	ROW SPACING	DAYS TO EMERGENCE	THIN TO	DAYS TO MATURITY	MATURE SIZE
After last frost	1 inch	3 inches, or 4 seeds per pole	4 feet	8–16	3 inches	55–80	6–8 feet

Note: Train pole beans up a trellis, large-mesh wire fencing, pole, tripods/teepees, cornstalks or sunflowers. Plant corn or sunflowers 2 weeks earlier to give the supports a head start.

PEAS

The charm of peas, apart from their sweet flavor, is their carefree manner. Easy-peasy, as the schoolyard slang has it. Garden peas, or English peas if you prefer, appreciate cool but bright weather, a crop of joyful spring. Peas also can be sown in late summer for a fall picking. The seeds are like dimpled ball bearings. You push them into the soil, one by one, with the tip of your index finger, about 1 inch deep. Soon they will jump up and be ready for picking within 60 days, alongside spring lamb and young pullet. Peas continue to bloom and set pods until early summer, when they peter out not long after new potatoes come in.

As with beans, green is a state of immaturity as well as a color. Setting aside dry field peas, which are particularly relished in Southern cooking, green peas are more or less of the same reliably English character—dignified and steady. Peas in a pod, as the saying goes, and the pod is discarded. Sugar peas are no less snugly packaged, but their container is edible. Upon release in 1979 by Gallatin Valley Seed Company in Twin Falls, Idaho, where it had been developed by plant breeder Calvin Lamborn, the Sugar Snap pea was greeted by my West grandparents as a kind of miracle—pea flavor without the shelling. Snow peas, often seen in Asian stir-fries, likewise have edible pods. They are always flat, unlike cigar-shaped snap peas, and sometimes yellow or purple. Pea sprouts are cut soon after emergence. Pea shoots are vine tips, and they can be sautéed like spinach. Tendril Pea and Dwarf Grey Sugar Pea are extra-tender. Other peas have edible shoots but they toughen as they grow; pick the freshest tips.

In the Garden

Peas are a traditional spring crop, among the earliest vegetables of the year, because they germinate in cold soil, up to 8 weeks before last frost. For the quickest start, soak seeds for several hours or overnight before planting. If the variety's mature size is 30 inches or more, provide a light support. I use apple tree prunings pushed into the ground and woven together into a twig trellis. Peas are planted very close and left unthinned.

Among English peas, the 1843 heirloom Champion of England (70–75 days) climbs high and bears heavily within puffy pods. Little Laxton's Progress #9 (68 days) stays under 20 inches. Tightly packed Strike matures early (55 days). Among the sugar peas, the original Sugar Snap (58 days) has recently been reintroduced by Johnny's Selected Seeds. I once made the mistake of interplanting shelling peas with sugar peas, figuring I could tell the difference as they matured. I couldn't. Keep your peas separate. Or color-code them. The sugar peas include purple Royal Snap II (its flat-podded sibling is Royal Snow) and the yellow Honey Snap—all Calvin Lamborn introductions, like Sugar Snap and Sugar Ann.

One ounce of seed will sow a 10-foot row. Peas should be picked every 3 to 5 days during their monthlong harvest. Expect 4 to 5 pounds of pods from a 10-foot row. A pound of pods shells out to a heaping cup.

GREEN PEAS

DIRECT SOW	DEPTH	SPACING	ROW SPACING	EMERGENCE	THIN TO	DAYS TO MATURITY	MATURE SIZE
Up to 8 weeks before last frost	1 inch	2 inches	2–5 feet, depending on height	7–14 days	n/a	55–70	20 inches to 10 feet

Note: Provide a light trellis of sticks or mesh for varieties over 30 inches. Sweeten the soil with a light dusting of lime or wood ash.

KEATS'S PEAS

John Keats marvels at pea flowers in the poem "I Stood Tip-Toe upon a Little Hill." His are *Lathyrus odoratus*, the inedible sweet pea. Still, any vegetable gardener who has admired English peas will recognize his description of delicate, grasping tendrils as they climb to release their winged flowers.

Here are sweet peas, on tip-toe for a flight:
With wings of gentle flush o'er delicate white,
And taper fingers catching at all things,
To bind them all about with tiny rings.

WAX BEANS • GARLIC • HERBS | SNACK | MAKES 2 PINTS | VEGAN, GLUTEN-FREE | SUMMER

MARINATED WAX BEANS

"I'm going to put out a couple rows of green beans," I told Del as we sat around the winter fireplace. "You'll be sorry," he said. Del had spent many teenaged hours in his mother's garden, and that summer he complained every time we picked beans. Toward the middle of August, I gave him a jar of marinated wax beans with his reward, which is what he calls his six o'clock beer. After that he didn't complain as much.

Any type of young, tender bean will work for this recipe. I like wax beans for their look, and usually add a few of another color for contrast. I "top" the beans—trim the stem end—but leave them whole. The marinade is a quick pickling brine; white wine and herbs give it a suave French-y vibe. Float a layer of good olive oil on top and the beans emerge predressed in a fragrant vinaigrette. My first cookbook, *Saving the Season,* includes more recipes for pickled green beans, alias Dilly Beans, as well as for fermenting green beans, a delicious technique for anyone who appreciates kosher dill pickles.

1 cup dry white wine
½ cup white wine vinegar
2 cloves garlic, lightly crushed and peeled
½ teaspoon dried savory or thyme, or several sprigs fresh
2 small bay leaves
1 chile de árbol or other dried red chile, split lengthwise
½ teaspoon yellow mustard seeds
¼ teaspoon black peppercorns
1 tablespoon plus ½ teaspoon salt
1¼ pounds tender wax beans or other beans, topped
4 tablespoons extra-virgin olive oil

1. Fill a large bowl with ice and water.
2. In a small saucepan, combine the wine, vinegar, ½ cup water, the garlic, savory, bay leaves, chile, mustard seeds, peppercorns, and ½ teaspoon of the salt. Bring to a boil over high heat, then remove from the heat and set aside.
3. In a medium saucepan, bring 2 quarts of water to a boil and season with the remaining 1 tablespoon of salt. Add the beans and blanch, uncovered, for 4 to 5 minutes, or until just cooked. Drain and shock in the ice-water bath, then drain the beans again.
4. Pack the beans into two clean 1-pint jars. Bring the marinade to a boil for a second time and pour into the jars. Top each jar with 2 tablespoons of olive oil. Seal and refrigerate overnight.
5. The beans will keep for 2 weeks in the refrigerator. Remove them 1 hour before serving to knock off the chill.

PEAS • HERBS | SIDE | SERVES 4 | VEGETARIAN, GLUTEN-FREE | SPRING, FREEZER

ALSO FOR: ASPARAGUS, CARROTS, RADISHES, SMALL TURNIPS, AND OTHER SPRING VEGETABLES

GREEN PEA BASICS

Peas should be picked young and at the last moment. They barely need cooking to release a flavor that is delicate but compelling. They don't go very far, and I am greedy, so I save peas for myself and very small dinners, setting the table with runcible spoons. If you can spare any during their brief season, peas freeze better than almost any vegetable.

Stretch green peas by combining them with other vegetables of the season, such as asparagus tips, thin carrots, tiny radishes, miniature turnips, pearl onions, infant potatoes, slim scallions. I'll blanch each separately for better control over the cooking time, then bring them together with a little stock or cream to lubricate the introduction. Any two of these will harmonize. Any three or more become a springtime ragout with the addition of minced shallot softened in butter. Or use the pale fat from a roasted chicken, or even the fatty drippings from a pork roast.

3 tablespoons unsalted butter or drippings
2 pounds green pea pods, shelled (about 2 cups)
1/2 teaspoon sugar
1/2 cup stock or heavy cream, plus more as needed
2 tablespoons minced fresh herbs, especially chervil, mint, tarragon, parsley, or chives
Fine sea salt

1. In a medium sauté pan, heat the butter over medium heat until foamy. Add the peas and toss a time or two to coat. Sprinkle them with the sugar.
2. Add the stock. Increase the heat to high, cover, and cook for 2 minutes, shaking frequently. Taste a spoonful of peas and cook for 1 minute longer if any are underdone, adding more liquid as needed to keep the pan from drying out.
3. Remove the peas from the heat. Add the herbs and toss again. Salt to taste and serve in a hot dish.

Note: To make a quick stock from the shelled pea pods, half-cover them with water in a small saucepan and simmer with a few herb stems, a slice of shallot, and a pinch of salt for 15 minutes, pressing down on the pods as they soften.

To adapt for asparagus tips, carrots, radishes, small turnips, fava beans, the first new potatoes, etc.:
The standard technique for nearly any spring vegetable is to blanch until tender in salted water and toss in a skillet with warm unsalted butter, cream or stock, and herbs. One pound of prepped vegetables will serve 3 or 4.

To make a spring vegetable ragout: Blanch each vegetable separately. Melt unsalted butter in a large sauté pan until foamy. Soften a bit of minced shallot or green garlic in the butter and a pinch of salt. Add the blanched vegetables, the stock or cream, and more salt to taste. Bring to a boil and simmer a few minutes for the flavors to blend. Scatter minced herbs over the top and toss a time or two to combine.

FILET BEANS, SUCH AS VELOUR • SHALLOTS • HERBS | SIDE | SERVES 6 TO 8 | VEGETARIAN, GLUTEN-FREE | SUMMER, FREEZER

GREEN BEANS IN WARM VINAIGRETTE
AND SIMILAR DISHES

In *The Everlasting Meal,* Tamar Adler says if you're unsure whether a vegetable is done, cook it longer. I remember that every time I blanch green beans. The little French filet beans such as Velour that I like for this recipe cook fast, but after 2 minutes—"crisp-tender"—they still have a grassy chlorophyll taste. It's okay but raw-ish. After 3 minutes, the beans are mild and sweet with a crunch at the center, like risotto rice. Four minutes and they're fully cooked. Five minutes won't hurt, but by then you've already drained them because you know it's time. (Large Romano beans take twice as long and some of the full-sized, thick-hulled beans like Kentucky Wonder improve with long, slow cooking.)

Beans prepared this way are best a la minute, made only when everyone is ready to sit down, as with pasta. They emerge from their hot bath limber and wreathed in fragrant steam when tossed with vinaigrette and herbs. Remember that blanched vegetables continue to cook from residual heat, just as resting steak does. Any beans left in the serving dish at the end of the meal will be limp and soaked with flavor.

For a recipe better suited to preparing in advance for a buffet or picnic, try Cool, Creamy Green Beans (page 125).

1 medium shallot, minced
1 tablespoon red wine vinegar
Fine sea salt
1 teaspoon sherry vinegar
1/4 teaspoon Dijon mustard
1/4 cup extra-virgin olive oil
1 tablespoon walnut oil
1 1/2 pounds filet beans, wax beans, or other young green beans, topped
2 tablespoons fresh savory or marjoram leaves, roughly chopped

1. In a small bowl, combine the shallot, vinegar, and 1/4 teaspoon salt. Stir and set aside to macerate for 15 minutes.
2. Whisk in the sherry vinegar and mustard. Stream in the oils, whisking to emulsify.
3. In a large pot, bring 4 quarts water seasoned with 2 tablespoons salt to a boil. Blanch the beans until just tender, 3 to 4 minutes. Drain. Do not rinse.
4. While the beans are still hot, toss with the vinaigrette and herbs. Serve immediately.

(CONTINUED)

BUTTERED GREEN BEANS WITH MARJORAM

For the earliest green beans of the season or any that come in looking particularly fine, such as the first picking of little Blue Lake, this is the simplest essence of cooking from a garden—vegetables picked that day, dressed with butter, finished with herbs. The method also suits frozen beans, which are blanched briefly before freezing.

Blanch 1 pound beans as directed on page 122 and drain. In a large skillet, soften a minced shallot and a large pinch of salt in 2 tablespoons foaming unsalted butter. Add the blanched beans and 2 tablespoons chopped marjoram or savory. Sauté for 2 minutes to just heat through. If using frozen beans, defrost but do not blanch before sautéing.

ROMANO BEANS WITH PESTO

This would work with any green bean, but Romano beans such as my favorite Northeaster pole bean (pictured on page 123) and the flat-podded Dragon's Tongue are particularly delicious.

Blanch 1½ pounds green beans as directed on page 122, adjusting the time to suit Romano beans, which typically cook in 8 to 10 minutes. Taste the beans and decide. Be ready with ¼ cup pesto in a large serving bowl. Drain the blanched beans in a colander and toss with the pesto. Add a large handful of mixed whole parsley and basil leaves. Season with ½ teaspoon flaky finishing salt and several grinds of black pepper. Add another ¼ cup of pesto and toss a final time or two.

If you can't serve them right away, shock the beans in ice water after blanching and dry them well, either in a salad spinner or with kitchen towels, before tossing with the pesto. You might need to loosen the pesto with some olive oil.

GREEN BEANS AND ZUCCHINI

The great Marcella Hazan published this pairing. In my version, whole green beans, such as flavorful midseason Maxibel, contrast perfectly in shape and texture with chunky zucchini rounds. Brought together only by parsley and lemon juice—no oil—the warm salad goes with any summer meal and helps absorb the annual glut of zucchini.

In a large pot of salted water, blanch 5 or 6 whole smallish zucchini (about 1 pound) until just tender when pierced with a knife, about 5 minutes. Remove with tongs and drain. Using the same water, blanch 2 handfuls of young green beans (about 10 ounces) until just tender, 3 to 5 minutes. Drain. Cut the zucchini into 1-inch-thick rounds. Toss together the beans and zucchini rounds while warm with the juice of 1 lemon and ¼ cup chopped parsley or mixed herbs. Finish at the last moment with flaky salt and freshly ground black pepper.

GREEN BEANS • HERBS | SIDE | SERVES 6 TO 8 | VEGETARIAN, GLUTEN-FREE | SUMMER

ALSO FOR: SNOW PEAS AND OTHER SUGAR PEAS

COOL, CREAMY GREEN BEANS

Prepped in advance and thrown together at the last minute, this easy side dish of bias-cut green beans and torn herbs is dressed for summer with bright, tangy crème fraîche. Opal basil and sesame seeds were the inspiration, but vary the herbs and sprinkles to suit what you have, such as:

Genoa basil + toasted pine nuts

Cilantro + Thai basil + sesame seeds

Dill + chopped walnuts

Cilantro + jalapeño rounds + toasted pumpkin seeds or deep-fried garbanzos

Sorrel + toasted almonds

1½ pounds green beans, topped
2 tablespoons marjoram or summer savory leaves, torn or roughly chopped
2 teaspoons grated lemon zest (from 1 lemon)
1 tablespoon fresh lemon juice, plus more to taste
⅓ cup crème fraîche
2 tablespoons extra-virgin olive oil
¼ teaspoon fine sea salt
¼ teaspoon freshly ground black pepper
1½ teaspoons flaky finishing salt, or 1 teaspoon fine sea salt, plus more to taste
1½ cups whole green and purple basil leaves (or other herbs from the headnote)
1 tablespoon sesame seeds, toasted (or other sprinkles from the headnote)

1. Fill a large bowl with ice and water.
2. In a large pot of boiling salted water, blanch the beans until tender, 3 to 5 minutes or more, depending on the variety. Drain and immediately plunge into the ice water bath. Drain and dry in a clean kitchen towel. Slice thinly on the bias.
3. In a small bowl, whisk together the marjoram, lemon zest, lemon juice, crème fraîche, olive oil, fine sea salt, and pepper.
4. Just before serving, in a large bowl, combine the beans, dressing, and flaky salt and toss until well coated. Taste and adjust the lemon juice and salt, if necessary. Add the basil leaves and turn over one time only to partially coat with the dressing. Garnish with the sesame seeds.

Note: To measure whole herb leaves, pack into a measuring cup firmly but without crushing them—a little (or a lot) more won't matter.

To adapt for snow peas or other sugar peas: Substitute 1½ pounds of trimmed, edible-hulled peas for the green beans. Blanch until tender, 2 to 3 minutes for snow peas, or slightly longer for sugar peas. Continue as above.

THICK-HULLED MATURE GREEN BEANS, SUCH AS KENTUCKY WONDER, RATTLESNAKE, OR GREASY TYPES • ONION

SIDE | SERVES 6 TO 8 | GLUTEN-FREE | SUMMER, FREEZER

COUNTRY STRING BEANS

This recipe from the farmhouse is intended for mature, thick-hulled green beans, the heirloom types harvested after the bean fills out—string beans, half-runners, pole beans, and greasy beans. You cook them slowly, like braising greens, and fortify the broth with smoked pork and black pepper. To have them ready for supper, get the pot going by late afternoon. The pods will turn khaki and split up the seam when ready. The beans spill out, almost shelling size. What you get is a simple ragout with a satisfying contrast between the bean's meaty substance and the silky pod's vegetal slipperiness. The pot liquor is heady stuff with an antique, out-of-time quality. You spoon it up like a tonic or soak your cornbread in it.

If you don't have the recommended old-fashioned bean varieties, try a big-podded Romano type.

1 tablespoon extra-virgin olive oil
1/4 pound slab bacon or other smoked pork, cut into 1-inch chunks
1/2 small onion, peeled and quartered
2 × 3-inch piece of kombu (optional)
1 teaspoon salt, plus more to taste
1/2 teaspoon freshly ground black pepper, plus more to taste
1 1/2 pounds beans, trimmed and snapped

1. Coat the bottom of a Dutch oven or other heavy pot with the olive oil. Set over medium heat. Add the bacon and lightly brown on all sides, about 5 minutes. Remove the meat from the pan. Lightly brown the onion in the rendered fat that remains, about 5 minutes.
2. Return the bacon to the pot with the onion. Add 5 quarts water, the kombu (if using), the salt, and pepper. Bring to a boil and cook, uncovered, for 15 minutes. Taste and adjust the broth—it should be highly seasoned.
3. Add the beans and reduce the heat to maintain a lively simmer for 1½ hours, or until the pods are silky and splitting along the seam. Stir in a lot of freshly ground black pepper before serving.

GREEN BEANS • TOMATOES • HERBS • GARLIC

ALSO FOR: CHARD STEMS, HEARTY GREENS

SIDE, MAIN

SERVES 6 TO 8

VEGETARIAN

SUMMER, EARLY FALL

GREEN BEAN GRATIN

Quick to assemble, this summertime gratin replaces béchamel with cream, which binds the elements together with a lighter presence. Thanks to the umami deliciousness of grated Parmesan and the crunch of bread crumbs, it is a side dish that could anchor the plate for a vegetarian meal. A big leafy salad or a plate of sliced tomatoes or a platter of corn on the cob would finish the picture.

1 tablespoon extra-virgin olive oil, plus more for the baking dish
1½ pounds tender green beans, trimmed but not snapped
2 cups diced seeded Roma tomatoes (4 to 5 large)
2 cloves garlic, thinly sliced
2 tablespoons savory or thyme leaves, roughly chopped
1 teaspoon grated lemon zest (from ½ lemon)
1½ teaspoons fine sea salt
¼ teaspoon freshly ground black pepper
1½ ounces Parmesan cheese, grated (¾ cup lightly packed)
1 cup organic heavy cream
1 cup fresh bread crumbs, lightly toasted

1. Preheat the oven to 350°F. Lightly oil a 12-inch oval gratin dish.
2. In a large bowl, toss together the green beans, tomatoes, garlic, savory, lemon zest, salt, pepper, Parmesan, and cream until coated. Separately, in a small bowl, toss the bread crumbs with the 1 tablespoon of olive oil.
3. Turn the green bean mixture into the gratin dish. Cover loosely with aluminum foil and bake for 20 minutes.
4. Remove the foil and scatter the bread crumbs over the top. Bake, uncovered, until the bread crumbs are browned and toasted, about 25 minutes.

To adapt for chard stems: After stripping the chard leaves for another use, trim the stems into 3-inch lengths and blanch in salted water until tender. Proceed with the recipe, substituting the blanched chard stems for the beans.

To adapt for hearty greens: Trim small-to-medium chard, kale, or other cooking green. Blanch in salted water for 2 minutes. Drain and allow to cool. Squeeze out the excess water. Proceed with the recipe, substituting the blanched greens for the beans.

GREEN BEANS • ONION • HERBS

ALSO FOR: PEAS, ZUCCHINI, WINTER SQUASH, CELERY

MAIN | SERVES 6 TO 8 | VEGETARIAN OPTION | SUMMER, FREEZER

GREEN BEAN RISOTTO

The carcass of a roast chicken, stripped clean after a meal, is good for about 1 quart of stock. In my kitchen, stock signals risotto, and risotto is a year-round vehicle for garden vegetables. This version uses green beans (fresh or frozen). In other seasons you can substitute peas, zucchini, or celery (see how to adjust on page 131). If you stir in what appears to be an excess of butter and the tangiest, oldest Parmesan you can find, you'll have it about right.

For a vegetarian version, cook the rice in vegetable stock or even plain salted water.

About salt: Remember that a meaningful quantity of salt might be hidden in the stock—or might be lacking from it altogether. My homemade stock has about 1 teaspoon fine sea salt per quart. If you use unsalted stock (or plain water), add extra salt when you bring the stock to a boil.

1 quart stock
Fine sea salt
$1/4$ cup extra-virgin olive oil
1 medium onion, finely diced (about $1\frac{1}{2}$ cups)
2 cups Arborio rice
1 cup dry white wine
12 ounces green beans, thinly sliced on the bias (about 4 cups)
$1/2$ cup minced fresh parsley
4 tablespoons unsalted butter
4 ounces Parmigiano-Reggiano cheese, grated (about $1\frac{1}{3}$ cups), plus more for serving
1 cup pesto
Freshly ground black pepper

1. In a medium saucepan, combine the stock, 4 cups water, and 1 teaspoon salt. (If using unsalted stock or all water, add an additional 1 teaspoon salt.) Bring to a boil, then reduce the heat to maintain a bare simmer.

2. In a very large straight-sided pan or a 6-quart Dutch oven, heat the olive oil over medium heat until it shimmers. Add the onion and 1 teaspoon salt. Cook gently, stirring constantly, until the onion is softened but not browned, about 5 minutes.

3. Set a kitchen timer to 30 minutes. Add the rice to the onion and cook, stirring constantly, for 5 minutes. Add the wine and stir until evaporated, a couple of minutes. Increase the heat to medium-high. Add a ladleful of stock—I use a ¾-cup ladle—and keep stirring. Adjust the heat as needed to maintain a gentle but steady boil. Once the liquid is fully absorbed, add another ladle of stock, continuing to stir and shake until the stock is absorbed. Each addition of stock should take several minutes.

4. When 15 minutes remain on the timer, add the beans. Keep adding stock and stirring it in. Take care to turn the beans under as you go to prevent them from riding to the top.

5. When 5 minutes remain, stir in the parsley. When the timer sounds (30 minutes after first adding the rice to the pan), the risotto should be ready—with a firm but not chalky center.
6. Remove from the heat. Stir in the butter and Parmigiano until melted. Taste and add another ¼ teaspoon salt only if needed. The finished risotto should be suave, on the loose side. If it stiffens, stir in the last of the hot stock or a splash of boiling water to loosen it.
7. Divide the risotto among warm dishes—I like it spread on a plate. Top with a generous dollop of pesto and serve with extra grated Parmesan and the remaining pesto to pass. Garnish with freshly ground black pepper, to taste.

To adapt for peas, zucchini, winter squash: Replace the beans with an equal volume of shelled peas, ½-inch diced zucchini, or ¾-inch diced winter squash that has been steamed or blanched for 5 minutes and drained. Omit the basil pesto or substitute another pesto or herb sauce.

To adapt for celery: Celery makes my favorite vegetable risotto. Use 2 cups of celery cut into ¼-inch dice. Add half to the pan when you sauté the onion and half when you would have added the beans. Include some of the fragrant chopped leaves with the second addition. Add more chopped leaves with the parsley in step 5.

ALSO FOR: PEAS, HEARTY GREENS, BROCCOLI, CAULIFLOWER | PANTRY | MAKES 3 POUNDS | VEGAN, GLUTEN-FREE | SUMMER

FREEZING GREEN BEANS

Green beans emerge from the freezer with their summertime flavor intact. Although filet beans are not quite as firm as when fresh, I still prefer them to winter grocery-store beans trucked in from who-knows-where. No complaints about sturdy Romano beans.

The universal rule for freezing vegetables is to use only your best. Nothing will improve with time. Blanching is not strictly necessary, but the flavor of unblanched vegetables degrades more quickly in the freezer. Frozen vegetables will keep for a year.

Vacuum sealers eliminate freezer burn, but ziplock freezer bags are good enough. Batch by serving size. For green beans, I'll pack small, medium, and large bags—up to 18 ounces. Thaw the unopened bag in lukewarm water for 10 minutes. Label everything.

3½ pounds tender young green beans, trimmed and snapped as you like

1. Fill a large bowl with ice and water.
2. Bring a large pot of water to a boil. Working in ½-pound batches, blanch the beans for 2 minutes. (Blanch Romano beans for 3 minutes.) Remove with a slotted spoon and cold shock in the ice bath for several minutes, or until completely cool. Drain and spread on a baking sheet or clean kitchen towels. Repeat with the remaining beans.
3. Still working in ½-pound batches, dry the beans in a vegetable spinner. Pat dry with kitchen towels. Pack the freezer bags. It's better to flat-pack—a thin edge-to-edge layer—rather than bundle the beans. Flat bags will freeze faster and stack neatly.
4. Seal the bags using a vacuum sealer, if you have one. For ziplock bags, my neighbor Del showed me a neat trick: Insert the tip of a drinking straw into the bag and close the zipper around it. Pinch tight and suck out the air. In one motion, slip out the straw and press the zipper shut.
5. To make tidy and stackable packages, lay the filled bags flat in a single layer on a rack or baking sheet. Place in the freezer for 2 hours, or until solid.

To adapt for other vegetables: The above method applies to all manner of vegetables. The basic steps—blanch, shock, drain, vacuum-seal—don't vary. The variables are how to prep the raw vegetable and how long to blanch.

To adapt for peas: Shell peas. Blanch quickly, 2 minutes.

To adapt for greens: Strip the greens off the midrib and cut the leaves into ribbons. Wash three times. Blanch 1 minute. Drain and spread on a sheet pan to cool. Place another sheet pan of the same size on top, so that the greens are sandwiched in an improvised press. Hold it by the long sides over the sink and squeeze to extract as

much water as possible. (Alternatively, press the drained greens into the bottom of a colander with the back of a wooden spoon or simply wring them out, one handful at a time.) Gently tease apart and pack flat.

To adapt for broccoli: Cut into florets. Blanch 3 minutes. Dry in a salad spinner or with a kitchen towel. Spread the loose florets on a baking sheet and place in the freezer, uncovered, until solid, 1 to 2 hours. Bag and seal.

To adapt for cauliflower: Follow the same steps for broccoli, except blanch for 5 minutes.

DRIED BEANS • AROMATICS • HERBS | MAIN, SIDE | SERVES 6 TO 8 | VEGETARIAN OPTION, GLUTEN-FREE | FOUR SEASONS

A POT OF BEANS

If you have considerable space in the garden and enjoy repetitive tasks, growing dried beans will connect you to a very long agricultural tradition. Beans have been farmed in the Americas for seven thousand years. The yield is modest compared to an equivalent planting of green beans, and the picking and shelling part will try one's patience. But a long-cooked pot of beans is a solace, ample with comfort and sustenance.

Coaxing dried beans to release their meaty flavor takes water, salt, and time. There are countless recipes and many deeply held beliefs, but the starting point is an overnight soak to reduce cooking time. (The exception is black beans.) Simmering just below the boil, such that you'll be tempted to check that the flame hasn't gone out, a pot of beans will need 2 hours or more. Embellishments abound, some of which purport to improve digestibility. I add less aromatics and smoked pork than I once did, instead choosing to salt earlier, right at the start, having unlearned the demonstrably false injunction against salting before the beans soften, lest they never do. I stir as little as possible to leave the beans intact, whereas as I once did the opposite to produce a creamier broth. Unstirred beans achieve their ideal consistency after a night under their own cooking liquid. If possible, make them the day before serving.

This particular pot of beans is vegan. Omnivores could add ¼ pound of slab bacon or ham hock. (In which case, reduce the salt.)

The speckled beans pictured opposite are an heirloom half-runner named after Robert Hazelwood from Clay County, Kentucky. The most flavorsome dried bean I've yet grown in my New England garden is the Kenearly Yellow Eye, developed in Nova Scotia, a very long way indeed from the bean's ancestral home in the tropics.

2 cups dried beans
2 teaspoons fine sea salt, plus more to taste
¼ teaspoon baking soda
1 small onion, peeled and halved
2 cloves garlic, lightly crushed
1 bay leaf
Several sprigs herbs, such as thyme, rosemary, sage, epazote, or parsley
2 × 3-inch piece of kombu (optional)
Avocado oil or extra-virgin olive oil (optional)
Minced fresh herbs (optional), such as parsley, cilantro, or rosemary

1. Spread the beans on a baking sheet and pick through them to remove stones, debris, and broken pieces. Rinse in a colander.
2. Transfer to a 6-quart Dutch oven or other covered pot. Add the salt and baking soda. Add water to cover by 2 inches. Place in the refrigerator to soak overnight. Discard any floaters.
3. The next day, bring the pot to a boil over high heat. Boil hard for 5 minutes. Skim. Add the onion, garlic, bay leaf, herb sprigs, and kombu (if using). Reduce the heat to maintain a bare simmer. Cook, partially covered, for 1 hour. Don't stir.

(CONTINUED)

4. After an hour, check the simmer and adjust the heat if necessary. Taste. The broth should be on the salty side, so add more salt if needed. Keep cooking until the beans are tender, another hour or so. (Black beans will require less cooking time.) If the first bean you taste seems ready, test four more—another bit of advice from Tamar Adler. My goal is to get the beans as soft as possible without collapsing.
5. Remove from the heat and allow them to steep for at least 1 hour before serving, preferably overnight in the refrigerator.
6. Before serving, gently reheat and again taste for salt. Serve in warm bowls with a glug of avocado oil and a scattering of fresh herbs, if you like.

SHELLING PEAS • SUGAR PEAS • PEA SHOOTS • GARLIC • HERBS | MAIN | SERVES 4 | GLUTEN-FREE | SPRING

PAN-SEARED LAMB WITH PEAS

This dish is what restaurant cooks call a fast pickup. You pan-sear lamb chops—or, in another mood, big scallops or inch-thick monkfish medallions—and as they rest on a warm platter, you use the same pan to toss together the three expressions of a pea's culinary potential: shelled green peas, sugar-pea pods, and pea shoots. It's easy but flashy, something for company dinner. Natural pairings include rich scalloped potatoes and a leafy salad bowl of spring lettuces, herbs, and shaved carrots.

- 8 bone-in lamb loin chops (about 2 pounds)
- 1½ teaspoons fine sea salt
- 1 tablespoon grapeseed oil
- 6 tablespoons unsalted butter
- 1 shallot, minced
- 1 clove garlic, crushed
- 2 or 3 sprigs thyme
- 2 pounds green pea pods, shelled (about 2 cups)
- ¾ pound snow peas or sugar peas (6 cups), trimmed, blanched, shocked, and sliced on the bias
- ½ cup chicken stock
- ¼ pound tender young pea shoots
- ⅔ cup Lucques olives, pitted
- Juice of ½ lemon
- ½ cup chopped mixed fresh herbs, such as chervil, tarragon, chives, parsley, and thyme

1. An hour before cooking, salt the lamb chops with 1 teaspoon of the salt.
2. In a large skillet, heat the grapeseed oil over high heat until nearly smoking. Add half the chops and sear, 2 to 3 minutes per side, depending on thickness. Don't overcook! Transfer to a warm platter and cover loosely. Repeat with the remaining chops, adding ½ tablespoon more grapeseed oil if the pan seems dry.
3. In the same pan, melt 2 tablespoons of the butter over medium heat. Add the shallot, garlic, and thyme sprigs. Cook until the shallot softens, about 1 minute. Add the peas and snow peas. Toss to coat. Add the stock and scrape the pan with a wooden spoon to lift any browned bits. When the stock has reduced by three-quarters, about 3 minutes, add the pea shoots, olives, and lemon juice. Toss to combine.
4. Remove from the heat. Add the remaining 4 tablespoons of butter and the chopped herbs. Toss to combine. Spoon over the chops to serve.

Note: You can add other spring vegetables if you have them, including small scallions or scallion-sized green garlic. Blanch each separately and add to the skillet with the snow peas.

CH.

13

BROCCOLI AND CAULIFLOWER

BUDDING BRASSICAS

Brassicaceae,
THE MUSTARD FAMILY,
ALSO KNOWN BY THE OLDER NAME
Cruciferae

B. oleracea var. *botrytis*
CAULIFLOWER AND ROMANESCO

Brassica oleracea var. *italica*
BROCCOLI

JUNE 14: Cousin Jack Carey with Green Magic broccoli

The mustard family, or brassicas, includes a very wide range of vegetables. This chapter covers broccoli and cauliflower, vegetables grown for their tightly clustered flower buds. The next chapter covers the heading brassicas: cabbage, Brussels sprouts, and petticoated collards, which are essentially scatterbrained cabbages that don't form heads. Brassicas pop up in other chapters, as well: Arugula and mizuna are in Tender Greens, kale and mustard greens in Hearty Greens, radishes and turnips in Root Vegetables. Still more brassicas lie beyond the scope of these pages: horseradish, wasabi, and rapeseed (the source of canola oil).

Brassicas are sometimes referred to by an older alias, cruciferous vegetables, because their four-petaled blossoms reminded the pious of a cross. Another archaic name for the sub-clan of collards, cauliflower, and kohlrabi is cole crops, derived from the Latin *caulis,* for stalk or stem. Kohlrabi also piggybacks on *kohl,* German for cabbage, a word that pops up elsewhere as coleslaw.

President George H. W. Bush, a diplomat on most counts, expressed his open hatred for broccoli, which he no doubt knew in its worst form—boiled until soggy and drab. Homegrown broccoli cooked with a defter touch would have changed his mind. Today's chefs are apt to scorch brassicas with fierce heat to bring out their sweetness, such as roasting broccoli in a hot oven until the florets crisp at the edges or searing cauliflower steaks over wood coals.

The familiar big-domed Italian broccoli, or crown broccoli, invests its effort in one main bud. Cutting or blooming broccoli has a smaller central head, and will resprout vigorously after harvesting, sending up shoots from the leaf unions along the stalk. Broccolini has a similar look but different parentage: It was developed by

JULY 22: Deeva Gupta next to cauliflower, its leaves clothes-pinned as a sun-shade to blanch the head

JUNE 14: Happy Rich sprouting broccoli

the Sakata Seed Company in Japan as a hybrid between Italian broccoli and gai lan, or Chinese broccoli (*B. oleracea* var. *alboglabra*).

Cauliflower and broccoli have similar growing requirements and, often, similar kitchen uses, as does the mesmerizing Romanesco, an alien-green cauliflower-shaped broccoli with fractal curds.

Skip ahead to the Hearty Greens chapter for broccoli rabe or rapini—*rape di broccoli* or *cime di rape* in Italian. The writer of a 1976 letter to *The New York Times* claimed that "hardly any non-Italians know of broccoli di rape," and advised gardeners to plant it in early spring and again in mid-July, where it might be squeezed in among eggplant and pepper plants. Sound advice to this day, when many more of us know what it is.

In the Garden

Broccoli and cauliflower are cool-weather crops. Heat runs them down. To give you an idea: Aroostook County in northern Maine, where summertime highs hover in the mid-seventies, is a center for East Coast commercial broccoli production. Both vegetables should be started indoors for an early-summer harvest or directly sown in midsummer, about 2 months before first frost, for a fall crop. Mature plants are somewhat cold hardy, although less so than other brassicas—the heads will survive light frost. Broccoli and cauliflower are big plants: Allow them as much as 2 feet of spacing. Make use of the bare ground between seedlings by interplanting with a jump-up crop, such as lettuce, spinach, radishes, Hakurei turnips, or cilantro.

The budding brassicas are beautiful as they grow, statuesque and unhurried. Truth be told, the return on time and space is modest: one head per plant of cauliflower or one central bud for broccoli—a meal per plant. In a small garden or a raised bed, sprouting broccoli with its repeat harvests will be a better choice. Clip secondary shoots at 6 to 8 inches long, before they fully bloom out, although a few blossoms won't hurt and pollinators will flock around.

When cauliflower begins to head up, it first appears as a small clot of curds on the central stem. Protect it from sunburn by blanching, which in the garden means to shield from direct sunlight. Gently enclose the head within four leaves fastened at the top with a clothespin—an auto-parasol. Cauliflower won't resprout after harvest. Romanesco, alike in most other respects, doesn't need blanching.

Pests come with brassicas. Protect broccoli and cauliflower transplants with a cutworm collar, a 2-inch-wide strip of newsprint wrapped around the stem at ground level. In summer, darling white butterflies will flit about, the adult phase of the jade-green cabbage worm. Inspect the underside of leaves for eggs to crush and follow caterpillar tracery to the culprit.

Delicious Green Magic broccoli (57 days) is an early variety and heat-tolerant—an example of plant breeders' work to develop crops for the new climate. The heirloom De Cicco (60–70 days) has a smaller central head, but resprouts vigorously *and* has edible leaves. Happy Rich is a delightful sprouting broccoli (55 days). For fall, Umpqua is slow to mature (95 days) but resprouts as well. The unusual broccoli spigarello is a wonderful leafy green, similar in stature to heading broccoli in the garden and less bitter than broccoli rabe—highly recommended.

Heat afflicts cauliflower more than broccoli. Amazing (75 days) and the 1947 heirloom Snowball (80 days), progenitor of the other wintry-named offspring, are hefty and beautiful in mild climates. Smaller varieties, such as Early Snowball (50–60 days), mature quicker, making them better choices in warm regions. Ghostly pallor is not the final word among cauliflowers. Flame Star is cheddar orange, and the heirloom Purple of Sicily is noted for its flavor and insect resistance.

Brassica seeds are tiny beads—think of the mustard seed in your spice cabinet—with 100–300 seeds per gram. A single packet of broccoli or cauliflower will oversupply most gardens.

BROCCOLI—SPRING PLANTING

START INDOORS	DEPTH	DAYS TO EMERGENCE	TRANSPLANT	SPACING	ROW SPACING	DAYS TO MATURITY	MATURE SIZE
4–6 weeks before last frost	1/4 inch	5–17	2 weeks before last frost	24 inches	24 inches	55–85	30–36 inches tall

BROCCOLI—FALL PLANTING

DIRECT SOW	DEPTH	DAYS TO EMERGENCE	SPACING	THIN TO	ROW SPACING	DAYS TO MATURITY	MATURE SIZE
2 months before first frost	1/4 inch	5–17	4-seed clusters every 24 inches	1 seedling per cluster	24 inches	55–85	30–36 inches tall

CAULIFLOWER—FALL PLANTING

DIRECT SOW	DEPTH	DAYS TO EMERGENCE	SPACING	THIN TO	ROW SPACING	DAYS TO MATURITY	MATURE SIZE
2–3 months before first frost	1/4 inch	5–17	4-seed clusters every 24 inches	1 seedling per cluster	24 inches	65–85 days	24–30 inches tall

Note: For spring planting, follow growing instructions for broccoli.

IN THE KITCHEN

Harvest broccoli while the heads are tight and blue-green, before they jaundice and bloom. Cut the stalk cleanly 8 inches below the crown and leave the plant in place for side shoots to emerge. Cut cauliflower while the head is tight and clear white. If it starts to spread like a slow-motion explosion, cut it immediately. Because the budding brassicas grow upright, facing the sky and protected by a layered ruff of spreading leaves, broccoli and cauliflower come in from the garden generally free of dirt and grit. But beware the bugs within, like soldiers in the Trojan Horse. Trim leaves and plunge the head upside down into a bowl of cold water to purge.

Broccoli wrapped in a kitchen towel will keep in the refrigerator's humidifier drawer for a short week. Cauliflower becomes stronger as the days pass. Both freeze well: See the note to Freezing Green Beans for instructions (page 133).

Broccoli and cauliflower are all delicious when simply blanched (or steamed) and tossed

JULY 22: Snowball

with melted butter, salt, and pepper. (Peel broccoli's sweet, nutty main stem and cut it into ½-inch rounds to blanch with the florets.) Both also go well with rich, old-fashioned sauces, such as hollandaise and mousseline. In a more modern vein, dress them with olive oil, strongly flavored ingredients such as anchovies, and grated Parmesan cheese (see Brassicas in a Hot Skillet, page 146). Quicker still, stir-fry smallish florets in healthy oil for several minutes until softened around the edges. Add a ½ cup water or light stock and cover to steam until tender. When the liquid has nearly evaporated, add whatever flavorful ingredients go with the rest of the meal—garlic and chile, garlic and ginger, cumin and coriander, etc.

Brassicas become sweetish and slightly crispy when cooked at high heat. Toss florets with oil and salt, and roast in a single layer in a 450°F oven or grill until the edges are all but charred.

A well-grown cauliflower is a backyard triumph. Celebrate it by roasting whole and serving with bagna cauda (see recipe, page 152).

CAULIFLOWER • BRUSSELS SPROUTS • CARROTS • SHALLOTS • HERBS

ALSO FOR: LEEKS, CELERY, FENNEL, GREEN BEANS, EGGPLANT

STARTER OR SIDE | SERVES 6 TO 8 | VEGAN, GLUTEN-FREE | EARLY SUMMER, FALL, WINTER

CAULIFLOWER À LA GRECQUE

In the canon of French cuisine, a vegetable prepared à la grecque is made "Greek" by the use of lemon juice, olive oil, and coriander. Julia Child's version, perfect in its way, must have seemed exotic in 1961. She opened her chapter on "cold buffet" with variations on the theme—mushrooms à la grecque, eggplant à la grecque, leeks à la grecque, artichokes à la grecque, et cetera à la grecque. The vegetable was poached in a court bouillon, then dressed in the reduced cooking liquid. The result is elegant, simple—and a wee bit underpowered by today's standards, a vintage Citroën on the modern culinary superhighway. Twenty-first–century eaters redline at a higher RPM, their palates having been calibrated to the spice and big flavors of Mexico, China, India, Korea, North Africa, the Caribbean, the American South, and Southeast Asia.

Punchier than the classic, this version is still French-ish. Elemental chunks of cauliflower soak up the tangy dressing. Brussels sprouts and carrots, two fall companions, add color. Trimmed leeks would fit right in, as would mushrooms. But really any vegetable firm enough to poach could join in, à la Julia Child.

Vegetables à la grecque are traditionally served cold, a piquant side dish less sharp than a pickle. They are best slightly warm.

2 cups dry white wine
1/3 cup fresh lemon juice
Bouquet garni: 8 to 10 parsley stems, 3 sprigs thyme, 1 bay leaf, 1 small stalk celery (with leaves), tied with string
1/4 cup diced shallot
2 cloves garlic, lightly crushed
1/3 cup extra-virgin olive oil, plus more for drizzling
1/2 teaspoon peppercorns, lightly crushed
1 teaspoon coriander seeds, lightly crushed
1/4 teaspoon fennel seeds, lightly crushed
1 pound cauliflower, broken into florets (about 6 cups)
1/4 pound small carrots or carrot batons
1/2 pound Brussels sprouts, halved or quartered, 2 cups
1 tablespoon minced fresh parsley
Flaky finishing salt
Aleppo or Marash pepper flakes or freshly ground black pepper

1. In a medium saucepan or Dutch oven, combine the wine, lemon juice, herb bundle, shallot, garlic, olive oil, peppercorns, coriander, and fennel. Bring to a boil over high heat. Reduce the heat to a simmer and cook uncovered for 10 minutes. Discard the herb bundle.
2. Add the cauliflower and increase the heat to maintain a slow boil. Poach until tender but not mushy, turning over occasionally, 10 to 15 minutes. Using a slotted spoon, transfer to a large bowl.
3. Add the carrots and poach for about 5 minutes. Don't overcook. Add the carrots to the cauliflower.
4. Poach the Brussels sprouts until the interior is slightly crunchy, about 10 minutes. Add the sprouts to the other vegetables.

5. Turn up the heat and reduce the poaching liquid to about ⅔ cup. Pour over the vegetables. Marinate for 1 hour or up to overnight.

6. To serve, drizzle with the olive oil. Sprinkle with the parsley. Add finishing salt and pepper to taste.

To adapt for leeks, celery, fennel, green beans, eggplant: Trim the vegetable into manageable but still chunky pieces. Make the poaching liquid per recipe. Adjust the cooking times to suit the vegetable's size and density.

BROCCOLI OR CAULIFLOWER • GARLIC • DRIED CHILES • PARSLEY | SIDE | SERVES 6 TO 8 | VEGAN OPTION, GLUTEN-FREE OPTION | FOUR SEASONS

ALSO FOR: SPROUTING BROCCOLI, BROCCOLINI, BROCCOLI RABE, OR BRUSSELS SPROUTS

BRASSICAS IN A HOT SKILLET

This is a universal method for cooking broccoli and cauliflower. Blanch or steam the trimmed florets, then dress them by tossing in a hot skillet with olive oil, butter, bacon fat, ghee, or any other fat you like. Seasonings can be subtle or aggressive. An Italian-ish take would be: olive oil + garlic + red chile flakes. Jane Grigson, the masterful English vegetable cook, adds a colorful touch: olive oil + 2 dried red chiles + ½ red bell pepper, chopped. But the possibilities go on:

Butter + lemon juice (+ grated zest + minced parsley)
Olive oil + lemon juice + dried oregano + fresh mint
Butter + black pepper + grated Parmigiano-Reggiano
Olive oil + garlic (+ chopped anchovies) + olives + capers
Ghee + ground cumin and coriander + garlic + fresh ginger
Bacon fat + lardons + raisins + pickled onions + almonds

Finish any version with a garnish of grated cheese or crunchy toasted bread crumbs if you like.

Fine sea salt, for the blanching water
1½ pounds broccoli or cauliflower, cut into florets, stalks peeled and cut into ½-inch-thick rounds
3 tablespoons extra-virgin olive oil
2 cloves garlic, sliced
Big pinch of red chile flakes
½ lemon
Chopped fresh parsley
Freshly ground black pepper
Parmigiano-Reggiano cheese, for grating (optional)
Toasted bread crumbs (optional)

1. Bring a large pot of generously salted water to a boil. (Salt the water fearlessly. Only a tiny fraction gets absorbed.) Add the broccoli and cook just until the florets are tender all the way through to purge the raw sulfurous edge, 3 to 5 minutes. (Alternatively, cook in a steamer basket, but note that steamed broccoli will fade and lack flavor until seasoned.) Drain in a colander.
2. In a large cast-iron skillet, heat the olive oil over high heat until it shimmers. Add the garlic and chile flakes and sizzle for a minute. Add the drained vegetables and toss until well coated. Remove from the heat. Squeeze over the lemon and sprinkle with the parsley. Add generous grinds of black pepper. Toss to combine.
3. Turn onto a warm serving platter. If desired, top with grated Parmigiano-Reggiano and toasted bread crumbs.

To adapt for sprouting broccoli, broccolini, broccoli rabe, or Brussels sprouts: Blooming broccoli should be cut into manageable lengths—stems, leaves, buds, blossoms, and all. For Brussels sprouts, remove the outermost leaves and cut in half or, if small, leave whole but slash an X in the base to insure the stem cooks through. Blanch each until tender—3 minutes for sprouting broccoli, 6 to 8 for sprouts, depending on size. Blanch broccoli rabe whole. Drain and squeeze out as much water as possible. Chop as coarsely or finely as you like.

BROCCOLI SAUCE

Broccoli cooked until the florets collapse becomes a fantastic rough sauce to spread on grilled bread or toss with pasta and grated cheese. It's also a satisfying, if drab, side dish for roasty meats.

Follow the main recipe, adding an extra glug of olive oil when tossing the broccoli with garlic and chiles in the skillet. Pour in 1 cup water, cover, and simmer, stirring occasionally and adding more water as needed, until the florets break down, 45 minutes to 1 hour. Mash chunks with the back of a wooden spoon. Taste as you go and be liberal with salt and pepper—remember the result is meant to be highly seasoned, like a sauce. At the end, loosen with a pour of olive oil or water to serve.

BROCCOLI • GARLIC • MINT • ROSEMARY

ALSO FOR: SPROUTING BROCCOLI, BROCCOLINI, CAULIFLOWER, ZUCCHINI

MAIN | SERVES 8 | PESCATARIAN | SUMMER, FALL

ROASTED BROCCOLI PASTA SALAD

My working life started on Madison Avenue in Manhattan, after I fell into an editorial assistant job at a glossy magazine. Lunch usually came from one of several indistinguishable Midtown delis that all had the same buffet table with the same pasta salad. You shoveled the stuff into a clear plastic clamshell to be weighed by a brusque checkout clerk. Lunch by the pound.

This is not that pasta salad, apart from the nostalgic element of bow tie pasta with broccoli florets (here roasted for the crispy charred bits) and cubed mozzarella. What's different are the powerful flavors of preserved lemon, anchovies, fresh mint, and minced rosemary. You can dial them back, if you prefer, but bear in mind that a pound of dried pasta needs a lot of oomph for lift off.

Pasta salad is best slightly warm or at room temperature. It doesn't lose too much overnight in the refrigerator provided you take it out an hour before serving to unchill. The layered flavors make this a satisfying meal-in-one and, because pasta salad travels well, it's good for a picnic, road trip, or carry-on bag.

1 pound bow tie pasta
1/2 cup extra-virgin olive oil
3 cloves garlic, sliced
1 tablespoon Aleppo pepper (or use paprika with a pinch of cayenne)
4-inch sprig rosemary
2 pounds broccoli, broken into florets (about 10 cups)
2 teaspoons fine sea salt
1 preserved lemon, pulp scraped out and discarded, peel chopped
6 salt-cured anchovies, rinsed, boned, and chopped, or 12 oil-cured anchovy fillets
1/4 cup coarsely chopped fresh mint or basil leaves
1 1/2 tablespoons minced fresh rosemary leaves
Grated zest of 1 lemon
1 lemon, halved
12 ounces mozzarella cheese, cut into 1/2-inch cubes
3 ounces Parmigiano-Reggiano cheese, grated (about 1 cup)

1. Preheat the oven to 400°F.

2. Bring a large pot of salted water to a boil. Set up a bowl of ice and water and have at the ready. Add the pasta to the boiling water and cook to al dente according to the package directions. Drain and cold-shock in the ice bath. Drain well.

3. In a small pot, combine ¼ cup of the olive oil, the garlic, Aleppo, and rosemary and warm over medium heat until the garlic sizzles. Remove from the heat and set aside to infuse for 20 minutes. Strain through a fine-mesh sieve and discard the solids.

4. In a large bowl, toss the broccoli florets with the remaining ¼ cup of olive oil and 1 teaspoon of the salt. Toss until coated. Spread on a sheet pan and transfer to the oven. Roast until cooked through and charred at the edges, about 15 minutes. With a spatula, return the cooked broccoli to the bowl.

5. Add the infused olive oil, preserved lemon, anchovies, mint, rosemary, and lemon zest. Squeeze in the juice from half of the lemon and mix well. Add the pasta, mozzarella, Parmigiano, and remaining salt and mix thoroughly. Taste and squeeze the other lemon half over the top, or else serve with lemon wedges.

To adapt for sprouting broccoli, broccolini, cauliflower, or zucchini: Trim the vegetables into manageable pieces. Substitute for the broccoli and adjust the roasting time, as needed.

BROCCOLI • GARLIC • DILL | SIDE | SERVES 6 TO 8 | VEGETARIAN OPTION | SUMMER, FALL

ALSO FOR: CAULIFLOWER, BRUSSELS SPROUTS

GRILLED BROCCOLI ADOBO

Just before the garden wallops us with squash and the suite of high-summer produce, the last spring brassicas bridge the gap to grilling season. Here, a lip-tingling marinade of canned chipotles in adobo lays a smoky foundation for big flavors, but honey is the secret ingredient from my kitchen collaborator Dash. Don't fear the umami-rich fish sauce: It slips past even a sensitive nose to add depth without a recognizable fishy essence. This recipe pairs well with Barely Cooked Corn (page 186) and grilled shrimp.

2 pounds broccoli
7½-ounce can chipotle peppers in adobo sauce, pureed with liquid
2 tablespoons honey
2 tablespoons minced garlic
1 tablespoon coriander seeds, toasted and ground
2 teaspoons fine sea salt
1 teaspoon Red Boat fish sauce (optional)
¼ cup extra-virgin olive oil
½ cup picked dill fronds, or whole leaves of other herbs, such as cilantro and mint
2 limes, cut into wedges
Grated zest of 2 limes

1. Prepare the grill or preheat the oven to 450°F with a rack in the top position.
2. Peel the broccoli stems and cut the heads into long florets (to lay lengthwise across the grill). Place in a large bowl.
3. In a small bowl, whisk together the pureed chipotles, honey, garlic, coriander, salt, fish sauce (if using), and olive oil. Pour over the broccoli and massage in until all the florets are coated.
4. On the grill: Grill the broccoli over hot coals until tender and charred, 6 to 8 minutes per side. In the oven: Spread on a baking sheet and roast in the oven on the top rack until browned and crispy at the edges, 10 to 12 minutes.
5. Arrange the broccoli on a serving platter and scatter the dill over. Spritz it with several lime wedges and garnish with the zest.

CAULIFLOWER • GARLIC • SHALLOTS • THYME | MAIN, SIDE | SERVES 6 TO 8 | PESCATARIAN, VEGETARIAN OPTION, VEGAN OPTION, GLUTEN-FREE | SUMMER, FALL

ROASTED CAULIFLOWER
WITH BAGNA CAUDA

A young cauliflower plant bides its time in the garden, attractive but vague. Nothing about it suggests flamboyant maturity. The faint powdery bloom on its glaucous leaves looks like blushing in reverse, the applied pallor of a ballerina's face. Then something unthinkable happens. Atop the apical bud, the plant forms a clenched fist of pure white, a strange color in the garden, and punches itself into existence. The curd swells into a head. Unless quickly snatched in at this ripe moment, it will burst open over several days, flying into pieces, an explosion of flowers. Such drama!

Before that happens, seize it and roast whole until browned and tender as flesh. The dressing is a matter of taste, but cauliflower favors the bold. Splashed with bagna cauda, a potent warm sauce flavored with garlic, anchovies, and garden thyme, the head is fit to serve on a platter as the centerpiece of a garden meal. For a vegetarian option, omit the anchovies. For a vegan alternative, serve instead with Universal Green Sauce (page 248).

Roasted cauliflower can play several roles. As an appetizer, serve it fondue-style with chunks of crusty bread, trimmed artichoke hearts, and raw fennel or celery for dipping in a communal dish of bagna cauda. As a vegetarian main course, the plate can be rounded out with little boiled potatoes tossed with pesto and simple green beans. A side of roasted cauliflower drizzled with bagna cauda goes particularly well with beef, whether grilled flank steak or beef stew.

1 large head cauliflower, trimmed
Fine sea salt
Extra-virgin olive oil
Bagna Cauda (recipe follows), for serving

1. Preheat the oven to 400°F.
2. Plunge the cauliflower into a large bowl of cold salted water for a few minutes to purge aphids.
3. Drain and place the head upside down on a cutting board. Remove the small leaves that cling at the base like damp curls. Carefully pare the thickest part of the core while leaving it intact.
4. Place the head right side up in a cast-iron skillet. Drizzle with olive oil and rub all over to coat. Sprinkle generously with salt, inside and out. Loosely cover with aluminum foil and transfer to the oven.
5. After 30 minutes, remove the foil and baste with olive oil. Continue to roast, uncovered, basting occasionally, until tender when poked with a skewer, 1 hour or more.
6. At the end, brown the head under the broiler for a few minutes, if you like. Bring to the table whole. To serve, cut into wedges and splash with the bagna cauda.

BAGNA CAUDA

MAKES ⅔ CUP

¼ cup extra-virgin olive oil
5 tablespoons unsalted butter
1 teaspoon minced garlic
2 tablespoons minced shallot
Pinch of red chile flakes
¼ teaspoon chopped fresh thyme
3 whole salt-cured anchovies, rinsed, cleaned, and chopped (2 tablespoons)
Grated zest of ½ lemon
2 tablespoons fresh lemon juice

1. In a small saucepan, combine the olive oil, butter, garlic, shallot, chile flakes, thyme, and anchovies. Warm over low heat, without sizzling, until the oil is fragrant and infused, about 10 minutes.
2. Remove from the heat. Stir in the lemon zest and lemon juice. Taste and adjust seasonings as needed.

CHICKEN DIVAN

FOR MY MOM

My mother imbued me with a love of vegetables and gardening. As a single working mom, she somehow managed to put real food on the table, for several years with the help of Mrs. Braswell, a retired neighbor who smoked like a chimney, loved *Days of Our Lives,* and attributed the vigor of her old age to car-free living—she walked everywhere. The combined efforts of these two women kept meat and potatoes on the table with plenty of vegetables on the side. Much of what we ate came from my West grandparents' farm in Tennessee, carried home in the rear seat of my mother's baby-blue Volkswagen Beetle. Homegrown food and weeknight cooking. This book doesn't attempt anything more.

Eating was a challenge for my mother at the end, after radiation treatment, but one thing she could manage was rich, saucy Chicken Divan, a broccoli-chicken casserole. It's satisfying in the way of one-dish meals, and the taste of Marsala cooking wine evokes 1950s "Continental" cuisine. The minor trick is to soften the broccoli without baking the succulence out of the chicken. In this version, I make a mornay sauce to bind the flavors—rather than the usual condensed soup—and the 10 extra minutes of prep makes a world of difference.

6 tablespoons extra-virgin olive oil, plus more for the baking dish
Fine sea salt, for the blanching water
2 pounds broccoli, cut into florets (about 10 cups)
1½ pounds boneless, skinless chicken thighs
2 medium onions, thinly sliced (about 2 cups)
3 cloves garlic, peeled and thinly sliced
1½ teaspoons mixed chopped fresh herbs, including rosemary, thyme, and savory
4 tablespoons unsalted butter
¼ cup all-purpose flour
2¼ cups whole milk
Bay leaf
2 sprigs fresh or dried thyme
Freshly ground white or black pepper
3 tablespoons Marsala or cooking sherry
3 ounces extra sharp cheddar cheese, grated (3 cups lightly packed)
1 ounce Parmigiano-Reggiano cheese, grated (⅓ cup lightly packed)
¼ cup minced fresh parsley
2 tablespoons fresh lemon juice
1 cup fine dried bread crumbs

1. Preheat the oven to 350°F. Lightly oil a 9 × 13-inch baking dish.
2. Fill a large bowl with ice and water. Bring a large pot of well-salted water to a boil. Working in batches, blanch the broccoli until just tender, about 5 minutes. Lift it out with a slotted spoon and cold-shock in the ice bath. Drain and spread it out on clean kitchen towels to dry.
3. In a large skillet, heat 2 tablespoons of the olive oil over high heat until it shimmers. Add the chicken in a single layer and cook until nearly cooked through, about 10 minutes, taking care not to overcook—it can be slightly pink at the center. Transfer the chicken to a plate to cool.
4. Add another 2 tablespoons of olive oil to the same skillet. When it shimmers, add the onions, garlic, and chopped herbs. Reduce the heat to medium and sauté, stirring frequently, until completely soft but not browned, about 10 minutes. Empty into a large bowl.
5. To make the mornay sauce, first make a béchamel. In the same skillet, melt the butter and sprinkle over the flour. Stir and let the paste bubble without browning, about 2 minutes. Whisk in the milk a little bit at a time and stir constantly until it thickens. Add the bay leaf, thyme sprigs, and pepper. Reduce the heat to the lowest setting to infuse for 10 minutes, stirring occasionally. Discard the herbs and stir in the Marsala, cheddar, and Parmigiano, which turns the béchamel into a mornay.
6. Add the mornay to the onion mixture, then add the parsley. Add the chicken with any accumulated juices. Add the broccoli. Sprinkle the lemon juice over top. Toss until the broccoli and chicken are thoroughly coated. Pour the mixture into the oiled baking dish.
7. In a small bowl, combine the bread crumbs and the remaining 2 tablespoons of olive oil and mix until coated. Sprinkle over the casserole.
8. Bake until the top is browned and crunchy, about 40 minutes.

CH.

14

CABBAGES, COLLARDS, AND BRUSSELS SPROUTS

HEADING BRASSICAS

Brassicaceae,
THE MUSTARD FAMILY,
ALSO KNOWN BY THE OLDER NAME
Cruciferae

Brassica oleracea var. *acephala*
COLLARDS

Brassica oleracea var. *capitata*
EUROPEAN CABBAGE

B. oleracea var. *gemmifera*
BRUSSELS SPROUTS

B. oleracea var. *saubada*
SAVOY CABBAGE

B. rapa subsp. *pekinensis*
NAPA CABBAGE

B. rapa subsp. *chinensis*
BOK CHOY

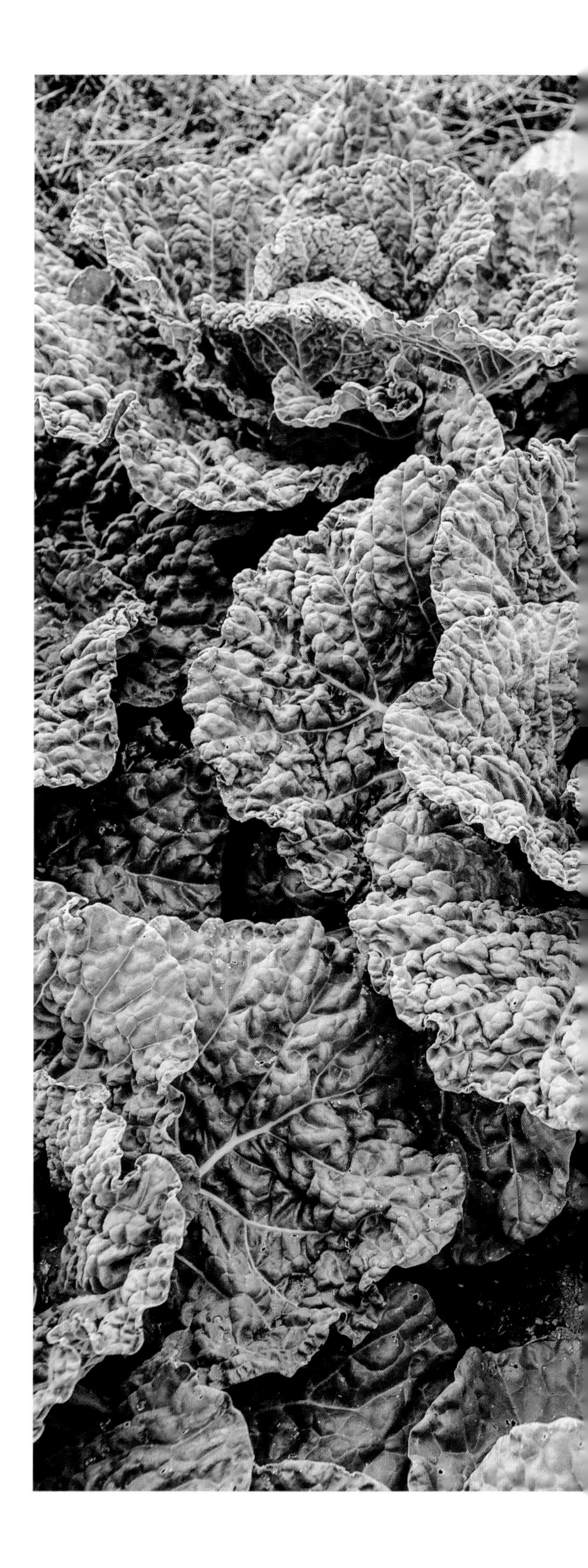

JUNE 24: Immature Savoy cabbage at Indian Line Farm

Cabbages belong to the protean species *Brassica oleracea*. Much of what I said in the previous chapter about the budding brassicas, broccoli and cauliflower, holds true here for the heading brassicas. They grow slowly and take up a lot of space. They attract squishy pests. And they produce a single harvest. But I couldn't have a garden without cabbages. Just look at them!

This chapter also covers the more compact Asian cabbages, mini-headed Brussels sprouts, and sibling collards, which usually drape themselves with leaves like layered petticoats but on occasion gather them overhead into a loose crown.

The European cabbages with their thick waxy layers and burly heft are biennials grown as annuals. They comprise the entire roster of the cultivar group Capitata, a reference to the noggin, as in Capitol and decapitate. European cabbage takes three forms: pale green with a whitish interior, anthocyanin-rich purple, and Savoy, which has thin leaves puckered like seersucker.

I grew up eating stewed green cabbage, which the cooks in my family seasoned with black pepper, salt, and butter, then placed on the back burner for a good long time. The cooking liquid was simple but profoundly flavorful, like the pot liquor from greens. I've never understood the snobbery against cabbage as an uncouth vegetable. In former centuries it was regarded as a delicacy, and choice varieties were esteemed for their refined flavor. Some were preferred for eating fresh, others for winter keeping. Particularly delicate are the thin-leafed Savoy varieties and a modern hybrid with a flattened head called Tendersweet.

Shredded and seasoned raw cabbage is coleslaw, from the German *kohl*. Salted and packed into crocks, shredded cabbage transforms with the assistance of lactic bacteria into sauerkraut. The venerable *Chez Panisse Vegetables* describes sauerkraut as "a new vegetable altogether." When I lived in Paris, I'd sometimes walk to an Alsatian brasserie on Île Saint-Louis for an eating spree. Heaps of choucroûte garnie—braised sauerkraut garnished with sausages, slab bacon, and potatoes—arrived with steins of beer. Like a python that swallowed a goat, I wouldn't eat again for days.

Asian cabbages mature more quickly and take less space, ideal for a small garden or raised bed. Fall-harvested Asian cabbage is preserved in Korea as kimchi; *The Cambridge World History of Food* supposes the practice is as old as agriculture. Crinkle-leafed napa is the standard kimchi-type cabbage, but juicy bok choy with its loosely bunched, thick-stalked leaves is also a cabbage, albeit of the Cheninsis cultivar group. The broader cohort of Asian greens spans a gamut from leafy mizuna to thick-stemmed gai lan. Bok choy is eager to grow. It self-seeded all over my garden for 2 years after I let a few plants bolt for pollinators.

Brussels sprouts are a strange vegetable. In the garden, they resemble collards with an unusual knack for setting buds at each leaf union along the stalk. If left to overwinter in a mild climate, each bud would unfurl a flower stalk. In the kitchen, Brussels sprouts can be shredded for slaw, a cabbage in microcosm, or steamed and tossed in a hot skillet with strong seasonings. See Brassicas in a Hot Skillet (page 146). They are an excellent vehicle for fish sauce and other umami-forward flavorings.

In the Garden

Cabbages look splendid as they grow, even after caterpillars, worms, snails, and slugs make cut-work of the wrapper leaves. The dense head escapes with only superficial damage.

Cabbage is a cool weather crop. Put it out well before last frost or plant in summer for a fall harvest. Late cabbages meant for keeping often have names such as Stonehead that indicate durability. Carefree and heat-tolerant Asian cabbages can be grown throughout the season. I sow and plant collards exactly as for heading cabbage and they provide repeat harvests into early winter. Brussels sprouts take up to

JUNE 28: Copenhagen Market cabbage

AUGUST 3: Churchill Brussels sprouts

AUGUST 13: Georgia Blue collards in need of picking

OCTOBER 31: Brussels sprouts hold steady through first snow.

4 months to mature; put out transplants before last frost and harvest after first frost has sweetened their flavors. Mature stalks hold well in the garden, extending the harvest period. I've picked them as late as January, after the onset of sub-zero nights.

Start seeds for European cabbage and collards 10 weeks before last frost, sowing 2 to 3 seeds per cell and selecting the strongest. Protect transplants against cutworms with a paper collar. To direct sow, up to 6 weeks before last frost, put out 3 or 4 seeds per hill, spacing hills 18 inches apart, with 2 feet between rows. Seedlings are slow to emerge. Sow fall cabbage starting late May and June. (Notice in the picture on pages 156 and 157 how offset rows maximize space.)

For napa cabbage and bok choy, sow directly in the garden at last frost. Napa matures in about 2 months, while "baby" bok choy will be ready in a month. Sow successions until 4 to 6 weeks before average first frost.

My recommendation for Brussels sprouts is to buy transplants and put them out as early as possible.

With their size and long growing season, European cabbages, collards, and Brussels sprouts are heavy feeders. Enrich the ground around them with compost and mulch seedlings. Stay vigilant against creepy-crawlies. Seek and destroy the inevitable slugs, snails, and caterpillars. Red cabbage is somewhat more resistant to pests in my experience.

The pointy heirloom Early Jersey Wakefield (63 days from transplanting) has set the standard for cabbage flavor in America since the 1840s. Copenhagen Market, another heirloom for fresh eating, produces crack-resistant small heads in 70 days. Ruby Perfection (85 days from emergence) is a standard late red cabbage, with 3-pound heads as solid as burl. It holds well in the garden and keeps in storage. The pale green Storage #4 is slower to mature (95 days), but likewise maintains its eating quality. For containers and small gardens, the compact variety Katarina matures in just 45 days from transplanting and produces 4-inch mini-heads on compact plants.

Cabbage seeds are small, with as many as 10,000 per ounce or 150 to 350 per gram. Expect a single head per plant—cabbage doesn't resprout. A 10-foot row will accommodate 7 plants, slightly squeezed. Mature heads typically weigh from 3 to 5 pounds—although Tribute produces basketball-sized whoppers of up to 10 pounds.

A recipe calling for 2½ pounds of cabbage will feed 6 to 8 people.

IN THE KITCHEN

Harvest European cabbage when the head is full and shiny. It should feel dense and compact when squeezed. (Savoy cabbage will be less shiny and less dense. See page 156.) Each variety has its average mature size—consult the seed packet. Sever the head at its base with a sharp knife and trim the outermost wrapper leaves before bringing it in from the garden.

Gather early varieties for fresh eating on the small side before hot weather. The early types don't hold well in the garden and will split after heavy rain. Large, slow-to-mature storage varieties will remain sound in the garden for weeks in fall. As with other brassicas, light frost improves their flavor. The plant's sugar content doubles within the first month of chilly weather. In mild climates, cabbage will overwinter.

Once inside, remove several more wrapper leaves and look closely around the base for bugs. Store an uncut head in the refrigerator loosely wrapped in a kitchen towel inside an open plastic bag. The summer cabbages for fresh eating will last a week. To store winter cabbage, harvest as late as possible before temperatures fall into the twenties. Don't wash. Keep in a cold dark place above freezing temperatures—the back of the refrigerator or an unheated garage—loosely covered with plastic. Storage quality depends on the variety, but most will last for a month and some will keep until spring.

Cabbage can be delicious raw, as in slaw and salad, or cooked through and through—stewed, braised, or stuffed. The cheffy approach these days is to sear, scorch, or grill cabbage.

The Asian cabbages, less sulfurous, have a delicate flavor best suited to stir-frying and quick blanching. Napa cabbage makes light, crisp slaw and can be skillet-braised quickly, like Savoy. Small varieties of bok choy grow tidy and come into the kitchen clean—easy prep. Store napa loosely wrapped in the refrigerator, as for European cabbages, up to a month. Bok choy should be eaten soon after harvest. It doesn't hold well in the garden, either. Rely on succession plantings for a steady supply.

Brussels sprouts keep best on the stalk. They mature sequentially from the bottom. Pick what you need. Peel away the thickest outer leaves from each bud until you get to a shiny tight layer. The smallest buds can be cooked whole; slice an X in the base for more even cooking. At Gjelina, the landmark vegetable-centric restaurant in Venice, California, founding chef Travis Lett scorched Brussels sprouts in a screaming-hot cast-iron skillet, then tossed them with bacon, dates, and vinegar.

CABBAGE AND COLLARDS—EARLY

START INDOORS	DEPTH	EMERGENCE	TRANSPLANT	SPACING	ROW SPACING	DAYS TO MATURITY	MATURE SIZE
8–10 weeks before last frost	1/4 inch	5–17 days	2 weeks before last frost	18 inches	24 inches	60–80 days from emergence (subtract 20 days for transplants)	15 inches tall, 18-inch spread

Note: Sow 2–3 seeds per cell and thin to strongest. For a fall harvest, direct sow in May or June.

CHINESE CABBAGE (NAPA) AND BOK CHOY—DIRECT SOWN

DIRECT SOW	DEPTH	EMERGENCE	SPACING	THIN TO	ROW SPACING	DAYS TO MATURITY	MATURE SIZE
Around last frost date and through midsummer	1/4 inch	2–15 days	2- to 3-seed clusters every 18 inches	1 seedling per cluster	24 inches	40–60 days	12–14 inches tall, 12- to 18-inch spread

GREEN OR RED CABBAGE | PANTRY | MAKES 3 QUARTS | VEGAN, GLUTEN-FREE | SUMMER, FALL

SAUERKRAUT

Biennial cabbage produces leafy growth in its first year before it hunkers down and waits to bloom in its second year. It's the hunkering phase that makes cabbage a good keeping vegetable. The greatest risk in storage is decay caused by microorganisms. One solution, a very ancient one, is to suppress microbial growth by salting. Salt also draws enough liquid from the leaves to submerge them beneath brine. Kimchi and sauerkraut are two expressions of fermented cabbage. The method given below for sauerkraut requires nothing beyond cabbage and salt. To explore the wide world of kimchi, consult the sixty recipes in Lauryn Chun's delightful *The Kimchi Cookbook*.

Among sauerkraut's many uses as a condiment, sandwich topper, and snack, it also makes a full meal when rinsed and stewed in hard cider, with or without a chunk of smoked pork and a few juniper berries. It is particularly good with steamed Carola potatoes.

5 pounds cabbage, quartered, cored, and sliced crosswise 1/8 inch thick

2½ to 3 tablespoons fine sea salt, plus more as needed

1. Put the shredded cabbage in a very large bowl, a few handfuls at a time, and lightly sprinkle with salt as you go. The cabbage should be salty in an agreeable, thirst-inducing way, like bar nuts. So start with 2½ tablespoons and add more to taste, if needed.

2. Bruise the cabbage with a wooden pounder or your fists. Continue to pound and knead until the cabbage is limp and drippy. Flatten in the bottom of the bowl and loosely cover. Set aside in a cool place overnight.

3. The next morning, knead the cabbage again to redistribute the juices. Pack it by handfuls into a clean, straight-sided gallon container. Compress each addition. The goal is to have a tightly compacted mass submerged in its own liquids. Top up the container with the remaining brine.

4. Place a weight on the cabbage to keep it submerged. Options include a heavy plate, a mason jar filled with water, or a ziplock plastic bag filled with brine rather than water, in case of leaks. For the brine, dissolve 2 teaspoons of salt in 2 cups of water.

5. Cover the container with a clean kitchen towel topped by a plate. Put in a cool, dark place, ideally between 55° and 70°F. Check the container daily. You will almost certainly find a speck of mold at some point. It's harmless—a superficial blemish—and can be removed with a spoon or paper towel and discarded. The cabbage will begin to ferment in several days. It will taste tangy within a week to ten days and be fully mature within about 2 weeks, or longer in a chilly setting. Whenever the taste is to your liking, put the sauerkraut in the refrigerator to slow further fermentation. It will keep for months.

Note: You can add to the shredded cabbage any combination of shredded carrots, turnips, or beets, if you like. Caraway seeds are a traditional flavoring.

BRUSSELS SPROUTS • SHALLOTS • PARSLEY

ALSO FOR: SAVOY CABBAGE

SIDE

SERVES 6 TO 8

VEGETARIAN OPTION

FALL, WINTER

WINTER SLAW

Coleslaw—which I thought was *cold slaw* until I was a teenager—is a creamy, tangy taste of summer. This wintery variation came together one night after my Churchill Brussels sprouts, like their namesake, refused to surrender. I was still picking them in January. In a rush one night, I sliced them thinly and threw them into a hot skillet with cream. The taste put me in mind of coleslaw from Scott Peacock, the foremost guardian of the South's home-cooking tradition and the former collaborator of the great Edna Lewis, god rest her soul. I dove into the spice cabinet for celery seed—the signature of coleslaw to my taste—and sharpened the flavor with vinegar. The result was warm slaw, suited to the winter months. Serve with seared lamb chops.

4 slices bacon, diced
4 tablespoons unsalted butter
2 large shallots, thickly sliced
3/4 pound Brussels sprouts, trimmed, halved, and thinly sliced
1/2 teaspoon fine sea salt, with more as needed
3/4 cup organic heavy cream
2 tablespoons minced fresh parsley
1 teaspoon celery seeds
2 tablespoons red wine vinegar or fresh lemon juice

1. In a large sauté pan, sweat the bacon over medium-high heat until it starts to render. Add the butter and shallots and sauté until the shallots soften, about 2 minutes. Add the Brussels sprouts and salt. Sauté until they soften, 5 to 7 minutes.
2. Add the cream and toss to combine. Reduce until the cream is mostly absorbed. Add the parsley and toss to combine. Remove from the heat. Taste and adjust the salt, as needed.
3. Turn into a warm serving dish. Scatter the celery seeds over the top and sprinkle with the vinegar.

Note: For a vegetarian option, omit the bacon and increase the butter by 2 tablespoons. Adjust salt to taste. Proceed with the recipe.

To adapt for Savoy cabbage: Replace the Brussels sprouts with thinly sliced Savoy cabbage. Adjust cooking time and cream, as needed—the cabbage should be softened but not mushy, slicked with cream but not dripping. Standard smooth-leafed cabbage requires longer cooking and is not right for this recipe.

RED CABBAGE • FENNEL | SIDE | SERVES 6 TO 8 | VEGETARIAN OPTION, GLUTEN-FREE | SUMMER, FALL, WINTER

MISO-BRAISED RED CABBAGE
WITH FENNEL

Braising red cabbage in wine preserves the color, but the flavor is reason enough. Here I use Riesling because its residual sugar balances the acid, but other wines would work as well, including light fruity reds. Or improvise a nonalcoholic braising liquid with a pint of water, a tablespoon of red wine vinegar, and a teaspoon of honey.

Red cabbage is spicier and firmer than green. The densest fall-harvested varieties will require considerable cooking time. Like potatoes, cabbage absorbs fat as it cooks, becoming silky and luxurious.

The sweet white miso stirred in at the end falls into the category of sneaky umami ingredients. You can do without, but it's an ingredient worth getting to know. Apple is a classic pairing with red cabbage. Here the Granny Smith isn't stewed but diced raw for a garnish.

3 slices bacon, cut into lardons (optional)
4 tablespoons unsalted butter
2½ pounds red cabbage, quartered, cored, and shredded (about 12 cups)
1 teaspoon salt
½ teaspoon freshly ground black pepper, plus more for grinding
2½ cups dry or off-dry Riesling
1 tablespoon white miso
1 tablespoon honey
1 Granny Smith apple, cut into ¼-inch dice
½ cup chopped fennel fronds

1. In a 6-quart Dutch oven, render the bacon over medium heat, if using. Remove the lardons and save for another use, such as snacking while the cabbage cooks.
2. Add the butter to the bacon fat. When foamy, add the cabbage, salt, and pepper. Stir to coat and cook down for 5 to 7 minutes. Add the wine. It should not quite submerge the cabbage. Cover the pot, reduce the heat to maintain a steady simmer, and braise for 1 hour.
3. Check the cabbage to gauge how much longer it will need, but don't rush it. Red cabbage cooked this way doesn't turn mushy but silky—eventually. Keep going until the texture relents, about 1 hour longer. Cooking time can vary depending on the variety, the season, and how the cabbage is cut. There should be a little braising liquid left at the end. Add water by the ¼ cup, if necessary.
4. Just before serving, stir in the miso and honey. Turn the cabbage over several times to coat. Transfer to a serving bowl. Top with the diced apple and fennel fronds. Finish with generous grinds of black pepper.

Note: For a vegetarian option, omit the bacon and increase the butter by 1½ tablespoons. Adjust salt to taste. Proceed with the recipe. Drizzle with walnut oil before serving.

BUTTERED GREEN CABBAGE

I cook buttered green cabbage—sometimes called white cabbage—weekly when cool weather arrives in September. It goes with everything and pairs unexpectedly well with pan-roasted salmon and steelhead trout. Edna Lewis suggests stirring in a handful of chopped scallion greens toward the end of cooking to brighten the drab color.

- 4 tablespoons unsalted butter
- 1 small onion, thinly sliced
- 1 clove garlic, lightly crushed and peeled
- 2 pounds green cabbage, cut into 1-inch-wide strips
- 1 teaspoon salt, plus more to taste
- 2 × 3-inch piece of kombu
- 2 ounces smoked pork (optional)
- Freshly cracked black pepper

1. In a large Dutch oven, melt 3 tablespoons of the butter over medium heat. Add the onion and garlic and cook for several minutes to soften.
2. Add the cabbage and turn it in the butter until it relaxes, 5 to 7 minutes. Half-cover with water. Add the salt, kombu, pork (if using), and lots of black pepper. Simmer until the cabbage is tender, about 30 minutes.
3. Stir in the remaining 1 tablespoon of butter and more salt and black pepper to taste. The broth should be highly seasoned.

SAUTÉED SAVOY CABBAGE

Savoy cabbage takes its name from the Savoy region of France. Its signature puckered thin leaves cook quickly. The classic pairing of sautéed cabbage with grilled sausages can't be beat.

- 3 slices bacon, chopped
- A handful of mirepoix: equal parts diced onion, diced carrot, and diced celery
- 1 small head Savoy cabbage (about 2 pounds), shredded
- White wine
- Bouquet garni (see page 144)
- 1 clove garlic, smashed and peeled
- Champagne vinegar

1. In a large skillet, brown the bacon. Add the mirepoix and cook to soften, 5 to 7 minutes.
2. Add the cabbage and turn to coat. Deglaze the pan with a small glass of white wine, then add water or stock to a depth of ½ inch. Tuck in a small bouquet garni and the garlic. Cover and cook at a lively simmer until the cabbage is tender, 15 to 20 minutes.
3. Finish with a splash of Champagne vinegar to taste.

AUGUST 21: Brussels sprouts in the garden with red and green cabbage in the background

CRISPY BRUSSELS SPROUTS
WITH PEAR AND CHÈVRE

This is a party dish, naughty but fun—flash-fried Brussels sprouts tumbled with the big fall flavors of pears, toasted nuts, and fresh goat cheese. Snacky, like an appetizer to scoop with crackers, it could also be served on a bed of dressed salad greens before a low-key main dish such as pork tenderloin.

Tweak the recipe to suit what you have on hand. Apples can replace pears. The balsamic doesn't have to be white. Hazelnuts have an affinity for Brussels sprouts; also try walnuts, pecans, or almonds. The fragrant herbs can vary.

2 red d'Anjou pears
Juice of 1/2 lemon
8 ounces fresh goat cheese
1 tablespoon minced mixed pungent herbs, such as thyme, rosemary, and savory
4 cups grapeseed oil
1 1/4 pounds Brussels sprouts, trimmed and halved or quartered
Flaky finishing salt
1 cup lightly packed fresh parsley leaves
1 clove garlic, minced
1/4 cup golden raisins
1/4 cup hazelnuts, chestnuts, or other nuts, toasted and chopped
3 tablespoons white balsamic or balsamic vinegar
Freshly ground black pepper

1. Quarter and core the pears. Cut lengthwise into 1/4-inch-thick slices. To prevent browning, plunge them into a medium bowl of cool water acidulated with the lemon juice.
2. Crumble the goat cheese into a small bowl. Add the minced herbs and mash together with a stiff spatula until combined. Spread in a layer on a serving platter large enough to hold all the Brussels sprouts. Cover loosely with parchment and set aside.
3. Line a wire rack with paper towels. In a medium saucepan, heat the grapeseed oil to 375°F.
4. Working in small batches, fry the sprouts until the outer leaves are browned and crispy, 2 to 3 minutes. Drain on the paper towels and salt lightly while hot.
5. In a large bowl, toss together the sprouts, parsley leaves, garlic, raisins, nuts, and vinegar. Season with flaky salt and black pepper to taste.
6. Transfer to the serving platter, mounding the dressed Brussels sprouts over the cheese mixture. Drain the pears, pat dry, and tuck them in among the Brussels sprouts.

NAPA CABBAGE • GARLIC • ONIONS • GREEN CORIANDER OR CILANTRO | MAIN | SERVES 4 TO 6 | OMNIVORE | SUMMER, FALL, WINTER

NAPA CABBAGE STIR-FRIED
WITH PORK AND VINEGAR

Chen Li and Deeva Gupta came to the Berkshires after I posted an advertisement for garden help at Deep Springs College, my alma mater, where rigorous academic work goes hand in hand with a daily schedule of physical labor. Both had garden experience as well as undeclared skills. Deeva was born to be a carpenter: She built a brawny, three-bay compost bin inspired by Donald Judd's Minimalist furniture. Chen could cook—really cook. She intuitively knew ingredients and how to bring them together. She sensed when vegetables were perfectly done. And she organized her work with an eye for beauty. Skills that can't be taught.

Chen originated this recipe, based on a favorite sweet-sour pork dish from her childhood. The first of several times she made it during our time together, she called her father in Inner Mongolia to check the proportions and method. The key ingredients that inspired her were fat napa cabbages from the garden and a piece of pork belly from the freezer—the last of a pig Del and I had butchered the year before. The stir-fry she perfected is tangy, spicy, filling, and brightened by the explosive flavor of green coriander—cilantro that's gone to seed. Once you try it—which is likely only if you grow cilantro—green coriander is an ingredient you can't get out of your mind.

Fine sea salt
2 tablespoons soy sauce
1/4 cup rice vinegar
2 tablespoons oyster sauce
1 tablespoon mirin
3 tablespoons sugar
2 tablespoons cornstarch
1 head napa cabbage (about 2 pounds), separated into leaves
2 tablespoons vegetable oil
1 pound pork belly (see Notes), sliced as thinly as possible
3 tablespoons chopped garlic
4 to 6 chiles de árbol
1 tablespoon grated fresh ginger
1 medium onion, diced
2 tablespoons Sichuan peppercorns (see Notes)
2 tablespoons green coriander berries (see Notes) or a dozen cilantro sprigs
Cooked rice (such as heirloom black or Carolina Gold), for serving

1. Bring a large pot of salted water to a boil over high heat.
2. Meanwhile, in a small bowl, whisk together the soy sauce, vinegar, oyster sauce, mirin, sugar, cornstarch, and 2 tablespoons salt.
3. Add the napa cabbage to the boiling water and blanch until softened, about 3 minutes. Drain well. Either spread it out on a baking sheet lined with a clean kitchen towel to dry or dry it in a salad spinner.
4. Heat a wok over high heat until it smokes. Add 1 tablespoon of the oil and when it shimmers, add the pork. Stir-fry the meat until no longer pink, about 3 minutes. Remove to a plate and set aside.

5. Add the remaining 1 tablespoon of oil to the wok. When it shimmers, add the garlic, chiles, ginger, onion, and peppercorns and stir-fry for 2 minutes, until fragrant. Add the cabbage and stir-fry for 2 minutes. Pour in the soy sauce mixture and cook for 3 to 4 minutes, to thicken the sauce. Return the pork to the wok and stir-fry until cooked through, 5 to 8 minutes.
6. Transfer to a serving platter. Sprinkle with green coriander berries or strew with cilantro leaves. Serve with rice.

Notes

The pork belly will be easier to slice if put in the freezer, unwrapped, for 15 minutes, until firm.

Other cuts of pork, such as shoulder, don't work for this dish, being too juicy.

The recipe calls for Sichuan peppercorns, which create the notorious tingly, mouth-numbing sensation called *ma* in Mandarin. Black or green peppercorns could be substituted, although they lack *ma*.

If green coriander berries are not available in the garden, use cilantro leaf instead.

COLLARDS • ALLIUMS • TOMATOES • HERBS | MAIN | SERVES 6 TO 8 | OMNIVORE | SUMMER, FALL, WINTER

SPICY COLLARD STEW
WITH CHICKEN THIGHS

This flavor-bucket meal was improvised one summer to make use of the endless collards in my garden. It's a Southern braise, a hearty one-pot supper. But that means it also has a deeper history. Groundbreaking culinary historian Michael Twitty's research and writing have recovered a core truth of American cooking. The foodways of the South—the basis of our national table and the stuff of my culinary upbringing—are directly tied to the cuisines of West Africa. This dish contains echoes from across an ocean. Serve in large bowls with rice or fonio, an ancient grain from West Africa.

2 tablespoons grapeseed oil
8 small bone-in, skin-on chicken thighs
1 teaspoon fine sea salt
2 cups chopped scallions
4 cloves garlic, peeled and chopped
3 dried cayenne chiles, split lengthwise
¼ pound slab bacon or other smoked meat, cut into 4 pieces
1 cup canned whole peeled tomatoes
1 tablespoon tomato paste
1 teaspoon fish sauce
2-inch piece fresh ginger (peeled and grated) or 1 tablespoon dried
½ cup chopped fresh parsley
½ cup creamy peanut butter
1½ pounds young collards, ribbed and cut into dollar-sized strips
¼ cup extra-virgin olive oil
Freshly ground black pepper
1 tablespoon molasses or dark honey
1 tablespoon apple cider vinegar

1. In a large Dutch oven, heat the grapeseed oil over high heat until it shimmers. Season the chicken with half the salt. Working in two batches, brown the chicken, 7 minutes per side. Remove and set aside.
2. Add the scallions, garlic, chiles, and bacon to the Dutch oven and sweat for 3 minutes. Add 4 cups water, the canned tomatoes, tomato paste, fish sauce, ginger, parsley, and peanut butter. Bring to a boil and cook for 5 minutes. Add the collards and return to a boil.
3. Tuck in the chicken and pour the olive oil on top. Return to a boil, then reduce the heat to maintain a steady simmer and cook, uncovered, until the chicken is done, about 45 minutes.
4. Taste and season with the remaining salt, as needed. Add quite a bit of freshly ground black pepper. Stir in the honey and vinegar. Taste again and adjust as needed.

Notes

"Spicy" is highly subjective. The three cayenne peppers called for in the recipe make the broth lively but not fiery. Consider "spicy" an invitation to add more chiles to taste.

Depending on the season, you could replace the scallions with 1 cup of chopped fresh spring onions or diced storage onions. Likewise, substitute chopped green garlic for mature cloves.

An equally delicious version could be made with bone-in pork, omitting the peanut butter.

CH.

15

CORN AND OKRA

THE TALL COHORT

Poaceae
THE GRASS FAMILY

Zea mays var. *saccharata*
SWEET CORN

Malvaceae
THE MALLOW FAMILY

Abeloschus esculentus
OKRA

AUGUST 28: Early Sunglow, a short-season hybrid sweet corn

What Americans call corn and other English speakers know as maize is a robust grass, distantly related to wheat, barley, and rice. (Beyond our shores, "corn" refers to any grain crop.) Thick-cobbed corn was developed in Mexico eons ago from spindly teosinte, *Zea mexicana*. Ears of corn as we know them didn't exist in the wild. They couldn't. The papery husk is too tightly wrapped to release the seeds. Corn can't shuck itself. It coevolved with farmers—in effect training us to cultivate it.

Okra hails from Africa and is kin to hibiscus, as its crepe-paper flower indicates. It lands in this chapter only because it is another tall, lanky, hot-weather crop. Okra is misunderstood and underappreciated except by Southerners and some non-European culinary traditions. Many gardeners will consider it a minor vegetable, but I love okra, so I give the basics for growing it on page 183 and the basics for preparing it starting on page 189.

CORN

Fresh sweet corn, the type you might want to grow at home, has nothing to do with dried commodity corn harvested by the billions of bushels to manufacture ethanol gasoline, high-fructose corn syrup, animal feed, and junk food. *That* corn covers more acreage in America than any other crop. That corn is not what you buy by the dozen at honor-system farm stands during summer vacation.

This corn is the result of a naturally occurring genetic variation that caused the kernel to store energy as sugar rather than starch. (See Types of Sweet Corn, page 180.) The ears are harvested young and eaten at the "milk" stage, when a pressed kernel will burst its thin skin. Cut from the cob, sweet corn is "creamed" if you scrape the cob with the dull edge of your knife.

The first picking of corn is a milestone, but the harvest date varies widely by region. Del, who grew up in Vermont, learned that a good stand of corn should be "knee high by the Fourth of July." My Tennessee relatives are eating early corn by then. In the Berkshires, river-bottom farms start picking in late July and the corn keeps coming until mid-September.

A genre of rural snobbery exists around corn's freshness. At the half-joking extreme, you first get your pot of water boiling, then you go out to pick. There is a kernel of truth, if you will, behind the call to last-minuteness. Corn's sugars quickly convert to starch after picking, although new varieties bred for long-distance shipping keep better.

Dried corn, also called field corn, is harvested in early fall, after the stalks have withered and the kernels have turned to stone. Indeed, one varietal group that includes popcorn and other indigenous types is called flint corn. Dent corn—named for a depression atop the dried kernel—is also widely cultivated for human consumption. A handful of water-powered grist mills still grind corn between spinning millstones in the southern Appalachians—an old-timey music soon to fade, I'm afraid. My mother shipped me stone-ground meal because finely ground commercial cornmeal doesn't make decent cornbread. Masa and grits are ground from hominy, corn that has been treated with alkali in a process called nixtamalization. So-called flour corn includes the soft blue corns.

Dried corn has still another traditional use, perhaps its highest and best. I once met a farmer in the southern Appalachians whose family has been growing the same variety of corn for over 150 years. He described it to me as the best bread corn there is, and ground some for me to take home. Then he led me to his shed to explain a local euphemism. The principal product of bread corn emerges not from the oven but from the still, he said as he tapped a barrel of family moonshine. It's called bread corn because purveyors of untaxed whiskey have cause to be circumspect in this choice of words.

AUGUST 25: Avalon, a "triplesweet" hybrid, in my father's Tennessee garden

TYPES OF SWEET CORN

According to Southern food historian David Shields, three strains of sweet corn arose independently among the indigenous agriculturalists of the New World: in South America, in Mexico, and among the Iroquois of New York State. The Iroquois called theirs Papoon, and it dates to about 1750. The seed was seized by colonists as a spoil of war in 1779 and improved through crossbreeding. The original red cob was bred out, and sweet descendants proliferated during the 1850s, among them Old Colony and Stowell's Evergreen. (The latter is still available.) Country Gentleman Shoepeg, also still available, joined them in 1890. It is an extra-tall white corn that sets kernels randomly, not in familiar rows. (See the photo on page 182.)

The hybrid Silver Queen, introduced in 1950, eclipsed all earlier varieties. It was the wonder corn of my childhood in the seventies and eighties but has since been displaced by more sugary varieties. Much of what is sold today as Silver Queen is likely a newer hybrid traveling under an assumed name—Silver Queen having become the Kleenex of white corn.

According to Shields, the pre-Columbian strains of sweet corn from Mexico and South America are still grown on their native soil. Papoon seems to be extinct, but Shields keeps looking. It is "inconceivable," he writes, that it has disappeared.

Sweet corn comes in yellow, white, and bicolor. Growers and seed catalogues have established a shorthand code for the genetic types of sweet corn.

- Standard sugary, or *su*, is original sweet corn. These varieties have traditional sweet-corn texture and flavor. The sugars convert to starch quickly, so *su* corn needs to be picked at peak maturity and eaten as soon as possible. The window for prime eating quality is a matter of days, and these varieties might not satisfy the modern sweet tooth. On the plus side, they germinate better in cool soil and are generally hardier. Some require a very long growing season. Proven *su* heirlooms include Stowell's Evergreen (95 days), Country Gentleman (92 days), Silver Queen (92 days), and the small, early yellow corn Golden Bantam (80 days), another survivor, introduced in 1902.
- Sugary enhanced, or *se*, has more sugar, tender kernels, and slightly longer storage time. It will retain sweetness for 2 to 4 days after picking. Varieties are sometimes sorted as heterozygous (only one parent contributes the *se* gene) or homozygous (both parents contribute). They are less hardy in the garden. Popular varieties include the bicolor Peaches and Cream (83 days) and yellow Kandy Korn (89 days).
- Supersweet types, *sh2*, have up to 10 times as much as much sugar and will stay sweet for as long as 10 days after harvest. Critics say these varieties have traded away corn flavor for sugar sweetness. *Sh2* varieties, named for their shriveled seeds, germinate only in very warm soil and must be isolated by hundreds of feet to avoid cross pollination.
- Synergistic, *sy*, has multiple genetics in the same ear, with varying mixes of homozygous or heterozygous *se*, as well as *sh2* kernels. The result is extended harvest and storage windows.
- Augmented supersweet, or sugary enhanced, has all homozygous *se* kernels with some kernels also having *sh2* genetics.

In the Garden

Corn is a warm-weather crop—the queen of the summer garden. It goes out late, after all threat of frost has passed, and comes in from midsummer through frost. Although not challenging to grow—it is grass—corn is the very definition of a heavy feeder: large, vigorous, long-lived. Some varieties reach 10 feet tall, and my father once grew antique Hickory King, a large-kernelled favorite of his, that topped out at 16 feet. Corn benefits from extra nitrogen, either in the form of compost or a nitrogen-rich organic fertilizer. Side-dress each stalk with more nitrogen midseason. Corn produces a single crop that matures simultaneously. Succession planting extends the harvest window.

Unlike insect-pollinated squash and tomatoes, corn relies on wind for pollination. Pollen rains down from overhead tassels, the male reproductive structure, and falls onto receptive silks emerging from ears at mid-stalk. A single pollen grain travels the length of the silk to fertilize one embryonic ovule, a future kernel. This synchronized timing—of pollen release and silk receptivity—goes by the nifty word *nick,* as in nick of time. Ears mature in 2 to 3 months. Poor pollination results in sparsely filled ears, a hockey player's gapped smile.

Corn is sown directly in the garden after the last frost date. Seed sold as "treated" has been sprayed with a fungicide to prevent rot in cold or wet ground. Corn seedlings are very tender. I sow about 2 weeks after last frost.

For better pollination, plant corn in blocks or in a 12 × 12-foot grid on 3-foot centers. Corn is not ideally suited to containers. Burpee offers a compact variety called On Deck, but to judge from the hundreds of reviews on the seed company's website—far and away the most discussed corn variety—real-world performance is mixed.

Sow corn rather thickly—3 or 4 inches apart—and 1 inch deep. The emerging cotyledons look like oversized grass. Thin to 1 foot apart as they grow. For hills, sow 3 seeds and select the strongest. Mulch with compost and side-dress with compost at midseason. Keep corn weeded.

When interplanting with beans, give the corn a 2-week head start. Sow 2 beans per stalk after thinning. For the Three Sisters, plant squash or pumpkin at the same time as the corn, leaving at least 6 feet between hills. Weave the growing vines among the stalks to shade out weeds.

If you have space, you can extend the corn harvest by planting successions at 2-week intervals. Or chose two varieties, one early and one late, with a 2-week span between the respective maturity dates, long enough to put the bloom out of sync and prevent cross-pollination.

Other creatures love corn as much as we do. A corn patch will attract earworms, cutworms, stinkbugs, stalk borers, blackbirds, crows, squirrels, raccoons, opossums, deer, bear, and probably Sasquatch, too, if he's around. Crows are notorious for picking corn seed from the ground. Tent the row with chicken wire, removing it before the corn grows through. An electric fence will discourage four-legged intruders. Raccoons are perhaps the most vexatious. They let you know when your corn reaches peak ripeness by eating the entire crop the night before you plan to pick it.

I gather from talking to commercial corn growers that it is nearly impossible to raise corn that is both unsprayed and worm-free. One organic option is Bt spray (see Organic Pesticides, page 77, for more). For the home gardener, another option is a carefully timed application of one part neem oil mixed with twenty parts vegetable oil. (Mineral oil has the same effect but is a petroleum product.) Apply 5 drops of oil to the tip of each ear when the silks begin to wilt and brown, or 5 to 6 days after 50 percent of the corn first shows silk. Earlier treatment can interfere with pollination and cause poorly filled ears. Personally, I opt for worms. The damage is usually limited to the very tip of the ear and can be snapped off with no great loss. A few varieties are marketed as having extra-tight wrappers to discourage pests.

See Types of Sweet Corn (page 180) for recommended varieties. Lady Finger popcorn is genetically closest to the first cultivated corn, according to archeologist Bruce D. Smith in *The Emergence of Agriculture*. Many old strains of flint, dent, and flour corn come from indigenous seed stock.

An ounce of corn will have 100 to 200 seeds and sow 25 feet. To calculate potential yield, estimate 1 stalk per foot and 1 or 2 ears per stalk. Some old varieties, notably Stowell's Evergreen, can produce 3, 4, or even 5 ears per stalk, but they are anomalies. To convert garden yield into meals, a typical serving of corn on the cob is 1 ear per person. (I can hear my father scoffing now. He'll eat 4 ears at a sitting.) For cut corn, 6 ears will yield nearly 5 cups, or enough to serve 8 as a vegetable side.

In the Kitchen

Harvest corn when the silks are dry and brown. Squeeze the tip of the ear lightly through the husk. It should feel full, plump, rounded. It's okay to peel back a strip for a peek. The kernels should be shiny as pearls, tight and bursting when pressed with a thumbnail. Pick the ear by gently twisting it as you steady the stalk with your other hand.

Most corn, and especially the old-fashioned favorites, don't hold well on the stalk or in the kitchen. Eat what you pick the same day and get your fill during the short harvest window. Invite friends. Freeze what you can't eat (see Freezing Corn, page 197).

Garden corn looks no different than ears brought home from a farmers' market. Shuck

SEPTEMBER 6: Country Gentleman, a "shoepeg" corn with randomly placed kernels, was introduced in 1890.

CORN—DIRECT SOWN

DIRECT SOW	DEPTH	EMERGENCE	SPACING	THIN TO	ROW SPACING	DAYS TO MATURITY	MATURE HEIGHT
After last frost	¾–1 inch	7–14 days	3–6 inches	12 inches	36 inches	70–90 days from emergence	5–10 feet

the husks and rub off the silks. It's a messy task better done outside. Snap off or cut away worm damage. Rinsing the ears in cold water and drying with a kitchen towel helps to remove persistent silks.

If you are lucky, you might find an ear infected by corn smut—*huitlacoche,* the Mexican word, is less off-putting. Like mushrooms and truffles, corn smut is the fruiting body of a fungus, in this case *Mycosarcoma maydis*. It causes infected kernels to swell grotesquely, but they become delicious. Cut the infected kernels from the cob along with the uninfected kernels, gently cook them in butter, and eat on tortillas.

To cut corn, place the cob, stem-end down, on the bottom of a large bowl. Use a short, sharp knife—I like a turning knife. Slicing away from yourself, adjust the angle of the knife until you feel it skimming along the cob to remove as much kernel as possible. For cream-style corn, milk the bare cob by scraping it with the dull edge of the knife. I prefer cream-style corn as a side dish and for soups and pasta. Clean-cut kernels work better in salads, relish, salsa, and succotash, that childhood friend.

The key to delicious corn is to undercook it. Cut corn is best when barely heated through, on the far side of raw. For corn on the cob, boil in unsalted water for 3 to 5 minutes.

Corn on the cob can also be grilled in its husks or roasted in the oven. Pull back the husks partway to remove the silks, then close them up again. Soak the ears a bowl of water for 15 minutes. Drain and, if you like, tuck a branch of herbs inside the husks at the last moment. (See the photo on page 448.) Grill for 15 minutes, turning several times. For darker grill marks, shuck the ears and grill for 8 to 10 minutes.

Corn's best condiments come from the cow: butter, cream, sour cream, grated Parmesan, and salty, crumbly cheese such as Cotija and feta. The summery garden flavors of basil, cilantro, minced rosemary, green coriander, and green chiles are complementary.

OKRA

Okra is a tender perennial grown as an annual. It belongs to the mallow family, related to cotton and cocoa, as well as hibiscus. Its origin is unclear, but okra was anciently cultivated in hot regions between Africa and Southeast Asia. Enslaved Africans brought the seed to America.

Okra demands heat. Typically, the seeds are sown directly after the soil has warmed, on the same schedule as corn. In northern climates, it can be started indoors 4 weeks before last frost and transplanted through black plastic mulch. Space okra as you would corn, with ample room between rows for the large, palmate leaves to reach for sunlight.

JULY 11: Young Clemson Spineless okra in my dad's Tennessee garden

AUGUST 22: Aunt Hettie's Red, Yalova Akköy, and Hill Country Red

Pick okra daily; the pods grow fast. Clemson Spineless (56 days from emergence), a 1939 introduction and a long-time garden favorite, is best when pods are harvested at 2 to 3 inches long, 2 or 3 days after blooming.

Two dozen heirloom varieties are available from Sow True Seeds and Southern Exposure Seed Exchange. Others are described in the book *The Whole Okra,* by Chris Smith. The so-called cowhorn types produce pods that remain tender at their mature length of 10 inches. Other heirlooms, such as the delicious Hill Country Red (70 days), are stout and stubby. Aunt Hettie's Red (60 days), from Tennessee, has a dandy name and terrific color.

Okra seed is ivory-white when fresh but dries down to look like small allspice berries. One ounce will plant a 100-foot row. In a hot climate, the plants are incessantly productive from the onset of bloom until cold weather. A 5 to 10-foot row would supply most families. Soak seeds overnight in tepid water to speed germination.

OKRA—DIRECT SOWN

DIRECT SOW	DEPTH	EMERGENCE	SPACING	THIN TO	ROW SPACING	DAYS TO MATURITY	MATURE HEIGHT
2 weeks after last frost	½–1 inch	7–15 days	3–4 inches	12 inches	36 inches	56–70 days	4–6 feet

Note: Soak seeds overnight in tepid water to speed germination.

PRIME FIRST-PICK CORN • BASIL | SIDE | SERVES 6 TO 8 | VEGETARIAN, GLUTEN-FREE | SUMMER

BARELY COOKED CORN

Corn at its most elemental is twisted from the stalk, shucked outside, plunged into boiling water, and eaten with your hands, all within about 15 minutes. But to distill fresh corn's flavor to its essence, there's no better technique than simple cut corn. Slice the kernels off the cob and gently heat through with butter. That's it. The corn is barely cooked. I serve some version of this dish every time company comes over in corn season. People go nuts for it. Whatever corn is freshest will be best.

Adapt the recipe to suit your taste. Mild as it is, corn nonetheless welcomes strong co-flavors. I like opal basil, plus enough freshly ground black pepper to notice it. Sometimes I'll add cream or crème fraîche for luxury and a few gratings of lime or lemon zest for fragrance. Or I'll stir in ¼ cup of grated Parmesan and some caramelized onions with a pinch of minced parsley. Experiment. A technique this simple can evolve with time into a complex, layered, personally meaningful family recipe. Or it can be corn, butter, and salt.

6 ears corn, shucked
5 tablespoons unsalted butter, plus more to taste
½ teaspoon fine sea salt
¼ teaspoon freshly ground black pepper
2 tablespoons crème fraiche (optional)
8 whole basil leaves, especially opal basil, chiffonade-cut
Grated zest of 1 lime

1. Cut the kernels off the cobs over a large bowl. Milk the cobs by scraping with the dull edge of your knife. (It should yield about 5 cups.)

2. In a large skillet, melt 3 tablespoons of the butter over low heat. Add the corn and sprinkle with the salt. Turn with a wooden spoon to coat. Reduce the heat to its lowest possible setting and gently cook the corn, uncovered, stirring every few minutes, until warmed through, as much as 20 minutes. The corn is done when it loses its raw look and releases plumes of steam when stirred. Don't overcook. The corn shouldn't caramelize in the least.

3. Remove from the heat and add the remaining 2 tablespoons of butter, plus more to taste, as well as the pepper and crème fraîche (if using). Stir to combine. Scatter the basil chiffonade and lime zest over the top and serve immediately.

Notes

For a Mexican variation, replace the basil with cilantro and add a finely minced jalapeño or serrano, then sprinkle green coriander over the top and hit it with a squirt of lime juice.

I've also dusted cut corn with ground New Mexico chiles, strewn it with roasted red bell peppers, splashed it with mole poblano, and wrapped it in a tortilla with habanero salsa and grilled summer squash to make a supreme vegetarian taco.

CORN • ZUCCHINI • PEPPERS • CILANTRO • BASIL • GREEN CORIANDER | SIDE | SERVES 6 TO 8 | VEGETARIAN, GLUTEN-FREE | SUMMER

ESQUITES

Esquites is Mexican grilled-corn salad, one step removed from elote, grilled corn on the cob. Both showcase corn's affinity for smoke and char. The flavors of chiles, herbs, salty cheese crumbles, and lime go with both, and elote is messy fun when you're eating outside with a plate balanced on your knees. A bowl of esquites looks more pulled together on the table—vegetable confetti. Serve esquites alongside watermelon salad and roasted pork or with any sort of taco.

6 ears corn, shucked
3 small zucchini, halved lengthwise
3 fresh mild chiles, such as poblano or cubanelle, or spicier jalapeños
1 medium red onion, peeled, trimmed, and sliced in half, with root plate left intact
1 tablespoon grapeseed oil
1½ teaspoons fine sea salt
1½ cups coarsely chopped fresh cilantro
¾ cup coarsely chopped fresh basil
2 tablespoons Herb Mayonnaise (page 250) or plain mayonnaise
⅓ cup crumbled Cotija or feta cheese
Grated zest and juice of 2 limes
Freshly ground black pepper
Green coriander berries (optional)
Chile powder, such as mild Aleppo, hot cayenne, etc., to taste

1. Prepare a gas or charcoal grill to medium-high heat.
2. Rub the corn, zucchini, chiles, and onion on all sides with the grapeseed oil. Season with ½ teaspoon of the salt. Grill the vegetables until tender and lightly charred, 5 to 7 minutes per side. (Because the vegetables are grilled in large pieces, a basket isn't essential.) Transfer to a baking sheet to cool.
3. When cool enough to handle, cut the corn off the cob into the bottom of a large bowl. Cut the zucchini into ½-inch cubes. Seed and dice the chiles. Chop the onion. Add them all to the corn.
4. Add the cilantro, basil, mayonnaise, Cotija, lime zest, lime juice, the remaining 1 teaspoon of salt, and the black pepper to taste. Toss to combine.
5. Sprinkle with green coriander berries (if using). Dust with pinches of chile powder before serving.

OKRA BASICS

Those who love okra need no convincing; others are mostly beyond reach. The chief complaint is sliminess. One solution is to steam or roast okra whole; it weeps only when cut. The tiniest pods, less than 2 inches, can even be eaten raw as part of a summertime crudités platter—see the photo on page 445. Whole okra also pickles well. It turns addictively crunchy when sliced into rounds and tossed with cornmeal to fry in a cast-iron skillet. At Chai Pani, a lively Indian restaurant in Asheville, North Carolina, the cooks slice it lengthwise to deep-fry sans batter for ultracrispy fries. In India okra is known as *bhindi* or "lady's fingers."

The other solution to okra's sliminess is to harness it for good. Sliced okra will thicken vegetable soup, most famously gumbo, which takes its name from a West African word for okra. In a hurry, I'll make a simplified gumbo-type vegetarian stew from sliced okra with tomatoes and onions. See also Nanny's Summer Vegetable Soup on page 430.

A pound of fresh young pods will yield 4 or 5 servings. The techniques that follow call for small okra pods, because they are tenderest, but heirloom "cowhorn" varieties can remain tender even at 8 to 10 inches long.

(CONTINUED)

OVEN-ROASTED OKRA

This sheet pan recipe will win over okra skeptics, if anything will. Small pods are oven-blasted with onion, herbs, and other vegetables. The drying heat crisps cut edges and concentrates flavors. Squash is the sweet element, and small tomatoes sauce everything with their savory juices. Go crazy with the fresh herbs. Mint is ace. Prep ingredients 30 minutes before cooking, to allow the flavors to blend.

1½ pounds okra pods, stems trimmed, halved lengthwise
2 medium yellow or green summer squash, roll-cut into ¾-inch chunks
2 or 3 small tomatoes, halved, or a large handful of cherry tomatoes
1 large onion, cut into ½-inch slices
3 cloves garlic, peeled and halved
½ cup packed fresh herb leaves, such as basil, mint, cilantro, parsley, etc.
¼ cup extra-virgin olive oil
1 teaspoon fine sea salt
Freshly ground black pepper

1. In a large bowl, combine the okra, squash, tomatoes, onion, garlic, herbs, olive oil, salt, and a generous amount of pepper. Toss to combine. Set aside for 30 minutes for the flavors to meld.
2. Set a rack in the top position of the oven and preheat the oven to 400°F.
3. Line a sheet pan with parchment or foil. Spread the vegetables in a single layer. Drizzle the juices over top.
4. Transfer to the oven and roast, uncovered, until the edges of the okra brown, the squash softens, and the tomato slumps, 15 to 20 minutes. The onion will remain half-crisp and juicy, with browned edges.
5. Transfer to a serving bowl and stir a time or two to coat with the scant pan juices. Good hot or at room temperature.

BUTTERED OKRA

This method from my father succeeds on the partnership of butter and Parmesan. The result is delicate in flavor if rustic in appearance. Select a pound of the most tender small pods, less than 3 inches long, by squeezing gently and discarding any that feel woody. Trim the stems to within ½ inch of the cap, otherwise leave the pods whole. In a medium saucepan fitted with a steamer basket and lid, steam until soft but not splitting, about 5 minutes. Immediately toss with 2 tablespoons softened unsalted butter, generous amounts of salt and black pepper, and ¼ cup grated Parmesan. Top with additional Parmesan to serve.

FRIED OKRA

Full flavored and crunchy, this method is approved by Southern gardeners who know what it means to "whup" okra, or strip the leaves from the lower stem to stimulate flowering. Slice 1½ pounds of tender okra into rounds ⅜ inch thick. In a large bowl or ziplock bag, toss with ½ cup cornmeal and ½ teaspoon salt. (For a lighter crust, replace half the cornmeal with flour.) Add corn oil, or another neutral oil, to a large skillet to a depth of ½ inch. Heat until nearly smoking; a test piece dropped in should sizzle violently. Carefully turn the okra into the skillet in a single layer. Do not stir. Adjust the heat to maintain a lively sizzle. Using a spatula, flip when the underside has achieved a dark gold crust, 8 to 10 minutes. Fry the second side until it is nearly browned, about 8 minutes. Transfer to paper towels to drain. Salt lightly while still hot.

An alternative comes from my father, who grew up on fried okra and in recent years has begun to dip whole pods in cornbread batter before frying. He prefers their more pronounced okra flavor.

STEWED OKRA AND TOMATOES

In a medium saucepan, heat 2 tablespoons olive oil over medium-high heat. Add a sliced onion and 2 crushed cloves garlic and sauté until softened, about 5 minutes. Adjust the seasoning to suit your meal, for example adding several fresh thyme branches or large pinches of dried savory or mint. In another vein, add 1 tablespoon black mustard seeds, 1 teaspoon cumin seeds, and 1 teaspoon coriander seeds. Once the onion has sweated, add 1½ pounds sliced okra, and sauté until the cut edges begin to soften, about 5 minutes. Season generously with salt and pepper. Add 2 cups canned whole tomatoes or the equivalent peeled fresh tomatoes. Bring to a boil, then reduce the heat to maintain a steady simmer and cook until the okra is completely soft, about 30 minutes. Good with pork chops.

CORN • SQUASH • EGGPLANT • TOMATOES • BELL PEPPER • ALLIUMS • HERBS

MAIN

SERVES 8

VEGETARIAN

SUMMER

SUMMER POT PIE

My kitchen collaborator Dash and I were talking pot pies. I had in mind a rich autumnal dish filled with curried cauliflower or winter squash. Dash flipped the script and suggested a light summery pie filled with corn and zucchini. It won't need a bottom crust, I said. Dash raised an eyebrow. "Is a pot pie with no bottom a pot pie?" he asked. In any case, *this* recipe doesn't have a bottom crust. The top is simplified puff pastry. You could replace it with Herb Biscuits (page 257) dropped by the spoonful, as for dropped biscuits.

The filling is a cousin to succotash. As with other recipes in this book, but even more so, the mix of ingredients will vary with the garden.

1 medium eggplant, cut into 3/4-inch cubes
2 medium summer squash, cut into 1/2-inch cubes
1 tablespoon plus 1/2 teaspoon fine sea salt
5 tablespoons unsalted butter
2 cloves garlic, thinly sliced
Large pinch of cayenne pepper
1 large red onion, diced
1 bell pepper, diced
2 cups chopped seeded tomatoes
2 cups corn cut from the cob
1 cup chopped fresh parsley
1 tablespoon chopped fresh thyme or savory
3 tablespoons all-purpose flour
1 1/3 cups whole-milk ricotta cheese
Rough Puff (recipe follows), Herb Biscuits dough (page 257), or pie dough

1. Preheat the oven to 400°F.
2. In a colander, sprinkle the diced eggplant and squash with ½ teaspoon of the fine sea salt to draw out water. Set in the sink to drip for 15 minutes.
3. In a 12-inch cast-iron skillet or other large ovenproof pan, melt the butter over medium heat. When it foams, add the garlic and cayenne and cook until the garlic is fragrant but not browned, about 1 minute. Add the onion and sweat it, stirring now and then, until softened, about 10 minutes.
4. Squeeze out the salted vegetables, either by the handful or by wringing them in a clean kitchen towel. Add to the pan with the onion. Increase the heat to high and cook, stirring frequently, until the squash is softened and slightly golden, about 10 minutes, Add the bell pepper and sauté for 5 minutes.
5. Remove the pan from the heat and stir in the tomatoes, corn, parsley, and thyme. Scatter the flour and the remaining 1 tablespoon of salt over everything and stir to combine. Fold in the ricotta and smooth the mixture.
6. Roll the puff pastry or other dough into a disc ¼-inch thick and approximately 1 inch wider than the cooking pan. Lift dough on the back of the rolling pin and center it on the pan as you drape it across the filling. Lightly press it into place and run a sharp knife around the inner edge to trim excess. Cut a vent in the middle.
7. Bake, uncovered, until the crust is golden, 40 to 45 minutes.

ROUGH PUFF

In this simplified version of puff pastry, you work cubes of butter into equal parts flour, much like making a pie crust, then fold the dough to build layers. Use a scale to simplify the measurements. The recipe makes one sheet of puff pastry large enough for a 12-inch pot pie lid. You can double the recipe and keep the extra in the freezer.

Save any trimmings to sprinkle with sugar and bake off as a breakfast treat. For a savory snack, top with grated cheese instead. Don't ball the scraps or you'll lose the layers.

2 cups (250 g) all-purpose flour
1 teaspoon (5 g) fine sea salt
2 sticks plus 2 tablespoons (9 ounces/250 g) cold unsalted butter, cut into small cubes
⅓ cup plus 1 tablespoon (100 g) ice water, plus more as needed

1. In a medium bowl, whisk together the flour and salt. Add the butter, toss it to coat, and smear the cubes between your fingers and thumbs until all the pieces are flattened and floury. Make a well in the middle of the mixture and add the water. Blend it in by using your fingers until the dough begins to come together. Add more ice water by the teaspoon, if needed, sprinkling it over the dry parts, until you have a shaggy dough.
2. Turn onto a lightly floured countertop and knead 3 or 4 times, no more than is necessary to bring the dough together into a compact, more or less consistent mass. It will still look a little rough. Pat it into a square 1 inch thick. Wrap in parchment and refrigerate for 20 minutes to hydrate and relax the gluten.
3. Remove from the refrigerator and unwrap. Roll out the dough, working in one direction only, until it is three times as long as it is wide. With a short side facing you, told into thirds from bottom and top, like a business letter. Turn the dough 90 degrees counterclockwise so that the top fold is facing left. Roll out and fold a second time. Turn 90 degrees counterclockwise again and repeat a third time. Mark the top fold by lightly incising an arrow with the tip of a knife. Press the edges of the dough together to make a tidy package. Wrap and refrigerate for 20 minutes.
4. Unwrap the dough and position it on the board with the incised arrow pointed left. Roll out and fold, as in the previous step, for a total of three more times, remembering to turn 90 degrees after each rolling.
5. Wrap the finished dough and refrigerate to rest for at least 20 minutes and up to 3 days. Take it out a few minutes before using so that it can warm slightly on the counter, making it more pliable for the final roll out to about ¼ inch thick.

SWEET CORN | PANTRY | MAKES 4 BAGS (4 SERVINGS EACH) | VEGAN, GLUTEN-FREE | SUMMER

FREEZING CORN

When your corn patch comes in quicker than you can eat it, freeze the surplus.

There used to be a clear rural preference for white sweet corn, perhaps because yellow field corn was grown for livestock. Golden Bantam, Burpee's first yellow sweet corn, introduced in 1902, helped turn the tide, and for generations children colored corn butter-yellow. Then Silver Queen appeared in 1955. It dominated mid-Atlantic cornfields by the late 1970s, according to reporting of the day in the *Washington Post*. Now bicolor is the thing.

Second-generation Berkshires corn grower Bruce Howden—whose honor-system farm stand near my house draws a crowd in August—says his customers asked for bicolor corn so consistently that he dropped other varieties.

Blanching vegetables before freezing, as I've mentioned elsewhere, locks in color and deactivates enzymes. Is it essential? In a head-to-head competition, nonblanched corn packed in vacuum-sealed bags lasted 6 months, then faded. Blanched corn lasted a solid year and—bonus—required almost no cooking time once thawed.

As a rough rule of thumb, a dozen ears will yield 10 cups of cut corn. Frozen corn on the cob is less satisfactory.

12 ears of corn, shucked

1. Fill a large bowl with ice and water. Bring a large pot of unsalted water to a boil.
2. Working in batches, blanch the ears for 4 minutes, then shock them in the ice bath. Cut the kernels from the cob with a sharp paring knife or turning knife. It's messy work. I cut in the bottom of my largest bowl to contain the splatters. For cream-style corn, milk the cob by scraping it with the dull edge of your knife.
3. Divide among 4 freezer bags. Seal with a vacuum sealer, or, if using ziplock bags, suck out the air with a straw as you pinch the zipper tight. Label the bags with the date and the name of corn used, so you'll know which to replant. Frozen corn keeps for 1 year.

CH.

16

CUCUMBERS AND GHERKINS

MELONLIKE CUCURBITS

Cucurbitaceae
THE GOURD FAMILY

Cucumis anguria
WEST INDIAN GHERKIN

C. sativus
CUCUMBER

Melothria scabra
MEXICAN SOUR GHERKIN
OR MOUSE MELON

JULY 13: Long Green Improved cucumber, an heirloom from 1842, at Thomas Jefferson's Tufton farm near Monticello

Its cells filled to capacity, a cucumber on the vine is 95 percent water. It holds the cool of nighttime into the heat of day, a reservoir of freshness. Thanks to its taut waxy skin, a raincoat in reverse, a picked cucumber will stay firm for a week or more as it chills in the refrigerator. (A pickled cucumber will stay crisp for many weeks.) And yet the finest virtue of a cucumber—its flavor—fades soon after its exit from the garden. Supermarket cucumbers are bland. If, on the other hand, you walk among the vines and take a cucumber while it's still on the small side, rub off its ⅛-inch pyramidical spines, and eat it where it grows, you might know the experience, not so frequent in adulthood, of having a preconception overturned. Cucumbers, it turns out, are full of flavor—focused, subtle, lingering flavor. The first year I grew the experimental 7082 Cucumber from Row 7 seeds, the vines ran like the dickens and by early July had set a heavy crop of 4-inch fruits. The taste was startling, like a reunion with a dull college acquaintance who in the intervening years has become magnetic. I couldn't get enough.

Cucumbers are a delight to grow, rambunctious and careening like kids on a playground, although they are subject to attack by cucumber beetles, a crafty nemesis. In the kitchen, cucumbers bring vitality but lack range. Their uses run the gamut from A to B, to borrow a phrase. Try as I might, I can't develop enthusiasm for grilled, sautéed, or stir-fried cucumbers. But taken as they are, sliced cucumbers offer a summertime pleasure as uncomplicated as line-dried sheets. Dust them with salt and pepper, or splash with yogurt, or feather with dill; they make a salad unto themselves. And there's always pickles.

Cucumbers belong to the cucurbit family, the gourds, and their close kin include melons and squash. (*Cucurbita* is Latin for gourd. Nero would have asked his chef for *cucumis* when he wanted a cucumber.) The rambling vines bloom prolifically and set fruit for a month. To save space, the vines can be trained up a trellis.

JULY 24: 7082 Cucumber overrunning its trellis, a tomato cage

Cucumber seeds are like carved ivory and so pleasingly smooth that I always overplant. Two or three hills would be reasonable for one household, with perhaps a midseason succession to extend the season.

There are many genres of cucumber, and they collect names like a dog collects ticks. Pickling cucumbers, also known as Kirbys, are thick-skinned, warty, and crunchy. The standard salad-bar cucumber is a slicing cucumber, and larger ones from a grocery store should always be peeled, because the skin bitters with age. (And also because long-haul cucumbers are waxed to extend shelf life.) A smoother, slimmer type of thin-skinned cucumber is sometimes marketed as "seedless," not true, or, worse, "burpless," a graceless name. Such specimens, sold in shrink-wrap, are often grown indoors, hence another of their names, "hothouse." Sometimes they bear

JUNE 23: A trellised hothouse cucumber at Husky Meadows

geographic monikers, such as English, Persian, or European cucumbers. There are also long, dark Asian cucumbers; round, yellow lemon cucumbers (which are picked an immature pale green), and striped Armenian cucumbers (in fact a melon). Two related cucumberlike cucurbits are Mexican sour gherkins (or mouse melons) and West Indian gherkins, which arrived in the Caribbean with the slave trade and were brought to America from Jamaica in 1792 by Richmond seed merchant Minton Collins, according to Monticello's Thomas Jefferson Center for Historic Plants. A gherkin can be any cucumber picked small and pickled; the word, coming straight from Dutch, meant precisely that in the original. The gherkin is defined by use rather than botanical taxonomy.

In the Garden

Like squash, melons, and other vegetables of high summer, cucumbers are very frost tender. Seeds are sown directly once the ground has warmed, typically a week or two after your average last-frost date. Cucumbers germinate quickly and grow fast on the way to an early harvest—less than 8 weeks from seed to salad bowl. If you're troubled by cutworms, start the seeds indoors and protect transplants with a cutworm collar. Succession plantings at 3-week intervals will give you cucumbers all the way to frost, although the plants will slow down and become listless as September temperatures retreat from their August highs.

Grow cucumbers in widely spaced hills, allowing them to sprawl, or train the vines up a trellis. See pages 198 and 199 for a twine trellis at Monticello's Tufton Farm and another at Husky Meadows at left. Space hills 4 feet apart, with 5 seeds per hill, sown ½ inch deep. Thin to the 3 best. For a trellised row, sow 2 seeds every 12 inches and pinch off the weaker. Space rows at least 4 feet apart. Train the young vines to find the trellis.

Cucumbers demand fertile soil and consistent watering. Dig in compost or lay down a 2-inch layer as the seedbed. Irrigate as needed to supplement rainfall. Erratic moisture during fruiting will lead to misshapen fruits, causing the water-starved blossom end to taper and curl like a horny toad's stub tail. Use mulch or black plastic to hold in moisture, suppress weeds—cucumbers are hard to weed once the vines get going—and keep grit off the fruit.

Cucumbers will attract cucumber beetles and squash bugs. Search and destroy in the early morning, when the culprits are sluggish. Check under leaves and inside blossoms. Barbara Damrosch goes after them with a vacuum cleaner fitted with the slot attachment. Cucumber beetles spread a sad disease called bacterial wilt. By the time you notice it, game over. The plants won't last more than a few days.

Salvage whatever fruit you can. Other diseases are reduced through good garden hygiene, such as crop rotation, disposing of plant residues, and spacing for ample sunlight and airflow. New varieties have been selected for disease resistance. As with any vegetable, an element of gardening success is learning which varieties do best in *your* garden, given the region, climate, microclimate, pest population, and local disease pressure.

There are many types to choose from. My quintessential cucumber, as mentioned prior, is 7082 Cucumber (58 days) from Row 7 Seed Company. Picked at 4 to 5 inches, it has exceptional flavor raw and pickles well. The heirloom slicer Straight 8 (58 days) matures at 8 inches, and the late-season Longfellow (70 days), another heirloom noted for flavor, keeps its eating quality all the way to its footlong maximum. Marketmore 97 (55 days) has disease resistance. Boston Pickling (55 days) and National from 1929 (52 days) are the standard American heirloom Kirbys. Parisian Pickling (50 days) dating to the late nineteenth century can be picked very small for cornichons. Many new options exist. Among the novelty cucumbers, Lemon (70 days) is yellow and round; Salt and Pepper (49 days) is cucumber-shaped and ghastly white; Mexican sour gherkins (75 days) look like inch-long watermelons, and spiked West Indian gherkins look like medieval weaponry from a gopher-sized battle royale.

Cucumbers with names that include words such as bush, miniature, or patio are suited to containers. Sow 5 seeds in a 24-inch container and thin to 2 or 3. Provide a trellis, water diligently, and fertilize every 3 weeks.

A gram of seed will number about 30, enough for a 15-foot trellised row or 6 hills. Given cucumber's productivity and the risk of bacterial wilt, it is prudent to sow succession plantings to extend the season and lessen the onslaught at harvest time. Otherwise, prepare in advance for a pickling campaign.

CUCUMBERS—HILLS

DIRECT SOW	DEPTH	EMERGENCE	SPACING	THIN TO	HILL SPACING	DAYS TO MATURITY	MATURE SIZE
After all threat of frost	½ inch	4–13 days	5 seeds per hill	3 per hill	4 feet	45–60 days from sowing	6–8 inches tall, 4–6-foot spread

Note: To start indoors, sow 1 seed per cell 2 to 3 weeks before transplanting. Harden off when seedlings have 2 true leaves. Don't postpone planting in the garden. Take care not to disturb roots.

CUCUMBERS—TRELLISED ROW

DIRECT SOW	DEPTH	EMERGENCE	SPACING	THIN TO	ROW SPACING	DAYS TO MATURITY	MATURE SIZE
After all threat of frost	½ inch	4–13 days	2 seeds every 12 inches	1 per 12 inches	4 feet	45–60 days from sowing	Will climb 3–5 feet

AUGUST 13: The high-season onslaught

SEPTEMBER 11: Cucumbers, like dahlias, thrive in hot weather.

IN THE KITCHEN

Unlike melons and winter squash, two kin species, cucumbers are harvested immature. If a specimen outgrows its taut green youth, the mature fruit becomes blimpy and yellow.

Pick over vines daily to harvest at prime eating size and to encourage further blossoming. Information on the seed packet will give the variety's ideal length, but when in doubt, smaller is better. If the vines produce more than you can use or give away, nonetheless continue picking at the proper size and compost the surplus. Monstrosities left on the vine will shut down production.

The vines, though vigorous, are fragile. Harvest with a clipper or use two hands to separate the cucumber without yanking the vine.

Short blunt spines stipple the fruit. Rub them off under running water. Eat cucumbers as soon as possible after picking. I leave them on the countertop covered with a cool damp towel for a few hours rather than dull their flavor by refrigerating.

The time-honored means to extend a cucumber's shelf life is pickling. Use fermented cucumbers for salads, as you would fresh. Or combine fresh and fermented, tossing with chopped dill fronds, thinly sliced red onion, and cream, sour cream, or yogurt. Fermented dill pickles will stay crunchy for a month or more and remain flavorful deep into winter, although the texture goes mushy with time. Two hundred years ago, very mature pickles were esteemed; to me they are an acquired taste.

A cucumber's character resides largely in its skin. If a large cucumber must be peeled, leave several thin green strips—a hint of bitterness—to quicken the flavor.

CUCUMBER PEELS • HERB STEMS • WHAT HAVE YOU | BEVERAGE | MAKES 1 QUART | VEGAN, GLUTEN-FREE | SUMMER

GARDEN REFRESHER

This recipe exists in the liminal space between the cutting board and the compost bucket, a boundary across which food becomes not-food, scraps become waste. Of such distinctions are taboos made. A parsley sprig on the counter is inarguably edible. Stripped of its leaves and tossed in the compost bucket, the stem is suddenly garbage, and a person who reaches in to eat it would be transgressive, perhaps insane. The parsley stem hasn't changed.

As a prompt to thinking about food waste, this infusion comes from the in-between. It is like a tisane but cold, a refresher to extract flavor from kitchen trimmings. Use it to hydrate when working in the garden. Or else drink it over ice when you come in. Add a splash of fizzy water or tonic if you like. A pour of dry vermouth makes it an aperitif and a little gin gives you a low-proof cocktail.

The ingredients will vary.

As you work in the summer kitchen, keep a quart jar on the counter to catch parsley stems, mint stems, juiced lemons and limes, and any edible plant matter that smells good. Add a sliced cucumber or just cucumber peels. Step outside to collect a few leaves of lemon verbena, lemon balm, basil, shiso, or another favorite herb. Look up at the sky. Bruise the leaves to release their aromatic oils and add to the jar. Top up with water and tighten the lid while saying to yourself the name of a person you love. Leave the jar on the countertop and go do your work. It will be ready when you are.

PICKLING CUCUMBERS • GARLIC • DILL

ALSO FOR: SMALL ZUCCHINI

PANTRY

MAKES 1 QUART

VEGAN, GLUTEN-FREE

SUMMER

REFRIGERATOR DILL PICKLES

The secret to snappy pickles is to not heat them, as you would when canning jars in a boiling-water bath. This recipe doesn't have to be canned, so the pickles retain their fresh-cucumber flavor and crunchy texture. The vinegar-based brine protects against botulism and other pathogenic bacteria. Store in the refrigerator and use within a month.

To scale up the recipe or otherwise adjust it, always maintain a ratio of 1 part vinegar to 1 part water. Otherwise, any other element can be adjusted to taste—or completely reimagined. You can safely change the quantity (or type) of fresh herbs, alliums, and spices. Adjust salt to taste, as well. From a food-safety perspective, you can also add sugar or honey to gentle the vinegar, but as a matter of taste, I prefer them tangy.

1½ pounds 4-inch pickling cucumbers or small zucchini
1½ teaspoons fine sea salt
4 bushy dill fronds
2 cloves garlic, lightly crushed
2 small dried red chiles
1 teaspoon mustard seeds
1 teaspoon dill seeds
½ teaspoon coriander seeds
½ teaspoon black peppercorns
1 cup distilled white or apple cider vinegar

1. Wash the cucumbers and rub off the spines. Slice ⅛ inch off the blossom end. Cut the cucumbers as you prefer. Options include halves or spears; ¼-inch-thick chips, ovals, or slabs; or chunky rounds. In a colander, sprinkle with ½ teaspoon of the fine sea salt and cover with a tray of ice. Leave the colander in the sink to drain for 1 hour.

2. In a clean (doesn't have to be sterilized) wide-mouthed 1-quart jar, combine the dill fronds, garlic, and dried chiles. Rinse the cucumbers quickly and drain. Pack into the jar, working in as many as will fit snugly without crushing. Leave ½ inch of headspace. Add the mustard seeds, dill seeds, coriander, and peppercorns.

3. In a small saucepan, heat the vinegar, 1 cup water, and the remaining 1 teaspoon of salt over medium heat just until it steams, stirring to dissolve the salt. It doesn't have to boil.

4. Pour the brine over the cucumbers to nearly fill the jar. Seal the jar fingertip tight. Allow to cool. Cure overnight in the refrigerator. Best if used within a month. Keep refrigerated.

SMALL PICKLING CUCUMBERS • GARLIC • DILL | PANTRY | MAKES 4 QUARTS | VEGAN, GLUTEN-FREE | SUMMER

FERMENTED KOSHER DILLS

- 1/3 cup (75 g) fine sea salt
- 2 quarts unchlorinated spring water
- 4 pounds 3- to 4-inch Kirby cucumbers
- 2 handfuls of dill fronds
- 2 large dill heads, or 2 tablespoons dill seeds
- 2 heads garlic, halved through the equator
- 1 tablespoon black peppercorns
- 2 dried red chiles
- Grape or cherry leaves (optional)

1. In a pitcher or large bowl, stir the salt into the spring water until it dissolves.
2. Slice 1/8 inch off the flower end of each cucumber.
3. In a 2-gallon crock or other suitable container, layer the cucumbers with the dill fronds, dill heads, garlic, black peppercorns, and dried chiles. Top with a double layer of grape or cherry leaves (if using). Pour over the brine to cover. Weight with a plate and a sealed quart jar of water or a ziplock plastic bag half-filled with brine. Cover the crock with a clean kitchen towel and top with a plate. Leave in a cool, dark place.
4. After 3 to 5 days, check the crock. Ensure that the cucumbers stay submerged. Skim the scum that forms on the surface of the brine and remove any specks of mold—they're nothing to worry about. Continue to check the crock daily. The pickles will start to sour within a week or 10 days, depending on the temperature. You can eat them at any time. When soured to your liking, pack pickles in 1-quart jars and cover with brine. Loosely close the lid: do not tighten. Store in the refrigerator for up to 4 months. Left in the crock, pickles will mature faster.

GHERKINS • PEARL ONIONS • TARRAGON | PANTRY | MAKES 1 QUART | VEGAN, GLUTEN-FREE | SUMMER

CORNICHONS

One year I trialed multiple cucumber varieties to find out which I liked best and thereby compounded my usual habit of overplanting. What followed was a glut. I ate cucumbers off the vine in the garden and brought them in for every meal. I served cucumbers small and large, whole and peeled, in slices and spears. I picked a 5-gallon bucket of Kirbys to ferment as Kosher Dills (page 208) and then another 5 gallons to make bread-and-butter pickles. Still the onslaught continued. I picked another 5 gallons for a family recipe known as Gran's 14-day sweet crunch pickles, and still the vines blossomed. At last my blood ran cold, and I went for the babies, picking mini-cucumbers an inch long. By the time the vines gave up, I had corralled their overwhelming productivity into four 1-quart jars of tiny cornichons, about a square foot of pantry space.

It takes 10 minutes to put up a jar of cornichons, and they retain their flavor and snap into the heart of winter, better than any other pickle, I think. Serve them with cheese or charcuterie, especially rillettes. For their highest and best use, place a hunk of raclette on a cutting board next to a roaring fireplace and scrape the melted edge to serve on little plates with boiled potatoes, cornichons, and bresaola. Drink an Alpine white wine. It's an indoor picnic.

1½ pounds 2-inch cucumbers
3 tablespoons plus 1 teaspoon fine sea salt
5-inch sprig tarragon
6 pearl onions, peeled
1 teaspoon yellow mustard seeds
½ teaspoon black peppercorns
1 bay leaf
2 cups best-quality white wine vinegar or Champagne vinegar

1. Wash the tiny cucumbers and rub between your palms to remove the spines. Using a sharp knife, nick off the blossom end. Leave the stem end intact. Toss with 3 tablespoons of the salt and leave in a colander placed in the sink to drip for 1 hour. Rinse and dry with a kitchen towel.
2. Pack the cucumbers into a wide-mouthed 1-quart jar, leaving ½-inch headspace. Layer the tarragon and onions as you go. Add the mustard seeds, peppercorns, bay leaf, and the remaining 1 teaspoon of salt at the end. Cover with room-temperature vinegar at full strength.
3. Cure for at least 1 week before using. Refrigerate or store in a cool, dark place. Keeps well.

Note: In place of fresh pearl onions, you can also use pickled cocktail onions. Or substitute one walnut-sized onion. But don't go overboard on either onions or tarragon.

SMALL CUCUMBERS
• SWEET ONION
• HERBS

STARTER

SERVERS 6 TO 8

VEGETARIAN

SUMMER

CUCUMBER SALAD

Dinners on the West family farm—dinner being the midday meal—lacked only in leafy salads, not that I cared much as a kid. Instead there were sliced tomatoes and trimmed raw scallions, and sometimes in summer a bowl of sliced cucumbers and Vidalia onions dressed in a vinaigrette of apple cider vinegar, Wesson oil, stale black pepper, salt, and a little too much sugar.

This version, while respectful of the old-school Southern flavor, skips the oil and uses honey to mellow the vinegar. Herbs and Niçoise olives provide the optional garnish.

Other vinegars will push the flavor profile toward other cuisines. For French-y or Italianate dishes, I replace onions with shallots macerated in red wine vinegar. White onions in lemon juice dusted with sumac take on Middle Eastern character. Rice vinegar, furikaki, and a drizzle of sesame oil skew it toward Japanese cooking, and so on.

- 1/2 teaspoon fine sea salt
- 1 teaspoon honey
- 3 tablespoons apple cider vinegar
- 1/2 medium sweet onion or a large shallot, thinly sliced
- 1/2 teaspoon freshly ground black pepper
- 4 or 5 small Kirby cucumbers, sliced into thin rounds or lengthwise spears
- 1/4 cup Niçoise olives, pitted
- 2 tablespoons packed chopped fresh dill, fennel fronds, carrot tops, or other herbs

1. In a large bowl, whisk together ¼ teaspoon of the salt, honey, and vinegar until dissolved. Add the onion and turn to coat. Add the black pepper and toss until distributed. Set aside for 15 minutes.
2. Add the cucumbers, the remaining ¼ teaspoon of salt, olives, and herbs. Toss to combine. Let marinate for at least 10 minutes for the flavors to blend, or up to 30 minutes before serving for a more pliable texture.

LARGE SLICING CUCUMBERS • RED ONION • PARSLEY • DILL • TARRAGON | STARTER | MAKES 6 CUPS | VEGETARIAN, GLUTEN-FREE | SUMMER

ICED CUCUMBER SOUP

This recipe is what you do with cucumbers that get ahead of you. Not yellow blimps bound for the compost pile, but the biggish ones, in cross section like a silver dollar, with seedy, jellied middles. Carve away the thickened peel, trough out the seeds with a spoon, and liquefy the rest in a blender for a mild, cooling soup.

Because peeled cucumbers will be mild to begin with, and because fat softens flavors, *and* because cold dampens volatile molecules that contribute to taste and smell, this cold soup requires all the boosters you can think of—alliums, fresh and dried herbs, olive oil, acid, salt. Season with a heavy hand. Use whatever herbs you have growing, as usual: fennel fronds, opal basil, shiso. Shallots or scallions could stand in for the red onion. Throw the spice rack at it.

The target consistency is pourable but thick enough to mound slightly in a spoon, like gazpacho. If you find it too thick after chilling, loosen with buttermilk. I like to serve cucumber soup on the rocks in a double highball glass with a sprinkle of za'atar and a side of cheese straws. It cools you off at the end of a sunburnt day when you're standing around the grill or just out there watching fireflies.

If you have an abundance of prime small cucumbers to spare for this recipe, all the better. There's no need to peel or seed them. The soup will have a deeper color and more cucumber flavor.

2 to 2½ pounds cucumbers
1 clove garlic, crushed and peeled
½ medium red onion, roughly chopped
¼ cup chopped fresh parsley
¼ cup chopped fresh dill
2 tablespoons chopped fresh tarragon
2 tablespoons fresh lime juice
1½ cups whole milk Greek yogurt
¼ cup extra-virgin olive oil
1 teaspoon sherry vinegar
2 teaspoons fine sea salt
1 teaspoon Aleppo or freshly ground black pepper
Finely diced cucumber, for sprinkling
Green Za'atar (page 254), for sprinkling

1. If peeling large cucumbers, leave on a few strips of skin for the color. Scoop out the seeds with a spoon and discard. Cut the flesh into 1-inch pieces. Keep going until you have 1½ pounds chopped cucumbers. If using small, prime cucumbers, cut up 1½ pounds, unpeeled.
2. In a blender, combine the cucumbers, garlic, onion, parsley, dill, tarragon, lime juice, yogurt, olive oil, sherry vinegar, salt, and pepper. Blend until smooth. Taste and adjust seasonings to taste. Cover and refrigerate for several hours or overnight for the flavors to blend.
3. Serve in chilled bowls or double highball glasses. I like to float an ice cube in each serving. Top with sprinkles of diced cucumber and a dusting of za'atar.

SLICING CUCUMBERS • RED ONION • CILANTRO • MINT

MAIN | SERVES 8 TO 10 | GLUTEN-FREE | SUMMER

SMASHED CUCUMBER CHICKEN SALAD
WITH SOUTHEAST ASIAN FLAVORS

At first glance, the ingredients list will seem wrong—so many herbs. Doesn't all that mint take over? It doesn't. Herbs appear in nearly every recipe in this book because, as discussed further in chapter 18, they grow aggressively but take the part of the second violin—essential, unmistakable, but scrupulously supportive. In pursuit of flavor, one can hardly have too many herbs.

Smashed cucumbers, borrowed from Asian cuisines, are more subdued in texture than un-smashed. Lightly whack each piece with the dull edge of your knife or your open palm. You're looking to crack the cucumber, bruising it at the cellular level, rather than smashing it to smithereens. The flavors here point toward Southeast Asia without being too specific and the chicken adds enough substance to make this hearty salad a meal. The ideal chicken is one poached with herbs, aromatics, and fresh ginger—the cooking broth will be delicious for pho or other soups—but roasted or grilled chicken can stand in. Try it also with grilled shrimp, or even sliced Thai-style grilled pork.

DRESSING

1 tablespoon grated or minced garlic
3 tablespoons grated fresh ginger
1 tablespoon chili garlic sauce or sriracha, preferably Huy Fong
2 tablespoons rice vinegar
1 teaspoon sugar
2 teaspoons fine sea salt
1/3 cup extra-virgin olive oil
1/2 teaspoon toasted sesame oil

SALAD

1 medium red onion, thinly sliced
1 1/2 pounds smallish cucumbers, sliced into 1-inch-thick rounds and lightly smashed (see headnote)
1 1/2 cups lightly packed fresh cilantro leaves and stems, chopped once or twice
1 1/2 cups lightly packed fresh mint leaves
3 1/2 cups shredded cold poached chicken or cold roasted chicken
2 tablespoons fine sea salt
1/2 cup peanuts, roughly chopped

1. **Make the dressing:** In a small bowl, whisk together the garlic, ginger, chili garlic sauce, vinegar, sugar, and salt until combined. Stream in the olive oil and sesame oil, whisking constantly. Set aside for 15 minutes for the flavors to meld.

2. **Assemble the salad:** Cover the sliced onion with cold water in a small bowl for 15 minutes. Drain and dry in a salad spinner.

3. In a large bowl, combine the onion, cucumbers, cilantro, mint, chicken, and salt. Toss to combine.

4. At the last moment before serving, pour in the dressing and toss to coat. Transfer to a serving platter and garnish with the peanuts.

Note: Dress the salad at the last moment. The generous yield on this recipe is enough for a table full of eaters, so if you find there is too much for a smaller group to eat right away, set aside a portion of it undressed and it will keep 2 days in a sealed container. Once dressed, the leftovers don't hold well.

CH.

17

HEARTY GREENS

THE COHORT OF COOKING GREENS

Brassicaceae
THE MUSTARD FAMILY,
ALSO KNOWN BY THE OLDER NAME
Cruciferae

Brassica oleracea var. *acephala*
KALE

B. napus
SIBERIAN KALE

B. rapa var. *ruvo*
BROCCOLI RABE OR RAPINI

B. rapa var. *rapa*
TURNIP GREENS

B. juncea
MUSTARD GREENS

Amaranthaceae,
THE AMARANTH FAMILY,
ALTHOUGH SOME GARDEN SPECIES
ARE PERHAPS BETTER KNOWN BY
Chenopodiaceae,
THE GOOSEFOOT SUB-FAMILY

Beta vulgaris subsp. *vulgaris*
CHARD, OFTEN CALLED SWISS CHARD

Spinacia oleracea
SPINACH

JUNE 6: Silverado chard

Robust, productive, beautiful to the eye, and nearly self-sufficient in the garden—the hearty greens generously unfurl new leaves throughout their long growing season. In my New England garden, the harvest stretches from late spring until the ground freezes solid in December. A 1915 catalogue from the John A. Salzer Seed Company of La Crosse, Wisconsin, active from 1868 to 1968, proposes "Swiss chard produces more food for the table than almost any other vegetable and it also requires less care."

Along with herbs and tender greens—the little salad bowl pickings—hearty greens are the novice's best introduction to cooking from a garden. This chapter focuses primarily on leafy vegetables from two botanical families, the goosefoots and brassicas. Chard (a goosefoot) and kale (brassica) are standard bearers for the larger cohort. As a group, the cooking greens are hearty, as in substantial, and many are also cold-tolerant, or hardy. Several less-familiar greens from warm regions of the world expand the summertime options.

Chard looks like a bouquet growing; its riblike stems come in ivory, orange, or red. It belongs to the species *Beta vulgaris* var. *vulgaris,* and botanists make a very fine distinction by further sorting two classes of chard into the cultivar groups Cicla and Flavecens. Chard is often assumed to have Helvetic ancestry—Swiss chard—but for no good reason. The wild proto-vegetable was Mediterranean. The most plausible explanation I've seen is that chard was once known in France as Swiss cardoon, i.e., a cool-climate analogue to an unrelated Italian vegetable noted for its edible central leaf rib. English, then, turned the French *carde* into chard by 1832, per Merriam-Webster, and apparently accepted the Swiss misnomer as part of the deal. The Cicla cultivar name is derived from the Arabic *al-silq,* which was once believed to suggest Sicilian origin. Not so. Better minds have since realized that the Arabic language received the word from the Babylonians, who reputedly cultivated the plant, known to them as *silqu,* in their famous hanging gardens. By rights, then, we should say Babylonian chard. The Romans called it *beta,* from a Celtic name, whence beet. Chard and beets are both *B. vulgaris* var. *vulgaris,* which makes sense when you examine the beet-colored stems of red chard and the chard-like leaves of beets. (Beet greens can be cooked like chard although they are less tender.) The leafy crop predates the root vegetable, which didn't develop until a late date, in the seventeenth century. The Queen's English preserves a faint memory of the old order in pairing the principal antique word for chard, silverbeet, with an amended term for the second crop, beetroot.

Kale takes many forms. Strappy Tuscan lacinato kale can be stewed to silky smoothness with long cooking. Portuguese kale (also known as beira or konchuda) is collard-shaped. Ultrafrilly ornamental kales are used in landscaping. Siberian kale, a separate species, has leaves that open in ruffles, like a poet's shirt. It is both pretty and tender. Kale's colors run deep for the most part, across a palette of jade, oceanic green, and Malbec-purple, with a few spooky albino varieties for sharp contrast. As with dogs and dating apps, there is a kale for every preference.

Bountiful mustard greens, turnip greens, broccoli rabe, spinach, and erbette are other leafy vegetables notable for flavor. Turnip and mustard greens can be sharp when raw but mellow with cooking, as Southerners will attest. Broccoli rabe, a leafier version of the familiar big-headed vegetable, has a pronounced character even when cooked, and its bitter edge might discourage greens skeptics. Use extra olive oil in cooking to soften the edge.

Spinach, another goosefoot, sometimes falls into the category of cooking greens—especially the meaty, flavorsome Bloomsdale types, introduced in 1925—and I'll touch on it in this chapter, but spinach is a small, fast-growing crop, in the garden more akin to lettuce. See the Tender Greens chapter (page 382) for cultivation tips. Spinach and chard are interchangeable in recipes. Biennial chard is less heat-sensitive than jumpy annual spinach and will not bolt in its first season. Delicate erbette, or Perpetual Spinach, will win over those who find chard too strong.

Because big-boned collards are in essence a non-heading form of cabbage, requiring more garden space and longer cooking times than most greens in this chapter, I've placed them in chapter 14, with their cabbage kin. That said, I wouldn't hesitate to pick a few tender, new collards to add to a pot of Young Greens Braised with Spring Onions (page 225).

Several antique species of greens are worth trialing as perennials—plant once, harvest for years to come. The potherb Good King Henry was a delicacy in the era from Chaucer and Jane Austen but is all but forgotten in the age of TikTok. Caucasian Mountain Spinach, *Hablitzia tamnoides* is a shade-tolerant perennial vine cultivated in Scandinavia. The common garden weed *Chenopodium album,* lamb's-quarter, is annual but freely self-sows and is equal in flavor to the tastiest cultivated vegetables. Finally, the world supplies a stewing pot full of greens celebrated in their home regions—African celosia, Asian Malabar, cosmopolitan amaranth, heat-tolerant orach, juicy New Zealand spinach, early-sprouting Belle Isle cress, twining morning glory, and the magical sounding Jewels of Opar, *Talinum paniculatum*. For the greens curious, Fedco Seeds in Maine and Southern Exposure Seed Exchange offer unusual varieties.

In the Garden

Most of the greens covered in this chapter are biennials grown as annuals, although kale will overwinter in mild climates. Most prefer cool weather, with the brassicas being somewhat more frost-tolerant and the goosefoots somewhat more heat tolerant. Look to tropical species for summer greens, such as Malabar, orach, and New Zealand spinach. All greens are suited to repeat harvests, putting out more leaves as you pick the largest.

The hearty greens don't need rich growing conditions. Chard and mustard greens will fit in small spaces, such as a raised bed or container garden, and many types of kale could also be grown successfully in containers, especially if you plan to cut them small. Chard and kale will tolerate some shade, unlike many vegetables.

Greens are sown early for spring harvest or in midsummer for fall harvest. Late plantings will be less subject to flea beetles. Frost improves the flavor of kale (and collards).

Grow greens in rows or beds. Sow directly once the threat of heavy frosts has passed, about 3 weeks before the average last frost date. Cover chard when temperatures fall below 30°F. Replant monthly until 2 months before first frost.

Sow greens seeds ¼ inch to ½ inch deep, depending on seed size. Space 4 inches apart in rows 18 to 24 inches apart. Chard seeds are fused seed clusters, and multiple seedlings will emerge. Pinch the less robust. When young greens reach the size of your hand, thin to 12 inches and cook the thinnings. Most greens will reach mature size in about 2 months, depending on type, and provide repeat harvests. I always start picking on the early side, after about 1 month, but at first take only a few leaves until the plants are well established. At maturity, cut or snap the largest outside leaves at the stalk, leaving those smaller than your hand to grow.

Common chard varieties include Fordhook Giant (50 days from emergence), Burpee's white-stemmed standard since 1934. It has large savoyed, or ruffled, leaves. Red-stemmed Red Rhubarb and yellow-stemmed Gold Silverbeet mature a week later. Any seed packet promising mixed colors, such as Bright Lights (56 days), contains a novelty mix of multiple single-colored varieties. Chard seeds are roughly BB-sized, 50 per gram (or 1,500 per ounce), and easy to handle. A gram will sow 6 feet.

Kale, the trendy green of the twenty-first-century healthy lifestyle, grows in many forms. My favorite type for stewing and freezing is Tuscan kale (60 days), also known as lacinato, black kale, and dinosaur kale. It produces steadily through the first snowfall. Curly kales, such as Winterbor and Redbor (both 60 days), have rigid, tightly ruffled leaves. Pretty as they are, I find them almost too textured, like eating vegetable crinolines. Ruffled Siberian kale, such as Red Russian (60 days), is more tender. Pick young for salads. Heirloom Portuguese kale (80 days) has very large nearly smooth leaves. Kale has tiny seeds, like other brassicas, approximately 300 per gram. One gram will sow a 25-foot row.

Mustard greens include the beloved heirloom Southern Giant Curled (45 days) and fancy newer types such as Golden Frills and Scarlet Frills (21 days to baby size), as well as Japanese Red Giant (43 days) and mild mizuna, which I pick small for salads. A unique Southern heirloom turnip, Seven Top, does not produce an edible root, just greens. Grow it like mustard greens, planting a bit more closely.

Ten feet of greens will feed a household, with some left over for the freezer. I put out plenty, but also take an opportunistic approach by sowing a few chard or kale seeds in empty spaces around the garden. Broccoli rabe grows well in beds.

A mixed bed of mustard, turnips, and collards is known by some as a greens patch. It will yield a fine mess of greens. A mess is enough to feed everyone at a Southern dinner table. Company might call for a big mess.

KALE AND CHARD—DIRECT SOWN

DIRECT SOW	DEPTH	SPACING	ROW SPACING	THIN TO	TIME TO HARVEST	SUCCESSION SLANTING	MATURE SIZE
3 weeks before last frost	1/2 inch	4 inches	24 inches	12 inches	1 month for baby, 2 months for mature	Monthly until 2 months before frost	Up to 3 feet tall, 2-foot spread

Note: The thinnings make prime eating. Start thinning when leaves are large enough for the salad bowl. You can also thin successively, leaving some to continue growing. For a single harvest of baby kale, sow like lettuce (see page 389).

MUSTARD GREENS, TURNIP GREENS, BROCCOLI RABE—DIRECT SOWN

DIRECT SOW	DEPTH	SPACING	ROW SPACING	THIN TO	TIME TO HARVEST	HEIGHT
3 weeks before last frost	1/4 inch	1 inch	18 inches	4–6 inches	45–60 days	16–24 inches

Note: For a greens bed, lightly broadcast mixed seed (mustard greens, turnip, cress, collards) over a prepared seedbed and cover with 1/4 inch soil. Thin to 6 inches in every direction. Pick across the patch to control crowding. Make room for the larger greens, such as collards, by strategically cutting out smaller, more ephemeral greens, such as cress, as they mature.

SEPTEMBER 7: Sautéed Silverado chard with gratinéed stems

JUNE 24: A bed of Winterbor kale at Indian Line Farm jazzed up by a stray Portuguese kale

IN THE KITCHEN

To manage greens plants for repeat harvests, take the largest outside leaves after they reach about 12 inches long. Some types of kale will get much larger, but mature leaves toughen and take longer to cook. Cut or snap each leaf at the base of the stem, along the stalk. Don't disturb the smaller central leaves. When thinning, sever at ground level.

Cut greens wilt quickly, especially in warm weather. Harvest in the morning, while the air is cool and leaves are fresh with dew. Immediately wash cut greens in a tub of cold water to clean and "remove the field heat," as commercial growers say. Adding ice to the rinse water would not be excessive. Once the leaves have cooled and firmed up, rinse in a second change of water, drain, and wrap in a clean kitchen towel to keep in the refrigerator's crisper drawer for up to several days. Thin-leafed chard and spinach are more perishable than kale and broccoli rabe.

In the kitchen, the greens combine and recombine endlessly, and pair with whatever else is growing. (Note that cooking times will vary by type, however.) In spring, I'll gently melt tender thinnings with scallions and scallion-sized green garlic. Summer chard gets sautéed with sunny herbs and tomatoes. For a fall lunch, I'll chop blanched mixed greens for a frittata or make a galette with the Sicilian flavors of dried currants and pine nuts. Two pots of stewed greens—one with traditional smoked pork, the other with a piece of kombu and a glug of olive oil for vegans—is a New Year's Day tradition at my house. (The greens represent greenbacks.) In the depths of winter, overcooked greens on toast is a humble but satisfying snack when drizzled with olive oil. A block of greens out of the freezer unfolds in hot soup, welcome sheets of chlorophyll.

The fine taste of greens blooms in fat—butter, olive oil, drippings. Slow cooking enriches the broth. Greens can also become a vehicle for strong flavors such as curry and Indian spices. When very young, almost any of them are welcome in green salads.

Some people find chard too strong. It can make your teeth squeaky—an odd sensation, but one smoothed out with the addition of cream, butter, stock, or olive oil.

When you're ready to cook your greens, wash them once again in a basin of cool water (the third rinse since picking—gritty greens are a downer). For chard, kale, and collards, first strip out the central rib (see Destemming Hearty Greens below) and discard, but reserve chard stems (see next paragraph). Stack the leaves and cut them crosswise into wide ribbons, like dollar bills. They are ready to cook. Greens prepped this way will keep in an airtight container in the fridge for 5 days.

Chard's rib-like stem, crisp and tangy, is almost a separate vegetable. Trim and blanch in salted water until tender, then cook with the greens. Or lay them in a gratin dish, cover with grated cheese, and brown in a hot oven (see the photo on page 219). Some people pull off the strings, but it's fussy work and generally unnecessary unless the stems are very large.

Greens freeze well. Blanch first in salted water, wring dry, and pack in vacuum-sealed bags. (See the Notes to Freezing Green Beans, page 132.) Canned greens are less appealing.

DESTEMMING HEARTY GREENS

For chard, kale, and collards, first strip out the central rib like this: Fold the two halves of the leaf together along the length of the rib. Firmly hold the stem in your left hand. With your right, pinch the junction of the folded leaf, just above where it joins the rib. Yank with your left hand and let the midrib run through the pinched fingers of your right hand. The leaf will unzip. Reserve chard stems to cook; compost kale and collard stems.

SIBERIAN KALE (BRASSICA NAPUS) • RADISHES • MINT • PARSLEY

ALSO FOR: CURLY KALE

SIDE | SERVES 6 TO 8 | VEGAN | SPRING, SUMMER, FALL

KALE AND DATE TABBOULEH

I grew up, kitchenwise, under the spell of Alice Waters and learned about the edible world through M. F. K. Fisher, Paula Wolfert, Elizabeth David, and Marcella Hazan. The foreign words I memorized were *mirepoix* and *aioli* and *bouillabaisse* and *prosciutto* and *tagliatelle* and *tagine*, words that no longer need to be italicized because they belong to the everyday kitchen patter of people who buy books like this one.

My kitchen collaborator Dash worked for Suzanne Goin, an Alice Waters protégée, but the culinary spellbinder of his generation was Yotam Ottolenghi. Dash's vocabulary spans the globe: harissa, za'atar, miso, yuzu koshō, leche de tigre, kombu, sumac, XO, masa, and a hundred other wonders.

Where our sensibilities converge is in the garden—around seasonality, vegetables, and salad. Which means our common vocabulary includes tabbouleh. In its original form, tabbouleh is a lemony herb salad. Veteran cookbook author Martha Rose Shulman describes it as "an edible garden." When Dash came to cook with me, a row of beautiful Siberian kale drew his eye, so this riff on tabbouleh is extra-leafy, a kale chopped salad. A bit of farro gives it chewy substance, and marinated dates make it opulent. The optional marigold petals don't add anything except color, but why not? Other edible flowers include peppery nasturtium, cucumber-y borage, perfumed violets, and Day Glo daylilies.

To round out a lunch menu, serve chunked tomatoes and feta. At suppertime, add grilled zucchini, roasted beets, or meaty skewers.

8 Medjool dates, pitted and chopped (3/4 cup)
Grated zest of 1 lemon
1/4 cup fresh lemon juice, plus more to taste
1/3 cup extra-virgin olive oil
1/4 teaspoon fine sea salt
1/2 pound tender, ruffly kale, such as Siberian or Red Russian
1/2 cup 1/4-inch diced radish or unpeeled seeded cucumber
1/2 cup almonds, toasted and coarsely chopped
1/2 cup lightly packed coarsely chopped fresh mint leaves
1/2 cup lightly packed coarsely chopped fresh parsley leaves
3 cups cooked and cooled farro or other grain
Freshly ground black pepper
Marigold petals or other edible blossoms (optional)

1. In a small bowl, combine the dates, lemon zest, lemon juice, olive oil, and 1/8 teaspoon of the salt. Mix to combine and set aside for 20 minutes.
2. Destem the kale (see Destemming Hearty Greens, page 221). Roughly chop the leaves into postage stamp–sized pieces. Transfer to a large bowl and sprinkle with the remaining 1/8 teaspoon of salt. Massage the leaves for 2 minutes or until tender.
3. Add the radish, almonds, herbs, and farro. Pour over the marinated dates and their liquid. Add 10 or 12 grinds of black pepper. Toss to combine. The tabbouleh should be dressed lightly. Garnish with marigold petals (if using).

To adapt for curly kale: Curly kale (*B. oleracea*) will need to be massaged more vigorously to make it supple and tender. Sprinkle it with salt in a large bowl and knead for several minutes until it relents. Alternatively, chop it more finely at the outset, as if mincing curly parsley.

MIXED GREENS • ALLIUMS • PARSLEY | SIDE | SERVES 6 TO 8 | VEGAN OPTION, GLUTEN-FREE | SPRING, SUMMER

YOUNG GREENS
BRAISED WITH SPRING ONIONS

After a mild winter of rain and mud—the new normal in New England's warming climate—I went out to the garden and found a surprise. Spinach had overwintered and collards were resprouting from the roots. I picked a mess of mixed greens, along with scallion-sized garlic, shallot sprouts, and tiny bunched clusters of dandelion leaves. In that way, I was able to cook the year's first garden meal before putting out a spring garden.

Fast-cooking young greens are full of vitality. They have a different character—a different vibe—than mature fall greens, which might need an hour or more on the stovetop. With young greens, it often isn't necessary to strip the central rib. Mix as many types as you can find, including weeds. My lucky 13 includes:

Kale, chard, spinach, mustard greens, turnip greens, young collards, radish tops, cress, stinging nettle, lamb's-quarter, dandelion, plantain, dock

The mix could also include early herbs such as parsley, chervil, and salad burnet, as well as any spring alliums, including chives and ramps.

This is a meat-free recipe. Kombu, dried kelp, is key to the broth. For omnivores, I add a few dashes of Red Boat fish sauce or colatura di alici, Italian anchovy sauce. See page 226 for a traditional long-cooked version of mature greens with smoked pork.

Thin-leafed chard and spinach can be cooked using this method all season long.

1 handful of scallions
2 tablespoons vegetable oil
2 tablespoons minced fresh parsley
Pinch of red chile flakes
¾ teaspoon fine sea salt, plus more to taste
4 cups light stock or water
2 × 3-inch piece kombu
½ teaspoon Red Boat fish sauce (optional)
½ teaspoon freshly ground black pepper, plus more to taste
6 quarts young greens, washed and run through with a knife once or twice
¼ cup extra-virgin olive oil
2 tablespoons red wine vinegar

1. Chop enough of the scallions to get 1 cup whites and ½ cup greens and set aside separately.
2. In a large pot, warm the vegetable oil over medium heat until it shimmers. Add the chopped scallion whites, parsley, chile flakes, and ¼ teaspoon of the salt. Sweat for 2 minutes until softened. Add the stock, kombu, fish sauce (if using), the black pepper, and the remaining ½ teaspoon of salt. Increase the heat to high and bring to a boil for 1 minute.
3. Add the greens. Return to a boil, then reduce the heat to maintain a steady simmer and cook, uncovered, for 8 to 10 minutes, until stems and midribs are tender. Don't overcook.
4. Remove from the heat. Add the olive oil, vinegar, and scallion greens. Taste and add more salt and pepper to taste.

LONG-COOKED TUSCAN KALE

ALSO FOR:
COLLARDS, MUSTARD, TURNIP GREENS, PORTUGUESE KALE

The cooking time for mature greens is quite a bit longer than for the tender young greens in the Young Greens Braised with Spring Onions (page 225). I plan for at least 1 hour and let myself be surprised if it takes less. The pot will hold at the back of the stove if they're done early. Greens lose nothing overnight in the refrigerator. Tradition holds that you never cover the pot during cooking or as it cools. Use as much salt and freshly ground black pepper as needed to make a highly seasoned pot liquor, adding a final grind just before serving. If using collards, mustard greens, or turnip greens instead of kale, I like to balance the broth at the end of cooking with molasses and apple cider vinegar—Appalachian style.

2 tablespoons extra-virgin olive oil, plus more for drizzling
4 ounces 1-inch cubes guanciale or pancetta
1 medium onion, sliced
1 clove garlic, peeled and halved
¼ teaspoon red chile flakes, or more to taste
1 teaspoon fine sea salt, plus more to taste
4 cups pork or chicken stock
2 × 3-inch piece of kombu
½ teaspoon Red Boat fish sauce
½ teaspoon freshly ground black pepper, plus more to taste
6 quarts Tuscan kale or other mature greens, destemmed (see page 221), leaves cut into 2-inch-wide ribbons
Freshly grated Parmesan cheese (optional)

1. In a large Dutch oven, warm the olive oil over medium heat. Add the guanciale and lightly brown on all sides, about 8 minutes total. Add the onion, garlic, chile flakes, and ¼ teaspoon of the salt. Sweat until the onion has softened and is pale gold, about 5 minutes.
2. Add the stock, 4 cups water, the kombu, fish sauce, black pepper, and the remaining ¾ teaspoon of salt. Increase the heat to high and bring to a boil for 5 minutes. Taste the stock and adjust if necessary. It should be salty and delicious.
3. Add the greens and return to a boil. Reduce the heat to maintain a lively simmer and cook, uncovered, until the greens are silky and slick, an hour or more.
4. Serve in bowls with a share of the pot liquor. Drizzle with olive oil and top with a few scrapings of Parmesan cheese (if using).

To adapt for collards, mustard greens, turnip greens, and Portuguese kale: Replace the guanciale with slab bacon and omit the olive oil when browning the bacon. Continue with the recipe as written and once the greens are silky, stir in ½ teaspoon molasses and 1 teaspoon apple cider vinegar, or more to taste. Serve with Fermented Hot Sauce (page 440) or pepper vinegar.

CHARD • PARSLEY • TOMATOES • DILL

ALSO FOR: SPINACH, LAMB'S-QUARTER, SIBERIAN KALE

SIDE

SERVES 4 TO 6

VEGAN, GLUTEN-FREE

SUMMER

SUMMER CHARD

WITH FRESH TOMATOES AND HERBS

Chard is a cool-weather crop that keeps going all summer long in milder regions. Picked in July and August, when heat-averse spinach has faded away, chard can be sautéed with juicy tomatoes to lubricate the leaves and keep them from squeaking against your teeth in that weird way. The green onions in this recipe could be replaced by diced bulbing onion and, as usual, feel free to swap out the herbs based on what's growing—cilantro or fennel in place of dill. Don't neglect to dry the chard leaves in a salad spinner before cooking to prevent a waterlogged pan.

2 tablespoons extra-virgin olive oil
6 scallions or small green onions, whites and half the greens, chopped
1 large clove garlic, crushed and peeled
2 pounds chard, destemmed (see page 221), rinsed, spun dry, and cut into wide ribbons
1/2 teaspoon fine sea salt, plus more to taste
1 large tomato, seeded and chopped
1/2 cup Niçoise olives, pitted
1/4 cup picked dill fronds, fennel fronds, or cilantro leaves, chopped
Freshly ground black pepper

1. In a large skillet, heat the olive oil over medium heat until it shimmers. Add the scallions and garlic and sweat until softened, 2 to 3 minutes. Add the chard and season with ¼ teaspoon of the salt. Increase the heat to high, cover, and cook for 5 minutes to wilt, turning the chard with tongs a couple of times to cook evenly.
2. Add the tomatoes and the remaining ¼ teaspoon of salt and sauté until soft and juicy, about 5 minutes.
3. Add the olives, herbs, and a few turns of freshly ground black pepper. Toss to combine. Taste and adjust the seasoning.

CHARD OR SPINACH • GREEN CHILES • ONION • GARLIC

ALSO FOR: OTHER HEARTY GREENS, GREEN BEANS, CAULIFLOWER, POTATOES, MIXED VEGETABLES

SIDE | SERVES 4 TO 6 | VEGAN, GLUTEN-FREE | SPRING, SUMMER, FALL, FREEZER

CURRIED GREENS

Kitchen English is a minor wrinkle in the grand tapestry of a global lingua franca, but interesting because it enfolds both the adaptability of the mother tongue and its failings.

The word *curry* came into use during the age of exploration. Originally from a South Indian language, it was applied to "exotic" dishes from multiple cuisines spanning maybe 50 degrees of longitude, a significant swath of Asia between the Indian subcontinent and Indonesia. As culinary vocabulary, the term is far too broad.

Not all curries have a common ancestral source. Ginger, garlic, turmeric, and chiles are shared themes but not bedrock requirements. Depending on how ingredients are apportioned, curry could be reddish, yellow/orange, or green. Pale vegetables (potatoes, cauliflower) soak up curry flavors, and bright ones (green beans, tomatoes, pumpkin) provide counterpoint. Chopped leafy greens bathed in curry are a homey flavor bomb.

The coconut milk used here skews the flavor palette toward Southeast Asia and makes the dish rich enough to anchor a vegetarian meal. Use whatever greens are available, and pair them with other garden dishes. In the lunch pictured opposite, curried chard sits alongside chunked Brandywine tomatoes, crookneck squash roasted with cumin and garlic, and brown rice from Et Cetera Farm in the Hudson Valley.

Depending on the season, you could instead use spinach, young collards shredded into thin ribbons, orach, amaranth, or wild lamb's-quarters. For a more Indian-inspired curry, sizzle fenugreek seeds with extra cumin and coriander in ghee, add a chopped tomato, and omit the coconut milk. The result is less rich, more biting.

Fine sea salt, for the blanching water
2 pounds white or yellow Swiss chard
3 tablespoons grapeseed oil
1 teaspoon cumin seeds
1 teaspoon coriander seed
1 small yellow onion, chopped
1 serrano chile, chopped
½ teaspoon fine sea salt
2 tablespoons grated or minced garlic
2 tablespoons grated or minced fresh ginger
1 tablespoon grated fresh turmeric
1 can (14 ounces) full-fat coconut milk
½ cup lightly packed fresh cilantro leaves

1. Bring a large pot of salted water to boil over high heat.

2. Destem the chard (see page 221). Trim the stems into 3-inch lengths. Blanch the stems for 5 minutes. Remove with a spider and drain in a colander. Chop into ½-inch pieces. Blanch the chard leaves for 1 minute. Drain. Once cool, squeeze the leaves very dry. Chop coarsely.

3. In a Dutch oven, heat the grapeseed oil over medium heat. Add the spices and sizzle for 30 seconds. Add the onion, serrano, and salt and sauté for a few minutes until softened. Add the garlic, ginger, and turmeric and cook for 1 minute. Add the chard leaves and stems. Stir a time or two to coat.

4. Pour in the coconut milk, reduce the heat, and cook at a brisk simmer until the chard is cooked through, about 20 minutes.
5. Transfer to a serving dish and shower with the cilantro leaves.

LACINATO KALE • WAXY POTATOES • ONION | MAIN | SERVES 8 | OMNIVORE | SPRING, SUMMER, FALL, FREEZER

GREENS AND POTATO PIE

Greens and potatoes are boon companions. This "pie" layers a filling of chopped lacinato kale and sausage between potato "crusts." As with other pies, the filling can be nearly anything: Replace sausage with browned mushrooms, diced roast pork, or leftover roasted carrots. In any case, the additions should be cooked through.

The best potato variety is the one you'd use for potatoes gratin. My favorite is Pinto Gold or the standby Yukon Gem, an improved version of moist-fleshed Yukon Gold.

- 1/4 cup plus 2 tablespoons extra-virgin olive oil, plus more for the baking dish
- 1 large onion, chopped (about 2 cups)
- 2 cloves garlic, chopped
- 1/2 teaspoon fennel seeds
- 1/2 teaspoon dried thyme or savory
- 1 bay leaf
- 1/4 teaspoon red chile flakes
- 1 teaspoon fine sea salt
- 1/2 cup dry vermouth or white wine
- 1 1/2 pounds kale or other greens, destemmed, blanched, squeezed dry, and chopped
- 2 links Italian sausage, cooked and sliced into rounds
- 1/4 cup all-purpose flour
- Freshly ground black pepper
- 1 1/2 pounds waxy potatoes, sliced 1/4 inch thick
- 1 1/2 cups chicken or vegetable stock

1. Preheat the oven to 350°F. Lightly coat a 10-inch round baking dish or 9-inch square baking dish with olive oil.
2. In a large skillet, heat 2 tablespoons of the olive oil over medium heat until it shimmers. Add the onion, garlic, fennel seeds, thyme, bay leaf, chile flakes, and ½ teaspoon of the salt. Sauté until the onion is golden, 7 to 9 minutes.
3. Deglaze the pan with the vermouth, scraping up all the bits. Stir until the liquid evaporates by half.
4. Transfer the onion mixture to a large bowl and add the chopped greens and sausage. Sprinkle with the remaining ½ teaspoon of salt, the flour, and ¼ teaspoon black pepper and toss to combine.
5. Cover the bottom of the baking dish with two layers of potato slices. Add the greens/sausage mixture in a single layer. Cover with overlapping potato slices. Pour the stock and the remaining ¼ cup of olive oil over everything. Sprinkle with a few grinds of freshly ground black pepper.
6. Loosely cover the baking dish with foil and bake for 30 minutes. Remove the foil and press the potatoes with a spatula into the cooking liquid. Bake uncovered until the potatoes are browned and crisped at the edges, about 45 minutes, pressing the top every 15 minutes so that the potatoes don't dry out.

SPINACH OR OTHER GREENS • SCALLIONS • GARLIC • PARSLEY • HERBS • POTATOES

MAIN

SERVES 8

VEGETARIAN, GLUTEN-FREE

SPRING, SUMMER, FALL

GREEN FRITTATA

This recipe accumulates chlorophyll. Dill and its cousin fennel are used in such quantity they almost become greens. I don't worry about cutting them back in the run-up to summer solstice, their annual growth spurt. The leafy vegetable called for is spinach, but it could be replaced with young chard, sorrel, or my favorite edible weed, lamb's-quarter.

My tip for bringing in the right quantity of greens is to measure by the bowl. I use a 6-quart stainless steel bowl. It's indestructible and holds about a pound of greens when heaped up.

- 3/4 pound medium waxy potatoes, such as Yukon Gem, peeled
- 2 tablespoons extra-virgin olive oil
- 1 1/2 cups chopped scallions or spring onions
- 1 clove garlic, minced
- 1/4 teaspoon fennel seeds
- 1/4 teaspoon red chile flakes
- Fine sea salt
- 1/2 cup minced fresh parsley
- 1/2 cup minced fresh leafy herbs, such as dill, fennel, chervil, tarragon, chives, etc.
- 1 pound spinach or lamb's-quarter, chopped (about 8 cups)
- 1/2 teaspoon baking powder
- 3/4 cup buttermilk
- 8 large eggs, lightly beaten
- 1/4 cup grated Parmesan cheese
- 4 tablespoons unsalted butter

1. Steam the potatoes in a steamer basket until just cooked, about 20 minutes. Remove from the heat. Once cool enough to touch, cut into a large dice and place in a large bowl.
2. Preheat the oven to 375°F.
3. In a 12-inch cast-iron skillet, heat the olive oil over medium heat until it shimmers. Add the scallions, garlic, fennel seeds, chile flakes, and a pinch of salt. Sauté until soft, about 5 minutes.
4. Add the herbs and another pinch of salt and wilt for 2 minutes. Add the spinach by handfuls and wilt before the next addition. Add two pinches of salt. Cook until soft, 6 to 8 minutes. Drain in a colander in the sink and press with the back of a spoon. When cool enough to handle, chop and add to the potatoes. Wipe out the skillet with a paper towel.
5. To the potatoes and greens, add the baking powder, buttermilk, eggs, and Parmesan and stir to combine.
6. In the skillet, melt the butter over medium heat. Pour in the egg mixture and transfer to the oven. Bake, uncovered, until a knife inserted in the middle comes out clean, about 30 minutes.
7. Remove from the oven and allow to cool for 5 minutes. Run a knife around the edge and shake the pan sharply to loosen. Cover with a plate and, without flinching, flip it. Allow to cool before serving.

CH.

18

HERBS PLUS CELERY AND FENNEL

THE FRAGRANT COHORT

Apiaceae* or *Umbelliferae
THE CARROT FAMILY OR UMBELLIFERS

Anethum graveolens
DILL

Anthriscus cerefolium
CHERVIL

Apium graveolens
CELERY

Coriandrum sativum
CILANTRO

Foeniculum vulgare
FENNEL

Levisticum officinale
LOVAGE

Petroselinum crispum
PARSLEY

Lamiaceae
THE MINT FAMILY

Alyosia citrodora
LEMON VERBENA

Lavandula spica
LAVENDER

Melissa officinalis
LEMON BALM

Mentha spicata
SPEARMINT

Mentha × piperita
PEPPERMINT

Ocimum basilicum
SWEET OR ITALIAN BASIL

O. basilicum var. *thyrsiflora*
THAI BASIL

O. basilicum var. *purpureum*
OPAL BASIL

Origanum majorana
MARJORAM,
ALSO CALLED SWEET MARJORAM

O. vulgare
OREGANO

Perilla frutescens
SHISO

Salvia officinalis
SAGE

S. rosmarinus
ROSEMARY

Satureja hortensis
SUMMER SAVORY

S. montana
WINTER SAVORY

Thymus citriodorus
LEMON THYME

T. vulgaris
THYME

As well as

Allium schoenoprasum (lily family)
CHIVES

Artemisia dracunculus (aster family)
TARRAGON

Borago officinalis (forget-me-not family)
BORAGE

Dysphania ambrosioides (amaranth family)
EPAZOTE

Laurus nobilis (laurel family)
BAY LAUREL

Pelargonium spp. (geranium family)
SCENTED GERANIUM

Man cannot live by herbs alone, it is true, but fresh herbs improve everything they touch. Nearly every recipe in this book makes use of them. Herbs bring to the table the most potent and distinctive flavors of the garden, making them a cornerstone of seasonal home cooking. If you grow nothing else, dare to plant a few herbs. You can buy most of them potted at a garden center, half-grown already, and a single thyme or basil plant will improve your kitchen life. From early spring until after Thanksgiving, my dinner prep begins with going out to pick herbs. I used to add pinches while cooking. Now I use handfuls. An herb garden makes one lavish.

HERBS

There are a dozen or more culinary herbs growing in my garden at any time, along with tansy, an insect repellent. There's always room for another because herbs are never out of place. Basil flourishes among the tomatoes, a reminder to gather them together. Put marjoram or summer savory in with the beans for the same reason. Clumps of chives, beautiful when they bloom, mark out the corners of beds. Herbs even can be released into the edges of the yard. Mint, a water-seeker, creeps along the stone foundation of my house, straying from where I put it beneath the hose spigot. Thyme, oregano, and winter savory spread along a gravel walk to the front door, where extra sunlight and droughty heat makes the Mediterranean natives feel at home. Salad burnet, a weedy cucumber-flavored Elizabethan potherb, grows under lilac bushes out back. I go around the yard by pinches and smells, in a true sense following my nose.

The herbs you grow determine the flavors of your cooking, a kind of culinary thumbprint. My favorite is quirky chervil, a souvenir from the time I lived in France. It is hardly grown commercially, perhaps because it has a short season, doesn't hold in the garden, and isn't worth drying. But from April until June, it flavors every meal I cook, as much a harbinger of spring as the robins' return. Its delicate anise flavor, complete by itself, also blends with tarragon, chives, and parsley to make up a fragrant mince aptly called *fines herbes*. Chervil is cold tolerant, springing up with the peas and ready for the earliest garden salad. Summer heat drives it out of business around the time cilantro goes rampant. Chervil self-sows as readily as fennel and dill. A late-sprouting fall crop might overwinter.

Many of the leafy herbs—parsley, basil, cilantro—are annuals. Rosemary, bay, lemon verbena, and rose geranium are tender perennials. I grew them outside in California; now my potted rosemary comes indoors before Halloween. Thyme, sage, chives, and mint are cold-hardy perennials. Over time, an herb garden becomes a collection of fragrances.

In the cold months, herbal flavors persist in the spice cabinet: dried savory, thyme, oregano, and bay leaves in repurposed jelly jars, quart jars for whole dried lemon verbena leaves and rose geranium to lift the spirits on gloomy days—rose-scented glasses. My mother used to send me packets of lavender to put among folded quilts. Dried tansy dispels moths from closets.

Other herbs hold medicinal properties, plant magick known to herbalists, curanderos, witches, and Friar Lawrence in *Romeo and Juliet*.

> *O, mickle is the powerful grace that lies*
> *In herbs, plants, stones, and their true qualities:*
> *For nought so vile that on the earth doth live*
> *But to the earth some special good doth give.*
>
> —ACT II, SCENE 3

In planning an herb garden, start with the lyrical bouquet from "Scarborough Fair"—parsley, sage, rosemary, and thyme. (An older version of the ballad proposed savory over parsley, a fair trade.) Basil and cilantro are essential to my summer cooking, but I'd also recommend Thai basil and shiso. Fennel and dill,

JUNE 22: Clockwise from top left: dill, chervil, celery, lovage, chives, sage, thyme, tarragon, and mint

AUGUST 23: Opal basil

AUGUST 13: Shiso

botanical cousins, have a hundred uses if they are within reach at the right moment, which they never are unless you grow them. Ditto chives and mint. Neither lovage nor lemon thyme seemed essential until I grew them. Lemon verbena, lemon balm, and lemon grass continue a theme and open the path to herbal desserts.

In the Garden

Keep your herbs close to the kitchen. They thrive in small spaces and pots. Herbs prefer full sun but don't require rich soil or any special treatment. The perennials are drought tolerant once established.

To start with transplants, buy potted herbs early in the season. Parsley and the pungent herbs can go out around last frost, or slightly before. The leafy herbs—basil, cilantro, shiso—go out after last frost, as do the tender perennials, such as lemon verbena and rose geranium. Refer back to chapter 6 for instructions in transplants (page 66).

To grow parsley, cilantro, chervil, and dill from seed, sow directly in the garden from midspring through late summer. Parsley can take 3 weeks to germinate. For a steady supply of cilantro and dill, plan succession plantings every 2 to 3 weeks.

Mulch around herbs with straw or shredded leaves for cleaner pickings.

HERBS—ANNUAL, GROWN FROM SEEDS OR STARTS

HERBS	SEED SPACING	DEPTH	MATURE SPACING	ROW SPACING
Basil *warm weather only*	1 per inch	1/4 inch	8 inches	18 inches
Chervil *cool weather only, tolerates light shade*	2 per inch	Barely covered	6 inches	12 inches
Cilantro	2 per inch	1/4 inch	2–4 inches	12 inches
Dill	2 per inch	1/4 inch	4 inches	18 inches
Epazote	1 per inch	Barely covered	12 inches	24 inches
Parsley *seeds very slow to germinate*	2 per inch	1/4 inch	12 inches	12 inches
Shiso	1 per inch	8 inches	12 inches	18 inches

TYPES OF HERBS

I sort herbs into three categories based on how I use them in the kitchen.

Parsley stands alone, a category of one. With its mild and accommodating flavor, flatleaf parsley is an indispensable ingredient for salads, sauces, stocks, pasta, soups, and stews. Chopped, it winds up everywhere, like storm-blown leaves. Combined with other herbs and bound with olive oil, it becomes the basis of a universal green sauce, alias salsa verde or chimichurri. Parsley finely minced with garlic and lemon zest is gremolata, a sprinkle for osso buco, bear stew, and other dark braises. Stripped from the stem, flat parsley leaves can be dressed and served as a little herb salad, a strong pinch of green next to an iron-rich steak. Chopped curly parsley gives bulk as well as flavor to tabbouleh. The seeds are slow to germinate but otherwise parsley grows itself and yields continuously for nonstop harvests.

Leafy herbs such as basil, cilantro, tarragon, dill, chervil, and mint should be used fresh, in combination, and with abandon. Green and tender like parsley, they exhibit the further virtue of fragrance. Each one is distinctive: think mint compared to cilantro.

So distinctive, in fact, that leafy herbs can create the aromatic hallmark for specific dishes (basil + tomatoes = Caprese salad) or an entire meal (tarragon + chervil = French bistro night). Their strong flavors harmonize surprisingly well, especially when joined by parsley, the uniter. Somehow, the greater the quantity of herbs in a recipe, the more harmonious the mix, as in the recipes for Smashed Cucumber Chicken Salad with Southeast Asian Flavors (page 212) and Green Frittata (page 233).

Except for mint, none of the aromatic herbs dries well.

Cilantro, dill, and chervil are easy to grow from seed, but are up and gone before you know it. Plan for succession plantings. Mint, the perennial outlier of the bunch, spreads aggressively by its roots and becomes invasive. Keep an eye on it or grow in a container. Tarragon is a more disciplined perennial. It will enlarge year by year but not take over.

Pungent herbs, such as thyme, winter savory, oregano, marjoram, and rosemary, have strong personalities. They produce small, oil-rich leaves on wiry stems or woody stalks. Most are perennial and get larger over time; some self-sow. All the pungent herbs dry well and can be used in spice blends such as herbes de Provence and za'atar (see recipe, page 254).

HERBS—PERENNIAL, USUALLY GROWN FROM STARTS

	MATURE HEIGHT
Bay laurel *tender*	35 feet
Chives	12 inches
Lavender	12–18 inches
Lemon balm	2 feet
Lemon verbena *tender*	5 feet
Lovage	4 feet
Mint *spreads aggressively*	1–2 feet
Oregano	12 inches
Rosemary *tender*	3–5 feet
Sage	2 feet
Scented geranium *tender*	2 feet
Tarragon	2 feet
Thyme	6 inches
Winter savory	6 inches

AUGUST 29: Borage flowers

In the Kitchen

The flavors of fresh herbs run the gamut from parsley's mild chlorophyll freshness to basil's sultry perfume, strong enough to smell at a distance on a hot day. The dusty musk of sage is the smell of Thanksgiving to me, as well as the smell of gnocchi in brown butter. The menthol trickery of peppermint creates a sensation of cool in the mouth, even when tasted in hot tea.

The uses of herbs vary according to type. Parsley can go into nearly any savory dish. The leafy herbs and pungent herbs have more targeted uses, but they are often interchangeable within their category, meaning you might well swap basil for cilantro (leafy for leafy) or savory for thyme (pungent for pungent) in many recipes. The results will differ, but isn't that the point? Less successful is trying to substitute across categories—basil in place of rosemary or vice versa. Chopped-herb sauces (see Green-Herb Sauces and Condiments, page 247) are arguably the most versatile in this book, as they go with many grilled or roasted dishes and can be fine-tuned by adjusting the herbs. What can't be helped by herbs? Even canned beans take a sure step toward home-cooked flavor when topped with minced cilantro or parsley.

The stage at which you introduce herbs to a dish will determine their influence. Added at the outset, herbs contribute to the foundation of a dish's flavor—thyme branches and bay leaves subtly infusing a long braise. Herbs added at the last moment are a garnishing flavor—the freshness of minced green flurries over the finished braise. Sometimes, as with the mint sprig in iced tea or a chiffonade of basil strewn over ratatouille, what the herb contributes is mostly fragrance, the harbinger of flavor, an airborne emissary of the substantive dish.

Gather herbs early in the day, before noon heat makes them droopy. Pinch or cut only what you need at the time. To harvest parsley, cilantro, chervil, and chives, cut the stem back to the ground. For basil, mint, tarragon, and rosemary,

SEPTEMBER 8: Dried lemon verbena and coriander (cilantro) seeds

JUNE 19: Bolted cilantro, with its special flavor, is not a missed opportunity but a second chance.

pinch from the tips to stimulate side shoots lower down the stem.

Recipes often call for sprigs, branches, stems, bunches, and other vague quantities of herbs. Rely on common sense, taste often, and adjust to your own liking. Err on the side of a little more, except with camphoric rosemary.

Harvesting herbs frequently and pinching out flower buds when they first appear will slow the growth cycle of annual herbs. Even so, the blossoms will get ahead of you at some point. Pick them for salads and garnish, just as you would use the leafy parts. Abundant chive blossoms have a light allium flavor. (Break apart the compound pom-pom flower heads and sprinkle the bell-shaped individual flowers.) Borage flowers have a faint but distinct flavor, usually likened to cucumber but to my taste an exact replica of Hood Canal oysters. In any case, the reason to grow borage is for the flower's gorgeous blue color.

Cilantro flowers are also edible. Even more intriguing are the glossy seeds known while still tender as green coriander berries, a special ingredient that pairs well with corn, tomatoes, Mexican dishes, and Chinese cooking. If you grow cilantro for its leaves, but sure to let at least one plant bolt. Once the seeds dry, they keep all winter—homegrown coriander. Dill seed and fennel seed are likewise easy to gather after they have dried on the stalk.

If herbs are gritty or muddy after rain, swish a few stems at a time in a basin of cool water and spin dry or lay them on kitchen towels to drain. Basil loses something if washed and refrigerated, like tomatoes. Stand a bouquet of basil branches in a jar of water in the windowsill for a day or two. Parsley and other herbs will keep for a week in the refrigerator in an airtight container lined with a kitchen towel or paper towel. See my herb box, page 237.

To dry pungent herbs: Gather long stems or full branches of oregano, marjoram, thyme, lemon verbena, etc., on a dry day in early summer before they bloom. Strip the lower stem. Gather into small bouquets and tie with string. Hang upside down in a warm, breezy place out of direct sun. Once fully dry to the touch and crackly, gently strip the leaves and store in an airtight container in a cool, dark place.

Save parsley stems and thyme branches for the stockpot. Other fragrant trimmings go into the compost bucket to recycle their herbal powers.

CELERY AND FENNEL

Fennel and celery are character actors, sharply drawn and individual, but too offbeat for marquee stardom. Both are highly fragrant umbellifers related to parsley, dill, and cilantro. I tend to use both as herbs, although eventually I'll get around to cutting the stalks—petioles, more properly—to prepare as vegetables. Most backyard growers won't displace their tomatoes and beans to make room for whole beds of celery or fennel, but surely there's room to stick in a few plants of each if you think of them not as minor vegetables but as major herbs.

CELERY

A kitchen staple, celery supplies the rib in a bouquet garni and the third element of mirepoix and Cajun/Creole trinity. Raw, it is a dieter's cliché because the juicy, salty crunch momentarily dulls a craving for caloric food. Homegrown celery is stronger, leafier, and much more fragrant than the wan "hearts" sold in grocery stores. The leaves smell of grass, cumin, cedar, musk, and damp sea salt, a clinging scent that stays on your hands like cologne. I add whole leaves to salads or chiffonade them to sprinkle, as I would cilantro, basil, or tarragon. At Sticky Rice, a Thai restaurant in Los Angeles, the daily specials sometimes include grilled-pork-and-celery-leaf salad, a flavor bomb. Celery risotto (discussed in Green Bean Risotto, page 130) tames the raw flavor. Braised celery is an excellent side dish for any meat, especially roast

OCTOBER 5: Celery and celeriac at full maturity

chicken. Liquefy braised celery with stock and a bit of cooked potato for garden bisque. It's October fare, lunch after raking leaves.

Celery needs to be sown indoors very early. Or buy transplants to put in the ground after frost with a later succession at midsummer. Its water-filled stems demand the equivalent of 2 inches of rainfall per week—twice what other vegetables need. Celery matures to knee height over 3 months.

Begin picking leaves as soon as the plants can spare a few, within weeks of transplanting. After the bunches are about a foot tall and shaggy headed, harvest a few stalks at a time. When you're ready for it, slice the entire bunch at ground level with a knife. Trimmed and wrapped in a towel inside a plastic bag, a bunch will keep in the refrigerator for weeks. As a biennial, celery also holds very well in the garden. Mature bunches will withstand a light frost, but harvest before sharp cold arrives.

The robust perennial herb lovage, another carrot relative, has a very strong celery scent. A single mature plant, 4 feet tall, would supply an entire neighborhood.

FENNEL

Fennel's dill-like fronds can be clipped as an herb or else whole stalks can be snapped into 10-inch lengths and laid on a sheet pan as a fragrant pallet upon which to bake fish. The flowers provide edible pollen—sweetly scented fairy dust. To gather it, shake fresh blossoms into a bag or onto a sheet of paper and dry for several days out of direct sun. Store in a sealed jar and sprinkle on pork, chicken, eggs, vegetables, or any dull person lacking vigor. Gathered in the fall, homegrown fennel seeds have a stronger, wilder flavor than the store-bought spice.

Two or three fennel plants used this way will carry you through the season. Bronze fennel is similar in all ways except color. Put out fennel transplants a week or two before last frost. Sow seeds as for parsley and thin seedlings to 8 inches.

To grow fennel as a vegetable, cut the flower stalk when it emerges to divert the plant's energy into thick stems, the "bulb."

CELERY AND FENNEL—TRANSPLANTS

TRANSPLANT	SPACING	ROW SPACING	TIME TO HARVEST
Spring or midsummer	8 inches	18–24 inches	2 weeks (leaves), 70–90 days (stalks)

Note: Fennel is put out before last frost, celery after. Thirsty celery requires the equivalent of 2 inches of rain per week. Fennel can be grown from seed more easily than celery; sow as for parsley.

JUNE 25: Fennel can be harvested for its vegetable bulb or grown out for abundant herblike greenery as well as for its aromatic pollen and flavorful seeds.

GREEN-HERB SAUCES AND CONDIMENTS

During afternoon video chats with my mother, a near-daily occurrence in the last years of her life, I always knew if she had had an appointment earlier—a church meeting, doctor's visit, or house call from a friend. The tell was her silk scarf. It indicated intention, a small gesture of respect for whomever it was she planned to see, her unfussy way of putting forward her best.

A quick herb sauce, any one of the several options that follow, is the silk scarf of the kitchen. It won't turn a sandwich into a banquet, to quote a line from Tom Waits, but it's a nice touch others will appreciate. They will think, "How nice," and know you've taken care with their supper. It shows panache.

UNIVERSAL GREEN SAUCE

MAKES ABOUT 1 CUP
VEGAN OPTION

This rustic green sauce will never bore you because it will never come out the same way twice. I often finish chopped-herb sauce with scrapings of lemon peel and a squeeze of lemon juice. Sometimes I add minced salt-cured anchovy, especially when the sauce is for lamb or beef, in which case a bit of chopped mint (for lamb) or minced rosemary (for beef) would also make sense. A quarter teaspoon of fish sauce or a pea-sized dab of anchovy paste introduces umami at a more background level. I sometimes throw in pitted olives—green or black or both—to go with grilled chicken. If I'm cooking fish, I chop in chervil and tarragon, and for summer vegetables I'll add basil and extra olive oil until it puddles in the mortar. Sometimes finely chopped jalapeño is the right extra something.

The following recipe provides general proportions and basic directions, but there's no danger in veering off course.

1 medium shallot, minced
2 to 3 tablespoons red wine vinegar
Fine sea salt and freshly ground black pepper
1 clove garlic, peeled but whole
1 salt-cured anchovy (rinsed, boned, and chopped), 2 oil-cured anchovy filets, a dab of anchovy paste, or a splash of fish sauce (optional)
1 tablespoon capers, preferably salt-cured, rinsed and chopped (optional)
1 cup chopped parsley, optionally substituting:
a portion of chopped leafy herbs
a portion of chopped fennel fronds, carrot tops, or celery leaf
Pinches of minced fresh pungent herbs, such as rosemary, thyme, oregano, or marjoram
Pinch of cayenne pepper
2/3 cup extra-virgin olive oil
Olives (optional), pitted and roughly chopped
Grated lemon zest and/or lemon juice

1. In a small dish, cover the minced shallot with the vinegar. Add 2 large pinches of salt and several grinds of black pepper. Set aside for 15 minutes.
2. Pound the garlic into to a paste in a mortar with 2 large pinches of salt. Add the anchovy (if using) and pound it to a paste. Add the capers (if using) and crush them lightly. Add the herbs and bruise them with the pestle. Add the shallots and half their soaking vinegar, as well the cayenne, another pinch of salt, and freshly cracked black pepper. Stream in the olive oil while stirring with the pestle. Taste. If it tastes a bit flat, add the remaining reserved vinegar.
3. To finish, stir in the olives (if using). Finish with a small amount of lemon zest and/or lemon juice to taste. Set aside for the flavors to blend for 30 minutes before serving. Store leftovers overnight in an airtight container.

GREMOLATA

MAKES 1/4 CUP
VEGAN, GLUTEN-FREE

Gremolata is potent—raw garlic and lemon zest minced with parsley. Traditionally sprinkled over osso buco, it will brighten any stew or braise. I also use it in less time-honored ways—on a roast chicken sandwich, or on leftover grilled vegetables, or sprinkled over soup or white beans or anything else that gets along with garlic, lemon, or parsley. Gremolata covers a lot of territory.

Gremolata differs from green sauce in that it uses no oil as a binder. It has an almost crumb-like texture—a sprinkle. Use your sharpest knife for a fine mince. The parsley must be completely dry before chopping. Whirl it through a salad spinner and dab it with a kitchen towel. Grate the lemon zest with a Microplane, if you have one. Otherwise, remove the zest in strips with a vegetable peeler, as thinly as possible, and mince.

1 cup lightly packed parsley leaves
2 cloves garlic, minced
Grated zest of 1 lemon

Chop the parsley medium fine, or until you have a compact pile. Add the garlic and lemon zest. Continue chopping until the ingredients are finely minced and blended. Spread the mixture in a thin layer on a saucer or soup plate to dry on the counter for a few minutes while you make dinner. Gremolata is best used the same day, but leftovers will keep overnight covered with olive oil in the refrigerator.

CHERMOULA

MAKES 1 CUP
VEGAN, GLUTEN-FREE

Cilantro and cumin are the foreground flavors of this Moroccan seasoning for fish. Its evocative aroma will lead you toward additional uses: on roasted vegetables, stewed chickpeas, braised greens, and so on.

2 cloves garlic, peeled but whole
Fine sea salt
1 cup lightly packed parsley leaves
2 cups lightly packed cilantro leaves
1 tablespoon cumin seeds, toasted and ground
1 teaspoon coriander seeds, toasted and ground
1 tablespoon paprika
Pinch of cayenne pepper
2 tablespoons fresh lemon juice
1 teaspoon rice vinegar
1/2 cup extra-virgin olive oil

Crush the garlic with a pinch of salt in a mortar or on a cutting board, using the dull edge of a chef's knife. Mince the herbs on a cutting board. Add about one-third of the herbs to the mortar and grind into a coarse paste. Repeat with the remaining herbs. Add the cumin, coriander, paprika, cayenne pepper, lemon juice, vinegar, and olive oil. Blend together using the pestle.

HERB MAYONNAISE

MAKES 1 CUP
VEGETARIAN, GLUTEN-FREE

Homemade mayonnaise, if you've never sweated to emulsify oil into an egg yolk drop by literal drop for 20 unceasing minutes, is less stiff in texture and finer in taste than store-bought. Its countless uses run beyond the utilitarian—sandwiches, egg salad—into the realm of the purely gustatory, for instance splashed on roasted La Ratte fingerling potatoes. It's a little bit naughty, like thick cream on strawberries. If you whisk a handful of chopped herbs into basic mayonnaise, it becomes a flavorful finishing sauce.

Grated zest of 1 lemon
Fine sea salt
1 clove garlic, peeled but whole
1 cup lightly packed parsley leaves
3 tablespoons lightly packed tarragon and chervil leaves
2 tablespoons finely chopped fresh chives
1 egg yolk
1 teaspoon Dijon mustard
3/4 cup neutral oil
1/4 cup extra-virgin olive oil
Juice of 1 lemon
Freshly ground black pepper

1. Rub together the zest and a pinch of salt in a medium bowl. On a cutting board, mince and mash the garlic with a pinch of salt using the dull edge of your knife. Add it to the bowl. Mince the parsley and other herbs, adding them to the bowl as well.
2. Whisk the egg yolk in the bottom of a large mortar or medium mixing bowl for 30 seconds. Whisk in the mustard. Combine the oils in a small pitcher or spouted measuring cup. Whisking all the while, slowly drizzle in the oil, drop by drop, emulsifying each drop. After one-quarter of the oil has been incorporated, squeeze in a few drops of lemon juice to loosen the consistency. You've passed the worst danger now. Continuing to whisk, incorporate the rest of the oil in a thin stream. Add drops of lemon juice as you go if the consistency gets too thick. At the end, add more lemon juice to taste. You can also adjust the consistency with drops of water. The finished mayonnaise should be smooth and billowy, not stiff.
3. Stir in the lemon zest, garlic, and chopped herbs. Add salt and pepper to taste. Store in a sealed container in the refrigerator for up to 3 days.

Note: To make mayonnaise in a stand mixer, use the whisk attachment. Add the egg yolk, 2 teaspoons of Dijon mustard, and the lemon juice to the bowl. Whisk at the highest setting until frothy. Add oil a few drops at a time until the mixture begins to thicken, then slowly stream in the remaining oil.

ZESTY HERB DIPPING SAUCE

MAKES 1½ CUPS
VEGETARIAN

This sauce moves toward the Eastern Mediterranean, the Levantine coastline of flatbreads, sumac, and kebabs, where the culinary imagination retains echoes of Byzantium, Jerusalem, and Babylon. Its mild yogurt tang is a vehicle for chopped herbs. Use over raw, blanched, or grilled vegetables. Or put it out as part of an array of small dishes in the spirit of mezze, perhaps alongside Kale and Date Tabbouleh (page 222), Smoky Eggplant Dip (page 444), and lamb kofta.

1 cup whole milk Greek yogurt
⅓ cup extra-virgin olive oil
1 teaspoon grated or minced garlic
½ cup chopped dill fronds
⅓ cup thinly sliced chives
Grated zest of 1 lemon
3 tablespoons fresh lemon juice
1 teaspoon fine sea salt
½ teaspoon freshly ground black pepper
Buttermilk or cream, as needed

In a medium bowl, whisk together the yogurt, olive oil, garlic, dill fronds, chives, lemon zest, lemon juice, salt, and pepper. The sauce should be on the loose side. Thin the consistency, if necessary, with buttermilk or cream by the tablespoon.

PUNGENT-HERB SAUCES AND CONDIMENTS

My kitchen collaborator Dash told a story about the time our chef friend Wes Whitsell went to a colleague's restaurant and got a full tour of the kitchen. Wes's favorite part was the pantry: shelves full of homemade pickles, hot sauces, ferments, spice blends, and all other manner of proprietary recipes—the tricks of the chef's trade. "It was a shelf of superpowers," Wes said. The following recipes are of the type.

BAGNA CAUDA

MAKES 1 CUP
PESCATARIAN, GLUTEN-FREE

Bagna cáuda, "hot sauce" in the Piedmontese dialect, is served warm, traditionally in a terra-cotta chafing dish over a candle. It is a potion of olive oil, butter, garlic, and minced anchovies. Thyme leaves and shallots expand upon its pungent flavors. Summer helpers Chen and Deeva had never tasted bagna cauda when we first had it together, spooned over roasted cauliflower. (Read about them in the headnote to Napa Cabbage Stir-Fried with Pork and Vinegar and see the photo on page 173.) I believe Chen would have drunk it straight from the dish if not for Deeva and me. Bagna cauda is a dunking sauce for any raw or blanched vegetable, especially fennel, celery, carrots, artichokes, and cardoons, should you ever find them at the farmers' market. It goes well with small boiled potatoes.

1 stick (4 ounces) unsalted butter
1/3 cup extra-virgin olive oil
3 tablespoons minced shallot
1 1/2 teaspoons minced garlic
Pinch of red chile flakes
1/2 teaspoon thyme leaves
3 tablespoons minced anchovy fillets
3 tablespoons fresh lemon juice, plus more to taste
Fine sea salt

In a small saucepan, heat the butter and olive oil over low heat. When the butter melts, add the shallot, garlic, chile flakes, thyme, and anchovies. Stir to combine. Simmer over lowest possible heat, uncovered, using a diffuser if it sizzles. After 10 minutes, remove from the heat and add the lemon juice. Taste and adjust with more lemon juice and salt, as needed. The sauce will separate in the dish.

GREEN ZA'ATAR

MAKES ABOUT ¼ CUP
VEGAN, GLUTEN-FREE

Za'atar is a Middle Eastern spice blend that includes dried herbs, sesame seeds, and sumac. My loose adaptation goes light on the red sumac in favor of green dried herbs. The Arabic word *za'atar* also applies to a wild herb, *Origanum syriacum,* that smells of sweet marjoram, thyme, and oregano, according to Johnny's, which sells the seed. This spice blend approximates the flavor profile by mixing dried marjoram, thyme, oregano, and savory, the last of which imparts a complex, resinous, wild quality. Fennel pollen brings in the fragrance of dust and enlightenment. Adapt the blend to what you have on hand. Sprinkle it liberally over raw, roasted, or steamed vegetables; on salads, soups, and grain bowls; on feta cheese and labneh; on eggs; on cold roast chicken, and on bread dipped in olive oil. Think of it as seasoned salt, elevated.

1 tablespoon dried thyme
1 tablespoon dried savory
1 tablespoon dried marjoram or oregano
½ teaspoon dried mint
¼ teaspoon fennel seeds
¼ teaspoon fennel pollen
½ teaspoon sumac
½ teaspoon Aleppo pepper or a pinch of cayenne pepper
1 tablespoon sesame seeds, toasted
½ teaspoon flaky finishing salt

In a mortar, combine the dried herbs, fennel seeds, fennel pollen, sumac, and Aleppo. Grind with the pestle a few times, just enough to bruise and blend everything. Stir in the sesame seeds and salt. Store in an airtight container, such as a 1-pint mason jar. It keeps for a month or more. Double the recipe if you'd like.

PRESERVED LEMONS

MAKES 1 PINT
VEGAN, GLUTEN-FREE

Salt-cured lemons are a sneaky umami booster. As they mature in the jar, they develop the mouth-filling complexity of sauerkraut brine. I like to tuck a few garden ingredients among the lemons, such as thyme, red chile, and coriander seeds. Meyer lemons, with their supple skins and whiff of mandarin, are my favorite to preserve.

Fine sea salt
5 small lemons or Meyer lemons
2 sprigs thyme
1 bay leaf
1 small dried red chile
6 coriander seeds
3 peppercorns

Wash and dry a wide-mouth pint jar. Add 1 tablespoon of salt to the bottom. Take one lemon and cut it into quarters lengthwise, stopping before you slice all the way through the base. Pry apart the quarters, opening it like a tulip, and rub a heaping teaspoon of salt onto the cut surfaces. Reclose the lemon and very firmly press it into the bottom of the pint jar, squeezing out its juice so that it is nearly submerged. Repeat with 3 more lemons. Sprinkle in salt as you layer the lemons into the jar and tuck in the thyme, bay leaf, chile, coriander, and peppercorns as you go. Press out air pockets. Juice the last lemon and pour into the jar to cover, leaving ½-inch headspace. Save any leftover juice. Loosely seal the jar and leave in a cool place. After 3 days, open it and push down the lemons. If necessary, add the reserved juice to cover. Leave in a cool place for 30 days to cure. Keep in the refrigerator for up to 1 year. To use, scrape and discard the mushy pulp and rinse the peel.

ONION • PARSLEY • LEAFY HERBS • CHIVES | STARTER, SIDE | MAKES 36 MINI-BISCUITS | VEGETARIAN | SPRING, SUMMER, FALL

HERB BISCUITS

Biscuits are quick—that's part of the appeal. Drop biscuits are quicker. You make a wet dough and drop it by the spoonful onto a baking sheet. The more peaks and crags, the better: they brown in the oven. I like small biscuits for the favorable ratio of crunchy exterior to tender crumb, which is fragrant with two handfuls of herbs and two types of cheese. Biscuits are not only for breakfast. Serve them hot as a predinner snack, in the manner of gougères, or put them on a buffet. The dough could also be used to top a pot pie or a savory cobbler.

To make biscuits in advance, portion out the dough on an ungreased baking sheet and put it in the freezer for 1 hour to harden. Transfer to an airtight container to keep for up to 1 month. Bake them off in a 425°F oven for 14 to 16 minutes, no need to thaw first.

- 2 cups (240 g) all-purpose flour
- 1 tablespoon baking powder
- 1/2 teaspoon baking soda
- 1/2 teaspoon fine sea salt
- 1/4 teaspoon freshly ground black pepper
- 6 tablespoons chilled unsalted butter or 1/3 cup lard, cut into small cubes
- 1 small shallot or yellow onion, diced and sautéed in unsalted butter until soft
- 1/3 cup finely chopped fresh parsley and mixed leafy herbs, such as tarragon, sage, dill, fennel fronds, celery leaf
- 1/4 cup chopped fresh chives
- 1 tablespoon minced mixed pungent herbs, such as thyme, savory, and rosemary
- 1/2 cup grated sharp cheddar cheese
- 1/4 cup grated Parmesan cheese
- 1 cup buttermilk

1. Preheat the oven to 425°F.
2. In a large bowl, whisk together the flour, baking powder, baking soda, salt, and pepper. Cut in the butter until the mixture has the texture of coarse cornmeal. Add the sautéed shallot, chopped herbs, cheddar, and Parmesan and stir to combine. Add the buttermilk all at once and stir just until a sticky dough comes together.
3. Drop the dough by the tablespoonful (about 1 ounce) 2 inches apart on an ungreased baking sheet optionally lined with parchment. Bake until browned and crispy at the edges, 12 to 15 minutes. Serve immediately.

TOMATOES • CUCUMBERS • RADISHES • EPAZOTE, MINT, OR OTHER LEAFY HERBS | STARTER, SIDE | SERVES 6 TO 8 | VEGAN, GLUTEN-FREE | HIGH SUMMER

WATERMELON-HERB SALAD

Fifty percent of home cooking is sourcing—having good ingredients. Another thirty percent is planning. The piddly rest is technique, the flashy knifework and pan tossing. This recipe for peak-of-summer produce shifts the balance even further, almost entirely onto the sourcing side of the equation. Start with a good watermelon from the farmers' market. Cut it up and arrange on a platter with the best of your garden—big glistening chunks of heirloom tomatoes, cucumber rounds, whole leaves of mint, basil, cilantro, or epazote, the Mexican bean herb. Squirt with a few of the lime wedges you cut for cocktails. A pinch of Aleppo pepper or spicier cayenne mimics the heat outside.

- 1½ pounds watermelon heart, broken into lime-sized chunks
- ¾ pound tomatoes, cut into similar-sized chunks
- ½ pound ¾-inch roll-cut small cucumbers
- ⅓ cup thinly sliced radishes
- ¼ cup lightly packed epazote, mint, or other herb leaves
- Grated zest and juice of 2 limes, plus wedges for serving
- 1 teaspoon fine sea salt
- 2 tablespoons extra-virgin olive oil
- ½ teaspoon Aleppo pepper, or 2 pinches of cayenne pepper

1. In a large bowl, combine the watermelon, tomatoes, cucumbers, radishes, and epazote. Add the lime zest, lime juice, salt, and olive oil. Gently turn over to coat.
2. Arrange the melon and vegetables on a serving platter. Sprinkle with the Aleppo. Serve with lime wedges.

CELERY AND CELERY LEAVES • SHALLOT • PARSLEY • MINT | SIDE | SERVES 4 TO 6 | VEGETARIAN, GLUTEN-FREE | SUMMER, FALL, WINTER

DOUBLE CELERY WITH MUSHROOMS

At Zuni Café in San Francisco, a starter that has never left the menu consists of scattered chevrons of sliced celery, a small shoal of house-cured anchovy fillets, five Coquillo olives (a Spanish version of the French Niçoise), and several thin slices of Parmesan. Olive oil and coarsely ground black pepper finish the plate. It's a sketch, the notation of a culinary mind in movement toward an idea, a glimpse of creative thought in action. It's also a touchstone for how not to do too much. Celery speaks for itself.

As it does in this recipe, mushrooms provide meaty substance and textural contrast, and the umami in Parmesan cheese helps round out its astringency. The result is a rather chewy salad best served in small portions alongside a rich dish, such as zucchini stuffed with ground meat, cheese, and pine nuts. (See recipe on page 362.) For a splurge, use wild porcini in season, perhaps topped with shaved black truffles.

1/4 teaspoon honey
4 tablespoons sherry vinegar
2 shallots, minced
1/2 teaspoon fine sea salt
Freshly ground black pepper
1 medium bunch celery, trimmed and thinly sliced on the bias (about 3 cups)
1/3 cup lightly packed pale inner celery leaves
1 1/2 cups lightly packed parsley leaves
1/2 cup lightly packed mint leaves
4 ounces brown mushrooms, thinly sliced (about 3 cups)
1/4 teaspoon Dijon mustard
1/2 cup extra-virgin olive oil
1/2 cup Parmesan curls, shaved with a vegetable peeler
1 teaspoon Aleppo pepper

1. In a small bowl, whisk together the honey and vinegar until the honey dissolves. Add the shallots, 1/4 teaspoon of the salt, and several grinds of black pepper. Set aside to macerate for 15 minutes.
2. In a large bowl, combine the celery, celery leaves, parsley, mint, and mushrooms and toss to combine.
3. Whisk the mustard into the shallot/vinegar mixture and stream in the olive oil, whisking constantly. Pour over the celery mixture and toss to coat. Add the remaining salt, Parmesan curls, and Aleppo. Toss a time or two to combine.

FENNEL MEATBALLS

Italian cuisine seasons pork sausage with fennel seed and rolls chopped fennel frond into stuffed pork roasts. When I lived in Los Angeles, I came across a similar flavor pairing at Beijing Pie House in the San Gabriel Valley east of downtown. The specialty was xian bing, an oversized dumpling filled with ground pork and fennel frond. Seared on a flattop griddle, it was large enough to eat with two hands, a burger-sized potsticker.

Meatballs are an easier play on the same concept. Garden intern Chen showed me her family's trick of using tofu to bind meatballs, making them both firm and juicy. We ate them with stir-fried vegetables and rice, but I've also served Fennel Meatballs with Italian-ish vegetables such as sautéed broccoli rabe. Simmered in tomato sauce, they would garnish a plate of spaghetti. Ground lamb could replace ground pork.

2 tablespoons extra-virgin olive oil
1 medium onion, diced
2 tablespoons minced garlic
1½ tablespoons grated or minced fresh ginger
2 teaspoons fennel seeds
¼ teaspoon cayenne pepper
3 cups chopped fennel fronds
1 pound ground pork
1 egg, beaten
½ pound firm tofu, crumbled and finely chopped
2½ teaspoons fine sea salt
Freshly ground black pepper

1. Preheat the oven to 400°F. Line a baking sheet with foil.
2. In a large skillet, heat 1 tablespoon of the olive oil over medium heat until it shimmers. Add the onion, garlic, ginger, fennel seeds, and cayenne and sauté until the onion has softened and is fragrant, about 5 minutes. Scrape into a large bowl.
3. Using the same skillet, heat the remaining 1 tablespoon of oil until it shimmers. Add the fennel fronds and sauté until wilted, about 3 minutes. Add to the onion.
4. Add the ground meat, egg, tofu, salt, and several grinds of black pepper to the sautéed vegetables. Mix thoroughly with your hands. Shape small meatballs, 1 tablespoon each, and place 1 inch apart on the lined baking sheet.
5. Bake until browned, 15 to 20 minutes.

LEMON THYME OR THYME • LEMON BALM OR LEMON VERBENA | SWEET | MAKES 1 LOAF | VEGETARIAN | SPRING, SUMMER, FALL

LEMON-HERB POUND CAKE

With a nose like a foxhound, my mother grew flowers and herbs as much for fragrance as for color—anise-scented roses, gardenias perfumed with boudoir drama, herbs as sharp as fresh ideas. I can't remember if she planted lemon thyme first and then went looking for a recipe, or if she found the recipe and grew her ingredients. Either way, she'd make lemon-herb loaf cake when I came home from college in California. It seemed very Berkeley in a way—herbs in cake!—but also easygoing and motherly. Her source was *Herbs: A Country Garden Cookbook* by Rosalind Creasy. I've fiddled with Creasy's tender, airy quick bread to make a denser pound cake. Lemon glaze drizzled across the still-warm surface cools as a thin and fragile sugar frost. Where it seeps into fissures, the interior crumb is seamed with sweet tang.

Regular garden thyme will do if you don't have lemon thyme. Serve the cake with lightly sweetened sour cream and red fruit in season—cherries, raspberries, currants—or else put it out for breakfast with cultured butter.

Softened butter and flour, for the pan
1 cup powdered sugar
1/4 cup fresh lemon juice (see Note, page 264)
6 lemon balm leaves or 12 lemon verbena leaves
3/4 cup organic heavy cream
2 tablespoons minced fresh lemon thyme
1 tablespoon minced fresh lemon balm or lemon verbena
1/4 cup sour cream
2 cups (240 g) all-purpose flour
1 1/2 teaspoons baking powder
1 teaspoon salt
1 stick (4 ounces/115 g) unsalted butter, at room temperature
1 cup granulated sugar
Grated zest of 1 lemon
2 large eggs
Whisked Sour Cream (recipe follows)

1. Preheat the oven to 325°F. Lightly butter and flour a 9 × 5-inch loaf pan.
2. In a small bowl, whisk the powdered sugar into the lemon juice. Pour the thick sauce through a fine-mesh sieve into another bowl to remove lumps. Bruise the herb leaves to release their fragrance and muddle them in the lemon glaze. Set aside.
3. In a small saucepan, warm the cream over medium heat without scalding. Stir in the minced herbs and sour cream. Set aside to cool.
4. In a medium bowl, combine the flour, baking powder, and salt. In the bowl of a stand mixer fitted with a paddle attachment, cream the butter, granulated sugar, and lemon zest until fluffy. Beat in one egg at a time. Incorporate the flour mixture in several additions, alternating with the cooled cream. Don't overmix. Pour the batter into the prepared loaf pan.

(CONTINUED)

5. Bake until a toothpick emerges clean, about 45 minutes.
6. Cool briefly and turn the loaf out onto a wire rack placed over a baking sheet. Pluck out and discard the spent herbs from the lemon glaze and drizzle the glaze over the cake while it is still warm. If you like, lay a few fresh herb leaves or sprigs of thyme across the top of the cake. Gently press into the glaze as you drizzle. Allow to cool completely before slicing. Serve with the Whisked Sour Cream.

Note: Zest the lemon before squeezing it for juice.

WHISKED SOUR CREAM

1 cup sour cream
1/4 cup whole milk
1 teaspoon powdered sugar
A few drops of vanilla extract

Vigorously whisk the sour cream, whole milk, powdered sugar, and vanilla in a medium bowl.

CH.

19

ONIONS AND GARLIC FAMILY

ALLIUMS

Amaryllidaceae
THE AMARYLLIS FAMILY,
FORMERLY ***Liliacea,*** THE LILY FAMILY

Allimum apeloprasum
LEEKS

A. cepa
ONIONS

A. cepa var. *aggregatum*
DUTCH RED SHALLOTS

A. fistulosum
SCALLIONS OR BUNCHING ONIONS

A. oschaninii
FRENCH GREY SHALLOTS

A. sativum
GARLIC

A. sativum var. *ophioscorodon*
HARDNECK GARLIC

A. tuberosum
CHIVES, GARLIC CHIVES

AUGUST 13: Cured Blush onions ready for storage

Ubiquitous, oddly glamorous thanks to their eye-watering potency, and underappreciated. It's a mixed fate for this family of bulbing plants. On the one hand, a recipe I yearned to make as an aspiring teenaged cook was Chicken with 40 Garlic Cloves from *The Wonderful Food of Provence*. You sealed the lidded baking-dish with flour paste to trap flavors. It sounded so daring—extreme cuisine. What I learned from making it, as any home cook knows, is that alliums are tamed by cooking. I moved on to jaunty leeks vinaigrette and, with time, matured into pissaladière, a caramelized-onion tart with black olives and anchovies. Today hardly a meal at my house begins without peeling an onion, mincing shallots, or crushing a clove of garlic. French onion soup, scallion pancake, and deep-fried onion rings all attest to alliums' global appeal.

And yet. Perhaps familiarity has dulled our appreciation and lulled us into accepting the mediocre onion and garlic varieties found at the grocery store. Growing the pungent clan set me straight. Onions are seasonal vegetables, variable and nuanced, and not just color-coded background aromatics in white, yellow, or red. Some garlic varieties are transformative, lifting recipes with their unexpected nuances. Spanish Roja garlic, one of the supremely delicious Rocambole types, is a secret ingredient during its too-brief prime season, whether grated raw and stirred into a pimento cheese or just rubbed over grilled toast. The flavor lingers, the long finish of good wine or erotic afterglow.

The name *onion* relates to union: It is the unified vegetable, layers wrapped within layers, a complex mystery resistant to superficial investigation. The Latin *cepa* turned into the Spanish *cebolla* and Italian *cipolla,* which gives us the name for a type of diminutive, flattened cooking onion known as cipollini, or small cipolla.

The annual allium cycle runs from spring to spring. It begins with scallions, also called bunching onions. Mild, green, almost grassy, they can be chopped and strewn like herbs over salads and soups. Simmered in cream, they are a vegetable side dish; as kimchee, fermented with chili paste and garlic, they are a favorite banchan alongside Korean barbecue.

The fat bulbing onion, also called a storage onion, matures midsummer, but I start pulling them as soon as they reach scallion size. A little larger, like a swollen thumb, they are what my grandfather called "spring onions," mild as scallions and crisp as apples. Certain varieties, such as the wonderful torpedo-shaped Rossa Lunga di Firenze and the cannonball-sized Alisa Craig, are meant to be eaten fresh, not cured for storage.

Garlic is ready to eat in its infant state, as soon as it emerges in spring, as garlic scallions. At its early-summer adolescence, a stage when the bulb has filled out but not yet developed strong character, "green" garlic invites flagrant excess. The Asian garlic Xian matures in June in my garden, 6 weeks ahead of some Rocambole varieties. Garlic is cured for storage by hanging until dry. Some softnecks—the less flavorful grocery-store types—will keep until spring.

Shallots once impressed me as high-toned, and they cost more than standard onions at the farmers' market. I don't understand why, though. In the garden, shallots yield prodigiously, half a dozen clumped heads developing from every bulblet-seed, hence the varietal epithet *aggregatum*. (My record is 16 heads—a pound and a half of shallots—from a single bulblet the size of a cocktail onion.) The refined flavor defies time, outlasting a full 12 months in storage.

Leeks are elegant in their stripling youth, no thicker than a walking cane. By early winter they become stately with age.

Chives embellish eggs and egg dishes, potatoes, fish, and anything made with dairy. The fairy-bell flowers are a pretty garnish and the most flavorsome of the edible blossoms.

The alliums accumulate sulfur from the soil, the factor that works in concert with oxygen to create acrid fumes. They are likewise filled with indigestible plant fiber called inulin, which reaches the lower gut undigested and in excess causes windiness.

In the Garden

Alliums like loose, fertile soil and regular water. The swollen bulb is the plant's energy storehouse to carry it through winter. A well-fed plant will pack away more sustenance—a larger bulb. When preparing beds for onions, shallots, garlics, or leeks, be generous with a 1- to 3-inch layer of compost or composted manure. Work it into the soil or, for no-till, plant directly into it. Mulch the row. Water between rains. Weeding is crucial.

With their slender, upright habit, alliums can be planted densely, making them a good option for small gardens and raised beds. Chives and quick-growing scallions are better suited to pots than larger onions and garlic.

Continue reading for species-specific planting and growing instructions.

"THE TRAVELING ONION"

by NAOMI SHIHAB NYE (1952-)

"It is believed that the onion
originally came from India.
In Egypt it was an object of
worship—why I haven't been able
to find out. From Egypt the onion
entered Greece and on to Italy,
thence into all of Europe."

—Better Living Cookbook

When I think how far the onion
has traveled
just to enter my stew today, I
could kneel and praise
all small forgotten miracles,
crackly paper peeling on the
drainboard,
pearly layers in smooth
agreement,
the way the knife enters onion
and onion falls apart on the
chopping block,
a history revealed.
And I would never scold the
onion
for causing tears.
It is right that tears fall
for something small and
forgotten.
How at meal, we sit to eat,
commenting on texture of meat
or herbal aroma
but never on the translucence
of onion,
now limp, now divided,
or its traditionally honorable
career:
For the sake of others,
disappear.

ONIONS

Onions (*Allium cepa*) include the regular grocery-store type of storage onions, as well as fresh "spring" onions pulled young. Specialty varieties include pearl onions for pickling. Among the varieties eaten fresh are the mild "sweet" onions often marketed with geographic placenames, such as Vidalia, Walla Walla, and Maui. They look much like storage onions but have a shorter shelf life.

Onions tolerate cold beginnings and are put out very early. They can be propagated by seeds, starts (seedlings started indoors), or bulblets called sets. Raising onions from seeds takes patience: They emerge slowly and mature over a 100-plus days after transplanting. Sow in trays or cells in late winter, then put in the garden as soon as you can work the soil, when the seedlings are 4 to 6 inches tall, like whips for a mouse-sized coachman. Commercial starts, or seedlings started in a professional greenhouse, are sold in bareroot bundles and look puny on arrival. Dibble them into the soil and they take off. Sets are pushed into a prepared seedbed, at least 1 inch deep, in the weeks before last frost.

Onions are photoperiodic plants, meaning their growth is regulated by seasonal changes in daylight hours. The peculiar consequence is that the variety you plant depends on your latitude. Long-day varieties need 14 to 16 hours of daylight to trigger bulbing. They are suited to regions north of 38° latitude, a line drawn from Richmond, Virginia, to Richmond, California, passing through Louisville, Wichita, Pueblo, and Yosemite. (Other sources say 36° north, the Kansas/Oklahoma border.) Short-day varieties, adapted to southerly regions, are triggered by as few as 10 to 12 daylight hours. Your local nursery will carry the right variety for your area. When ordering from a seed supplier, pay attention to the distinction.

Harvest onions when you're ready for them. To cure for storage, wait until half the tops in the row have flopped over. Knock over the rest and harvest within a week. On a dry day, gently lift the bulbs with a fork (rather than yanking out of the ground). Leaving the tops intact, spread the onions in a single layer in an airy spot out of direct sun, such as on screens under a picnic canopy or beneath a shade tree. Allow to air-dry for a week, covering at night to keep off the dew and bringing inside in case of rain. Finish curing in a warm, dry, covered place out of direct sun, such as on a covered porch or in a garage, for several weeks longer. In very humid weather or an enclosed space, use a rotating fan to stir the air.

When the tops have withered, trim them to 1 inch above the bulbs and store the cured onions in net bags. Alternatively, braid the tops into bunches, incorporating a strand of twine for added strength. Store in a cool, dry place. I keep onions in an unheated back bedroom, the coldest corner of the house. Do not store onions near potatoes; they ruin each other like delinquent friends.

Of the many onion varieties, Patterson is a storage champion, the slightly flattened Stuttgarter a runner-up. Alisa Craig is a big mild fresh-eating onion for burger-sized slices. Blush has a lovely pink color, and blocky Rosso di Milano is a gorgeous red, a pleasure to cook with. Walla Walla is sweet. Italian Torpedo and others of similar name have a beguiling shape. Many notable heirlooms are described in the charming book *Onions and Garlic Forever* by Louis Van Deven.

A 10-foot row of some 30 onions will yield 5 to 10 pounds.

ONION SETS

PLANT IN GARDEN	DEPTH	SPACING	ROW SPACING	THIN TO	MATURITY	MATURE SIZE
As early as possible	2–3 inches	4 inches	10–12 inches	n/a	70–80 days, but varies by variety	Varies

Note: Prepare a seedbed and push each set into the soft ground. Or poke a hole with your finger or a dibble. The flattened basal plate, or root end, faces down.

ONION TRANSPLANTS

TRANSPLANT TO GARDEN	DEPTH	SPACING	ROW SPACING	THIN TO	MATURITY	MATURE SIZE
At last frost	1–2 inches	4 inches	10–12 inches	n/a	70–80 days, but varies by variety	Varies

Note: Poke a hole for each seedling with your finger or a stick.
Water well and keep watered for 2 weeks while the transplants settle in.

ONION SEEDS—STARTED INDOORS (TRAYS)

START INDOORS	DEPTH	SPACING	ROW SPACING	DAYS TO EMERGENCE	THIN TO	TRANSPLANT TO GARDEN	SPACING IN GARDEN
8–10 weeks before transplanting	1/4 inch	1/2 inch	2 inches	6–16 days	n/a	When 4–6 inches high	4 inches

Note: When transplanting small onions such as cippolini space 2 inches.

ONION SEEDS—STARTED INDOORS (CELLS)

START INDOORS	DEPTH	SPACING	ROW SPACING	THIN TO	TRANSPLANT TO GARDEN	SPACING IN GARDEN
8–10 weeks before transplanting	1/4 inch	5 seeds per cell	n/a	3 per cell	When 4–6 inches high	6 inches

Note: Transplant each cell's three-onion bunch as a unit.
The growing bulbs will push each other aside and develop their normal shape.

SCALLIONS

Perky scallions or bunching onions, *A. fistulosum*, have a fresh, lively flavor that goes particularly well with springtime cooking. In May and June, I'll use them in place of storage onions, chopping up every bit of the greens along with the tidy whites. Scallions are an easy, enjoyable crop. They mature quickly and can be sown in clusters of 9 to 11 seeds—a ready-made bunch. (Alternatively, sow closely in rows or in a 2-inch-thick band.) To get longer whites, "blanch" the lower stems by mounding soil around the base as the tops grow.

Because scallions don't hold well, either in the garden or in the kitchen, divvy up your planting. Sow every 2 weeks from spring through late summer.

Tokyo White is a classic scallion. Nabechan has good flavor, and Deep Purple adds color.

Pearl onions are sown directly in the garden, like scallions. Put out early, on the same schedule as peas, and sow thickly in rows or bands. Thin to ½ inch apart. Harvest when marble-sized for cocktail onions or the size of a Ping-Pong ball for boiling onions. The classic open-pollinated heirloom is Crystal White Wax.

SEPTEMBER 14: Scallions, also known as bunching onions, ready for harvest

SCALLIONS—DIRECT SOWN

DIRECT SOW	DEPTH	SPACING	ROW SPACING	DAYS TO EMERGENCE	MATURITY	MATURE SIZE
When the ground can be worked	¼ inch	9–11 seeds per bunch	18 inches	6–16	50–60 days	12–14 inches

Note: To start indoors, follow the directions prior for planting onions in cells, except sow 9–11 seeds per cell. Space transplants 8 inches.

LEEKS

The long-shanked leek, *A. apeloprasum,* has the nonchalant elegance of a feather stuck in a hiker's cap. My friend Tom Hudgens, a meticulous cookbook author, gifted cook, and retired chef, taught me that leeks are known in the French kitchen as "the soup onion." (Shallots, discussed on page 276, are called "the sauce onion.") Grown from seeds or starts, leeks are heavy feeders. Go all-in on compost or composted manure. Leeks can be blanched, either by planting in a trench, which you backfill in in several stages, or by successively hoeing up soil around the base as they grow. Eliot Coleman developed a practical technique: Dibble a 6-inch-deep hole and drop in a start without refilling the hole. Rain will wash in soil over time. The blue-green top growth is a called the "flag."

Begin to pull leeks at scallion size or whenever you're ready for them. With regular watering through the summer, they will gain girth until cold weather. In mild climates, they will overwinter and can be dug as needed.

King Richard is the standard, although there are many heirlooms. A 10-foot row will yield 20 leeks. If planning to pull them young, as for leeks vinaigrette, sow densely, every 2 inches, and thin in stages.

AUGUST 24: King Richard leeks

LEEKS—TRANSPLANTS

TRANSPLANT TO GARDEN	DEPTH	SPACING	ROW SPACING	THIN TO	DAYS TO MATURITY	MATURE SIZE
Around last frost	Dibble 6 inches	6 inches	18 inches	n/a	75–90+	2 feet tall, 2-foot spread

Note: To grow your own transplants indoors, sow seeds indoors in trays 6 weeks before last frost, following the directions prior for planting onions in trays. Add 30 days to maturity.

GARLIC

Central Asia is the global hot spot for garlic biodiversity and the likely origin of the wild progenitor of *A. sativum*. Central Asia was also the center of global trade during the thousand years of the Silk Road, the trade routes that connected China, India, Persia, Greece, and Egypt. Luxurious trade goods such as silk and spice made it even as far as Venice, and garlic traveled with the caravans. When European explorers sailed west in search of Asia and bumped into the New World, they carried garlic to the Americas, where native alliums such as ramps, *A. tricoccum,* were already being used by indigenous peoples. Today garlic is eaten in huge quantities around the world. The 2018 harvest reached 28.5 million tons.

So-called softneck garlic has a stalk that flops over when mature. The tops can be braided for storage. Heads are composed of a layer of medium-sized cloves around an inner layer of small cloves. Softneck garlic is valued for its storage quality, up to 10 months compared to 6 months or less for hardneck. Most commercially grown garlic is softneck, and cultivars are grouped into three categories: silverskin, artichoke, and creole.

Hardneck garlic, the cool-weather subspecies *A. sativum* var. *ophioscorodon,* sends up a flower stalk, or scape, which is cut to force the plant's energy into producing larger bulbs. The scape is mild and edible. It can be snapped, blanched in salted water, and tossed with unsalted butter and herbs like green beans. Hardneck categories include large-cloved porcelain, superbly flavorful Rocambole, and purple stripe, which is nearly as delicious and longer storing. All have relatively large cloves, usually arrayed in a single layer. Some dried hardnecks are beautiful, with the sculpted form and subtly toned wrappers of hand-painted ornaments.

Garlic is planted in the fall; the seed comes from heads harvested that summer. Break heads into individual cloves. Plant cloves 2 inches deep and 5 inches apart. In cold climates, mulch to reduce frost heaves. In warm climates, garlic is planted in the spring after chilling for 3 months in the refrigerator in a paper bag—a false winter.

JUNE 29: Xian garlic hanging to dry

Harvest garlic at any point, starting at scallion-size. Bulbs are mature when the bottom one-third of the leaves turn brown. Pull back mulch and stop watering for a week. Gently lift the bulbs with a garden fork. Shake off clinging dirt but don't wash them. Spread in a shaded, airy place to dry, such as a shed or porch, or tie into bunches of 5 to 8 to hang out of direct sunlight. Cure for about a month. In a wet or very humid summer, it may be necessary to remove a layer or two of outer wrappers for adequate drying. Feel the bulbs after 2 weeks of curing and, if they are damp or there is any hint of mustiness, shuck off the dried outer skin until you reach the moist innermost delicate membrane enclosing the cloves. With very large-cloved varieties, such as Music, I might even expose the tips of the cloves. (Don't damage the cloves' individual tight-fitting jackets.) When the bulbs are cured, trim the stalks and roots with a pair of pruning shears and store them in net bags in a cool, dry, dark place. The ideal temperature is 50° to 60°F, not in the refrigerator, which will prompt sprouting.

A 10-foot row of garlic, 4 to 8 ounces of seed, will yield about 25 heads. To make the most early-season green garlic, space cloves every 3 inches and thin at scallion size.

Among desirable hardneck varieties, the best for flavor and worst for storage are the Rocamboles, notably Spanish Roja and Killarney Red. Eat them first. Chesnok Red, also known as Shvelisi, is a flavorful purple stripe with better storage qualities, up to 6 months. Music is a porcelain type with extra-large cloves, sometimes 4 per head. German Extra Hardy is reliable to grow and easy to find. Softnecks include Inchelium Red and Italian Softneck, while Lorz Italian, an artichoke type, has good flavor for a softneck. There are many more varieties to test; results will vary by region. Filaree Farm and The Garlic Store are reliable seed sources. Remember to save a portion of your first-year harvest to replant.

Despite its name and appearance, elephant garlic is a leek that produces weakly flavored jumbo cloves. Cultivate like garlic—as a curiosity.

GARLIC

PLANT IN GARDEN	DEPTH	SPACING	ROW SPACING	THIN TO	MATURITY	MATURE SIZE
From first frost to hard freeze (mid-October to mid-December)	2–3 inches	6 inches	10–12 inches	n/a	240–290 days from fall planting	18–30 inches tall, 18-inch spread

Note: Break apart heads only when ready to plant; don't peel cloves.
Plant individual cloves by pushing into a prepared bed or dibble in. The root end goes down, pointy end up.

SHALLOTS

The copper-skinned red shallot is a type of onion, *Allium cepa* var. *aggregatum,* and the variety name tells you how they grow, in aggregated clusters. Shallots are sometimes called multiplier onions.

A staple of the French kitchen, shallots have never fully conquered America. Perhaps the French association makes them seems suspiciously gourmet. The flavor is both more distinctive and subtler than that of regular onions. They can be used raw if subdued for a few minutes in vinegar, as when minced and whisked into vinaigrette. Cooked, they bloom in warm unsalted butter, making beurre blanc a very easy sauce for fish and blanched vegetables. Despite their delicacy, shallots also stand up to long cooking, as in braises, and slices can be flash-fried for an Asian-style crispy topping that beats fried onions as an all-American casserole topping.

Shallots are fall-planted in most regions, like garlic, or else as early as possible in the spring. Care for them like onions. You only buy shallots once. Each bunch will produce several multilobed bulbs as well as miniature "offsets." Replant the offsets ad infinitum.

A half-pound of seed shallots will plant a 10-foot row, which is quite a lot considering the yield. (Harvest yields in my garden average 25 pounds per pound of seed.) Cure and store shallots like onions. They will keep for a year.

Dutch Red and French Red are the common types. The shallot esteemed above others, French Grey, is a distinct species, *A. oschaninii.* It has an elongated teardrop form and is less productive than the reds. Despite its thick skin, it stores poorly, lasting only 3 to 4 months. The French flavor, however, is *superbe.*

CHIVES AND GARLIC CHIVES

Chives (*A. schoenoprasum*) and garlic chives (*A. tuberosum*) are perennial alliums grown for their foliage, like herbs. They readily sprout from seeds and are widely sold in little pots in springtime. Diminutive *A. schoenoprasum* is native to North America, Europe, *and* Asia—a unique distinction among alliums. The common North American lawn weeds *A. canadense* and *A. vineale,* which resemble chives, are edible. Keep chives handy for making omelets.

SHALLOTS

PLANT IN GARDEN	DEPTH	SPACING	ROW SPACING	THIN TO	MATURITY	MATURE SIZE
Fall or very early spring	1–3 inches, depending on winter cold	8 inches	18 inches	n/a	120 days from spring planting	8–10 inches tall

Note: Gently separate lobes before planting, but don't peel.
Mulch through the winter, then pull back mulch in the spring, as for garlic.

AUGUST 13: Cortland, an "improved" hybrid storage onion

SCALLIONS • GARLIC OR GREEN GARLIC • PARSLEY | SIDE | SERVES 4 | VEGETARIAN, GLUTEN-FREE OPTION | SPRING THROUGH FALL

SILKY SCALLIONS

When I learned to cook, recipes instructed me to chop scallions into rounds, using the whites and half the greens. I would wonder: What about the rest of the greens? Years later, a recipe from Edna Lewis brought me full circle, and not for the first time in my kitchen life. Her skillet scallions from *The Taste of Country Cooking* consists of whole scallions—greens and all—cooked in foaming butter with nothing more than the rinse water clinging to them. "No salt or pepper will be needed," Miss Lewis notes.

My variation finishes the dish with cream, grated garlic, herbs and, with apologies to my betters, a little salt to season the cream. The pan sauce is richly flavored, while the scallions have chlorophyll vitality. The combination is a vegetable condiment. In spring, I serve it alongside green peas and new potatoes. In summer: crookneck squash and roasted okra. In fall: winter squash and baked tomatoes. If a hunk of pan-seared halibut is part of the meal, I'll dribble the cream over it.

Ramps and other skinny alliums—such as the garden mix of spring garlic, onions, and shallots shown in the photograph opposite—can be cooked the same way.

12 scallions or pencil-thin spring onions, trimmed (2 bunches)
1/2 cup organic heavy cream
1 clove garlic, minced, or 1 scallion-sized green garlic, chopped
1/2 teaspoon fine sea salt
1/4 cup fresh bread crumbs, toasted (optional)
2 tablespoons unsalted butter, melted (optional)
1 tablespoon minced fresh parsley
Lemon zest, to garnish

1. Heat a large skillet over high heat until a drop of water sizzles. Add the scallions and 2 tablespoons water. Cover, reduce the heat to medium, and cook, shaking frequently, until the whites are nearly tender and the greens still bright, 5 to 7 minutes. Add more water by the teaspoon if the pan dries out.
2. Add the cream, garlic, and salt and shake to combine. Simmer uncovered for 5 minutes to thicken somewhat.
3. Meanwhile, if making the crunchy topping, in a small bowl, combine the bread crumbs and melted butter and set aside.
4. Add the parsley to the skillet and shake to combine. Transfer to a serving platter and sprinkle with the buttered bread crumbs. Finish with a few scrapings of lemon zest.

GARLIC SCAPES • HERBS • HERB BLOSSOMS | SIDE | SERVES 4 | VEGETARIAN | EARLY SUMMER

BUTTERED GARLIC SCAPES

A garlic scape is the flower stalk, but it's an odd one that curls back on itself as it grows, a curlicue topped with a pointy bud. The archaic word, heard only where garlic growers convene, comes from the Latin *scapus* for "stalk" and the still-earlier Greek for "staff" or "scepter." It does indeed look like a scepter from a drunken realm. Snapped into short lengths and cooked like green beans or asparagus, scapes are tender, delicious, and only faintly garlicky.

Each plant produces a single scape midway through its growth cycle. It points straight up at first, but soon seems to lose its way, tracing circles in the air. Eventually the bud will curl a full 360° and point skyward again. Harvest now to force the plant's energy into producing a larger bulb.

Snap or cut just above the top pair of leaves. You can loop scapes around your wrist and add more like bangles as you work down the row. Thirty to fifty scapes make a pound. They will last in the refrigerator for a week.

Fine sea salt, for the blanching water
1 pound garlic scapes, snapped into bite-sized pieces
2 tablespoons unsalted butter
1/4 teaspoon fine sea salt
1 tablespoon chopped fresh chives, parsley, or thyme
1/2 lemon
Chive blossoms or thyme blossoms (optional)

1. Bring a large pot of salted water to a boil over high heat. Blanch the scapes for 2 minutes, or until tender. Drain.
2. In a large skillet, melt the butter over medium heat until it foams. Add the scapes and toss to coat. Add the salt and chives and toss again.
3. Transfer to a serving dish and squeeze the lemon over. If desired, sprinkle with chive blossoms.

GARLIC SCAPE PESTO

It was black raspberry season when I went to see Talea and Doug Taylor at Montgomery Place Orchards. They are fruit people, and the extraordinary apples they grow read like poetry: Hidden Rose, Belle de Boskoop, Ashmead's Kernel, Duchess of Oldenberg, Black Twig, Cox's Orange Pippin, and dozens more. But Talea and Doug also grow vegetables to stock their farm stand until apples come in. Garlic is a reliable crop for them, and their daughter Caroline Olivia upcycles the scapes to make pesto, the recipe for which she generously shared. You use this flavorful sauce as you would herb pesto—tossed with pasta as a first course, tossed with blanched green beans as a side dish, or tossed with pasta, blanched green beans, and sliced grilled Italian sausage for a square meal. I make scape pesto in large quantities to freeze. On many, many nights it has proven to be the solution to the urgent question, what's for dinner?

- 1/2 pound garlic scapes (15 to 25), buds trimmed and discarded, stalks roughly chopped
- 1/2 cup pine nuts or coarsely chopped cashews
- 2/3 cup extra-virgin olive oil
- 1/4 cup fresh lemon juice
- 1/2 teaspoon fine sea salt
- 1/4 teaspoon freshly ground black pepper
- 1/3 cup freshly grated Parmigiano-Reggiano cheese

In a food processor, combine the scapes, nuts, olive oil, lemon juice, salt, and pepper and pulse until smooth. Transfer to a medium bowl and stir in the Parmigiano. Jar unused pesto and refrigerate for up to 1 week or freeze in half-pint jars.

Note: To use the pesto, toss ⅓ to a scant ½ cup with 8 ounces dried pasta, cooked al dente, or 1 pound of green beans that have been blanched and drained. Reserve ½ cup of pasta water or blanching water to loosen the sauce, if necessary, adding a tablespoon at a time. Finish with a drizzle of olive oil and a squeeze of lemon.

SHALLOTS • THYME • BAY | SIDE | SERVES 4 TO 6 | VEGETARIAN, GLUTEN-FREE | EARLY SUMMER

ALSO FOR: SMALL ONIONS

BRAISED SHALLOTS

The natural sweetness of alliums is concentrated by long cooking. Here sweetness is balanced by vinegar for a version of Italian agrodolce (literally sour-sweet). The result is rich, tangy, almost jammy, and delicious as a condiment. It transforms plain chicken or pork, and it adds savory depth to vegetarian meals, for example alongside a slice of stuffed Koginut squash. A slight variation makes a superb condiment for rare beef: Replace the sherry vinegar, vermouth, and thyme with balsamic vinegar, red wine, and rosemary.

3 tablespoons unsalted butter
2 tablespoons neutral oil
1½ pounds shallots or small onions, peeled
1 teaspoon fine sea salt
2 tablespoons sugar
2 teaspoons dried thyme or savory, or 1½ teaspoon fresh leaves
2 bay leaves
½ cup dry vermouth
1 tablespoon sherry vinegar

1. In a large skillet, heat the butter and oil over medium heat until foamy. Add the shallots and stir or toss to coat. Add the salt, sugar, thyme, and bay leaves and sauté gently until lightly caramelized, 7 to 9 minutes.
2. Add the vermouth and reduce by half. Add the vinegar and 1 cup water, cover, reduce to a steady simmer, and cook until the shallots are soft, about 15 minutes. Small onions will take longer, up to 30 minutes.
3. Uncover, turn up the heat, and reduce the cooking liquids to a syrup.

FRESH RED ONIONS • GARLIC • TOMATOES • BASIL | STARTER, MAIN | SERVES 2 TO 4 | VEGAN | SUMMER

SUMMERY BUCATINI
WITH NEW RED ONIONS AND TOMATOES

Fresh red onions come in from the summer garden attached to their spry tubular tops. Italian torpedo varieties are grown to be eaten fresh, but storage onions such as Blush, Rosso di Milano, and Redwing (see photo opposite) are also ideal for fresh eating at this stage—the latter half of summer, around late July in my region. Nearly full-sized and juicy, they still have the mild flavor and crisp texture of spring onions.

This dish comes together fast, speed being a virtue in the summer kitchen, when there are more interesting things to do outside. First up, put pasta water on to boil. Prep everything while it heats. This is what restaurant chefs called mise en place—roughly "everything in place"—and it's the key to efficient cooking. Once you throw the bucatini into the boiling water, you'll have 10 minutes, just enough time to get the sauce together.

Fine sea salt, for the pasta water
8 ounces bucatini or spaghetti
1/4 cup extra-virgin olive oil
1/2 medium fresh red onion, sliced from stem to root (see Notes)
1/2 teaspoon fine sea salt `
2 cloves garlic, minced
1 pound Roma tomatoes, seeded and chopped (about 3 cups)
8 large basil leaves, chiffonade-cut
Freshly ground black pepper

1. Place a large serving bowl in a 200°F oven to warm.
2. Bring a large pot of salted water to a boil. Add the pasta and cook to al dente according to the package directions. Reserving ½ cup of the cooking water, drain the pasta.
3. Meanwhile, in a large sauté pan, heat the olive oil over high heat until it shimmers. Add the onion and ¼ teaspoon of the fine salt and sauté until the onion sweats but is still slightly crisp, 3 to 5 minutes.
4. Add the garlic and sauté until fragrant, about 30 seconds. Add the tomatoes and remaining salt and cook until the tomatoes release their juices, about 3 minutes. Remove from the heat. Add the basil and toss to combine.
5. Transfer to the serving bowl. Add the drained pasta on top and toss to coat. Add tablespoons of cooking water, as needed, to loosen. Top with freshly ground black pepper.

Notes

I slice the onions lengthwise, from stem to root, for a slivery texture, rather than crosswise into half-rings. First trim the root end to release the slivers.

To make a winter variation, use a medium red storage onion and replace fresh Roma tomatoes with a pint of drained canned tomatoes. (Save the juice for soup or add to a stew or braise.) Simmer for 20 minutes.

STORAGE ONIONS • GARLIC • DRIED HERBS | STARTER, MAIN | SERVES 4 TO 6 | PESCATARIAN | WINTER

WINTRY SPAGHETTI
WITH SOFTENED ONIONS, SARDINES, AND FENNEL SEED

This recipe is the cook's version of a rabbit pulled out of a hat—a meal pulled out of a pantry. Storage onions wait out the winter in a dark, unheated space. When you fetch one out, it is as cool to the touch as a billiard ball. The cupped outer skin slips off with the rustle of fresh bank notes and wobbles on the counter, a vellum rotundity. The innermost peel clings like paint. When you start cutting, does it matter if the slices are even? "True shooting, certain hitting," says the Zen archer. The layers cleave into pristine curves. Some people openly weep. Inner sweetness is the secret of the onion, a vegetable for all seasons.

This recipe takes half an hour, so the moment you realize you're famished, put the pasta water on the stove. Get the sliced onions into a pan to sweat and relax. When the water boils, the spaghetti will need 10 minutes to cook, long enough to finish the sauce. Place serving bowls in the oven to warm. Call the other eaters to their tasks: One sets the table, another opens wine, a third grates Parmesan. The onions turn sweet, and the sardines melt away. Fennel seed and parsley freshen the mouth and the mood. Don't mention the anchovy. A magician never explains the trick.

Fine sea salt, for the pasta water
2 tablespoons extra-virgin olive oil
3 tablespoons unsalted butter
2 cloves garlic, minced
3 medium onions (1 pound total), cut in 1/4-inch-thick slices
1/2 teaspoon fine sea salt
1/2 teaspoon dried thyme or savory
1/4 teaspoon fennel seeds
Pinch of red chile flakes
1 (4-ounce) can sardines, drained and boned
4 fillets salt-packed anchovies, rinsed and chopped (about 1 tablespoon)
12 ounces spaghetti
2 tablespoons minced fresh parsley
1/2 cup grated Parmigiano-Reggiano cheese
Freshly ground black pepper
Lemon wedges, for squeezing

1. Place a large serving bowl and individual bowls in a 200°F oven to warm.
2. Bring a large pot of salted water to a boil for the pasta.
3. Meanwhile, in a large sauté pan, combine the olive oil, 1 tablespoon of the butter, and the garlic. Heat over medium heat until the butter melts and the garlic is fragrant. Add the onions, fine sea salt, thyme, fennel seeds, and chile flakes and toss to combine. Cook slowly, uncovered and tossing occasionally, until the onions are pale gold, about 20 minutes. They should still have some texture.
4. As the onions cook, in a small bowl, mash the sardines and anchovies with a fork.
5. Add the spaghetti to the boiling water and cook to al dente according to package directions. When the pasta is almost ready, add the sardine mixture to the onions. Stir to combine.
6. Reserving 1 cup of the cooking water, drain the spaghetti and add to the pan with the onions. Moisten with some of the pasta water and toss to combine.
7. Remove from the heat. Add the remaining 2 tablespoons of butter, the parsley, and half the Parmigiano. Toss again, adding more pasta water as needed to keep the mixture loose.
8. Turn the pasta into the warmed serving bowl. Top with the remaining cheese and a generous amount of freshly ground black pepper. Squeeze a lemon wedge or two over the top and serve the remaining wedges on the side.

ROAST CHICKEN
WITH BURNT SHALLOT JUS

Like the previous two pasta recipes, this one is a lifesaver, a meal for those in need—in need of sustenance, of flavor, of comfort, of time, of ideas. Roasted chicken is one of the very easiest main courses to prepare, and yet there's an air of ceremony about serving the whole bird as a centerpiece. It's fit for company, fit even for Sunday dinner with the preacher man, an offering to the gods. Onions and herbs create the incense, a wisp of fragrant steam.

The secret to this roast chicken—to any roast chicken—is to cook a small bird fast in a hot oven. Three pounds or a little more is enough if what you want is flavor. Four pounds is the upper limit. An organic pastured bird, although more expensive, will have the right stature. Remember that the price equals two meals, because the stripped carcass will provide a quart of stock for soup or risotto.

The method here is to stuff the bird with garlic and herbs and roast it on a bed of shallots and onions. Preheat a cast-iron skillet to blazing hot, and the chicken will be done in 45 to 50 minutes. While it rests, deglaze the pan with vermouth and press the flavor out of the oven-charred shallots, which are sweet in their depths and faintly bitter from their burnt edges. Strain the reduction for a richly colored, richly flavored jus.

1 whole chicken (3 to 3½ pounds), generously salted in advance (see Note)
1 small head garlic, halved horizontally
4-inch sprig rosemary or several thyme sprigs
Pinches of dried savory or thyme
Freshly ground black pepper
½ pound whole shallots, larger ones halved
1 small onion, quartered
1 tablespoon extra-virgin olive oil
½ cup vermouth
½ cup chicken stock or water
1 tablespoon unsalted butter, cut into small cubes

1. Preheat the oven to 450°F. Set a cast-iron skillet in the oven to preheat.
2. Stuff the chicken with the garlic and fresh herbs. Sprinkle with pinches of dried herbs and black pepper. In a bowl, toss the shallots and onion with the olive oil to coat.
3. Transfer the chicken to the hot skillet and surround with the shallots and onion. Roast until browned and a thermometer inserted in the thickest part of the thigh reads 180°F, 45 to 50 minutes. The leg will wiggle but not be loose. Transfer the chicken to a warm platter to rest.
4. Transfer the skillet to the stovetop over medium heat and deglaze with the vermouth. Scrape up the browned bits and mash the shallots and onions to extract their flavor. When the vermouth is nearly evaporated, add the stock and cook to reduce by half. Strain through a fine-mesh sieve. Quickly return it to the warm skillet and add any liquids released by the resting chicken. Turn the heat to the lowest setting, tilt the pan to pool the liquids, and whisk in the butter a few bits at a time. Serve immediately.

Note: As soon as you bring the chicken home, salt it generously inside and out and put in the fridge until you need it. For crisper skin, I leave the chicken uncovered to dry. The benefits of salting in advance—a tastier, juicier bird—are beyond dispute at this point. (See Judy Roger's *The Zuni Café Cookbook* and Samin Nosrat's *Salt Fat Acid Heat.*) Salting a day or two in advance is ideal. The morning before cooking is terrific. An hour before cooking is great. Salt with authority. It's the single biggest factor separating accomplished cooks from decent ones.

CH.

20

POTATOES

PLUS

SWEET POTATOES

THE TUBEROUS COHORT

Solanaceae
THE NIGHTSHADE FAMILY

Solanum tuberosum
POTATO

Convulvulaceae
THE MORNING GLORY FAMILY

Ipomoea batatas
SWEET POTATO

JULY 3: First new potatoes, Red Gold

AUGUST 3: Red Gold tubers (note attached root), foliage, and tomato-like fruit

POTATOES

The long potato season begins in early summer, a bare 2 months after planting. You slip your hand into the soil beneath the bushy vines to check if they're ready, like feeling for eggs beneath a broody hen. The new potatoes will be smooth, as cool as glacial pebbles. A half-dozen will fit in your cupped hand. Carry them gingerly inside to boil in their jackets and roll in unsalted butter with chopped chervil, chives, tarragon, and parsley. Served 30 minutes out of the ground, new potatoes taste like living roots and mineral earth, a nearly literal expression of what the French call terroir. Too young at this stage to keep for later, new potatoes belong to the carpe diem season—enjoy them now.

Smash cut to February, when ice grips the ground a foot deep, and hefty russet potatoes wait in my cellar, mute as a frozen pond, their slow thoughts bound within leathery skin. "Arshtaters" is what Pappaw called them, and years later, when I heard about the Irish potato famine, I understood that he had been saying "Irish." Baked, boiled, or mashed, storage potatoes ballast many winter meals.

Schoolchildren of my day learned that Andean farmers cultivate 4,000-some potato varieties. Less told was the story of how conquistadors looted the New World's agricultural hoard during the Columbian Exchange. The spud—a nickname by analogy to an archaic digging spade—made its way from Spain to India, China, and, of course, Ireland. The astonishing variety of potatoes speaks to its adaptability and food value in centuries when calories were scarcer than they are today.

Potatoes vary in skin tone from nearly translucent to opaque buff, barn red, beaten gold, and unearthly blue, the rarest edible color. Some types are "painted," or pinto, like a horse. The inside doesn't always match its cover—a red coat might conceal a rose quartz interior or Carrera marble. The mottled beauty of Masquerade is skin deep, but Purple Majesty's ink goes clear through. The peel contains most of the nutrition and fiber, so a peeled potato is a surfeit of starch, an intemperance of comfort.

Potatoes vary in shape and size from cutesy nuggets to two-handed loaves. At the diminutive end of the spectrum, Upstate Abundance grows to the size of shooter marbles with the occasional Ping-Pong ball prodigy. At the other extreme, I've had Kennebec weigh a full pound. New potatoes are dug soon after the vines flower. Potatoes for storage are dug after the vines wither, and late varieties can take 110 days to mature. Cured and kept in a cool, dark place, the most lasting storage varieties will keep for many months. Eventually, they wither like shrunken heads and sprout Medusa hair but if deeply peeled they remain sound at the core. My notes from one year record that Green Mountain planted on April 25 and dug in September kept until the following May. Less than 2 months later, on July 3, I dug the first new Red Gold (see the photo on pages 290 and 291). There is hardly a potato-free gap long enough to miss them.

Potatoes are propagated from seed potatoes. The term is something of a misnomer. A potato's dimpled "eyes" enclose not seeds but buds, incipient vines. The sprout is a vegetative clone of the mother plant. Abundant flowers flash white and pale purple against the verdant leaves but rarely produce fruit. When they do, it looks like a runty green tomato, a failed attempt, and the seeds are of interest only to plant breeders. (See opposite.)

Anyone can grow potatoes. I personally think anyone should for the food security they provide. Neighbor Del and I once harvested 92½ pounds of La Ratte fingerling potatoes from 3 pounds of seed, a gigantic return on investment. Be aware that potatoes require extra chores—hilling and mulching—and inevitably attract the Colorado potato beetle, a squishy foe.

Given their allegiance to the unknowable vastness of the subterranean realm, it may come as a surprise that potatoes do well in containers. In my experiment one year, Upstate Abundance grown in a compost-filled burlap sack yielded nearly as much as field-grown vines. (See the bottom right photo on page 295.)

In the Garden

To start with, what not to do: Don't plant potatoes from the grocery store. They likely have been treated to delay sprouting. Start instead with seed potatoes, ordered early in the new year. Maine Potato Lady, Fedco, and Johnny's have wide selections. Local nurseries and garden centers stock varieties suited to the region. Keep seed potatoes in a cool (45° to 55°F), dark, and slightly humid place (such as a closed cardboard box) until 2 weeks before planting. Then wake them up with warmth and light, a process known as presprouting, green-sprouting, or chitting. Spread in a single layer on a tray and place in a bright spot out of direct sunlight. The eyes will swell and turn green.

Potatoes are planted in midspring, after the soil has warmed to above 50°F, or 2 to 3 weeks before last frost. A traditional indicator in New England is dandelions in full bloom. Online suppliers will ship to you at the right time. There's no advantage to planting early, because the seed might rot in wet, cold soil. Don't plant in muddy ground after heavy rain; wait several days for the soil to drain.

Potatoes like rich, open soil, the same as most vegetables, although I've found they produce a moderate crop even in unimproved average soil. Drought will cause smaller tubers and a lower overall yield; heavy mulch conserves moisture. Do not amend the soil with lime or wood ash the year you plant potatoes; both encourage black scab, an aesthetic blemish that makes the peels hard to clean.

Two days before planting, cut the chitted seed into chunks and leave out to callus. Each piece should get three or more eyes. Small potatoes, the size of a pullet egg, can be left whole. The sprouts sometimes break off with rough handling, but usually to no ill effect.

Plant potatoes in rows 3 feet apart. In a tilled garden, dig a furrow 6 inches deep and mound the soil to one side. Space seed 12 to 15 inches apart, with sprouts facing up. Fill the furrow

TYPES OF POTATOES

Potato varieties are sorted into three types based on texture and use:

Starchy potatoes are dry and fluffy when cooked, a texture sometimes described as mealy or grainy. They are low in moisture and tend to fall apart when released from their skins—think of how a russet baked potato will billow up through its slashed top when squeezed from the sides. Starchy potatoes can absorb nearly their own weight in butterfat or liquid—a large bowl of mashed potatoes will take a reckless expense of cream.

Waxy potatoes have a denser, moisture texture that holds together when boiled, roasted, sliced for gratins, or cut up for potato salad. Skins tend to be thinner and smoother, as with Pinto Gold, although some picky eaters will peel them on principal. Fingerlings are waxy types, as are many red-skinned new potatoes.

All-purpose potatoes explain themselves. Yukon Gold and Kennebec, two named varieties that broke through grocery store anonymity, are all-purpose, although the former has become a quasi-generic marketing name since the original cultivar was replaced by newer, more reliable Gold-type varieties.

I'd recommend growing two or three or a dozen kinds of potatoes as space allows. Not all will thrive equally in every location and every growing season. Diversity offsets potential loses. See the Recommended Potato Varieties by Criteria chart (page 298) for some of my favorites.

halfway. When the emerging plants reach 6 to 8 inches tall, top up the furrow. (See the top left photo on page 295.) When the plants grow another 6 to 8 inches, mound up soil from between the rows, again nearly covering the plants. (See the top right photo on page 295.) Keep potatoes well weeded until the vines grow enough to shade out competition. Cultivate shallowly around the hill to avoid damaging roots and young tubers.

In a no-till garden, deeply aerate the row or bed where you intend to plant potatoes with a garden fork. Use a dibble or a hand trowel to open a narrow hole 6 to 8 inches deep. Bury the seed and lightly cover with only an inch or two of loose soil. Space holes every 12 to 15 inches. When the potato sprouts reach ground level, top up the hole. After another 6 to 8 inches of growth, mulch deeply with an 8- to 10-inch layer of leaves or straw. You may have to refresh the mulch midseason. In my experiments—and to my surprise—no-till potatoes thrive.

You can also grow potatoes in a 5-gallon bucket with drainage holes, a burlap sack, or a "tower" made from a 16-inch circle of woven-wire fencing. (See the bottom right photo on page 295.) Partially fill the container with 10 to 12 inches of well-aged compost or potting mixture. Plant 1 seed potato per 5-gallon container. When the plant reaches 6 to 8 inches, infill with enough compost or potting mix to nearly cover the leaves. Repeat when the plant has grown another 6 inches. Keep watered.

Diligent pest control is essential. In June and July, the first thing I do in the garden is to walk the potato patch with an eye for Colorado potato beetles. If you see one, crush it or knock it into a yogurt tub half-filled with soapy water. Subject the vine and its neighbors to a thorough and intimate inspection, like checking a dog for ticks. Egg clusters underneath leaves are bright orange. Crush them. Ruthless vigilance will prevent a population boom. If the pest situation gets out of hand, the organic pesticide bT is a last-ditch remedy.

A fungal disease called late blight can lay waste to potatoes and other nightshades—tomatoes, eggplants, peppers—at the end of summer, on the eve of harvest. It is encouraged by cool weather, damp days, and cloudy skies. Prevention includes disease-free seed from reputable suppliers and widely spaced rows to invite the cleansing benefits of sunshine and breeze. According to the Maine Potato Lady, the most effective commercial organic sprays are Serenade and Actinovate, which contain beneficial bacteria that colonize leaf surfaces and prevent the disease organism from taking hold.

Dig new potatoes starting 2 weeks after blooms appear. (See the bottom left photo on page 295.) Harvest potatoes for storage 2 weeks after the vines wither. See In the Kitchen, page 299, for more details. In my trials, harvest yields across all varieties average 2½ pounds per hill, or 20 to 25 pounds per 10-foot row.

You could spend years sampling potato varieties. My favorites include the beautiful Red Gold (early-season), with its shimmery skin and yellow interior, and the buttery Upstate Abundance (early-season), a newer variety with culinary appeal but limited commercial value due to its small size. Green Mountain (late-season) is an 1885 heirloom out of Vermont with a pronounced earthy flavor and impressive storage qualities. It makes the most delicious baked potatoes. Carola (midseason) is another flavor champion, very refined, with the taste of European haute cuisine. Yellow-fleshed Yukon Gem (midseason) and the white-fleshed Kennebec (midseason) are all-purpose standbys and excellent keepers. La Ratte (late-season) is a high-yielding heirloom French fingerling with a dense texture and a nutty flavor when roasted. My desert island potato, the one I couldn't do without, is Pinto Gold (late-season), a small-to-medium, smooth-skinned yellow-and-red-mottled waxy type with a dense texture and terrific flavor. It's great for roasting, potato salad, and Shaken Potatoes (page 306).

POTATO VARIETY COMPARISON

	EARLY 65–80 DAYS	MIDSEASON 80–90 DAYS	LATE 90–110 DAYS
STARCHY Baking Mashing Fries	Early Ohio*	Caribou Russet Goldrush Yellow Finn*	German Butterball* Green Mountain*
ALL-PURPOSE	Yukon Gold	Carola* Kennebec* Soraya Yukon Gem	Bintje* Katahdin*
WAXY Roasting Boiling Salads	Dark Red Norland Natascha Red Gold	Chieftain (red skin)	La Ratte* (fingerling) Pinto Gold Rose Finn Apple (fingerling) Russian Banana* (fingerling)

**heirloom*

HOW MUCH TO PLANT

POUNDS OF SEED POTATOES*	ROW FEET
1 pound	5–8 feet
5 pounds	25 feet
20 pounds	100 feet

*For fingerlings, use half the amount

POTATOES

DIRECT SOW	DEPTH	SPACING	ROW SPACING	HILLING	TIME TO HARVEST	MATURE SIZE
2–3 weeks before last frost	Plant in a 6-inch furrow and cover with 2 inches of soil. Reserve the remaining soil.	12 to 15 inches (wider spacing yields larger potatoes)	36 inches	When plants are 6 inches high, backfill the furrow with reserved soil. When plants have grown a further 6 inches, hill up soil to nearly cover or else mulch with 8 to 10 inches of leaf mold.	60 days for new potatoes and early varieties, up to 110 days for late varieties.	2 feet tall, 2-foot spread

RECOMMENDED POTATO VARIETIES BY CRITERIA

Easy to grow	Adirondack Red Dark Red Norland Kennebec Upstate Abundance Yukon Gem
Great flavor	Adirondack Blue Bintje Carola Green Mountain Pinto Gold Red Gold Soraya Yellow Finn
Clay soil	Dark Red Norland Red Chieftain Yukon Gem
Drought tolerant	Baltic Rose Belmonda Désirée Rose Finn Apple Adirondack Blue
Late blight resistant	Belmonda Burbank Russet German Butterball Nicola Soraya Strawberry Paw Upstate Abundance Yukon Gem
Scab resistant	Dark Red Norland German Butterball Red Chieftain Russian Banana
Containers	Bintje Yukon Gem Various fingerlings
New potatoes	Dark Red Norland Caribe Red Gold Upstate Abundance
Fingerlings	French Fingerling La Ratte Magic Molly (blue) Red Thumb Russian Banana
Keepers	Caribou Russet German Butterball Green Mountain Kennebec Pinto Gold Strawberry Paw Yellow Finn Yukon Gold / Yukon Gem
High yields	Adirondack Blue Kennebec La Ratte (and other fingerlings) Red Gold Soraya Strawberry Paw Upstate Abundance

Note: Charts compiled from information supplied by Maine Potato Lady, Johnny's, and High Mowing, as well as personal experience.

In the Kitchen

You begin to harvest new potatoes soon after the plants bloom. Reach into the soil to gather larger tubers selectively, leaving the smaller ones to grow, or dig the entire vine. New potatoes are immature, so digging early will reduce your overall yield. Storage potatoes are fully matured after the vines have died back to the ground. If the weather isn't wet, you can leave potatoes in the ground for up to 6 weeks, or until the temperature drops off in the fall. At my dad's house in Tennessee, I once dug potatoes at Christmas, but that's extreme even by his lackadaisical standards. Potatoes that overwinter in the ground will appear sound the next spring but will taste disgustingly sweet. By the same token, don't store potatoes in the refrigerator.

Digging potatoes is sheer delight. Harvest on a sunny day, preferably after a run of dry weather. The cool, damp tubers seem to pop out of the ground. I suppose it's an optical illusion rather than a trick of physics, but it tickles me every year. Children have fun on potato digging day. Dogs bark their excitement.

A garden fork is more effective than a shovel for digging potatoes and less damaging. Until you get a feel for the set, or how the potatoes grow under the vine, start by digging well away from the center of the hill. Plunge the tines into the soil and lever back the handle. If the dried vine is still visible, tug it at the same time. Some varieties set close, nested directly under the vine, others set wide, growing even beyond the shade of the mother foliage—way out in the row. Set is consistent by variety, so once you understand the growth habit, you can tighten your focus. Likewise, some potatoes lie close to the surface while others grow deep, as much as a foot down. Fingerlings often seem to burrow through the soil, like stubborn lizards, but La Ratte gathers in vertical clusters like fingers emerging from the deeper earth. Creepy. I always turn a dug hill a second time and run my hands through the loosened soil. You rarely get everything on the

AUGUST 13: Digging Green Mountain potatoes. This one hill produced 3½ pounds—exceptional.

first pass. Or the second. The next heavy rain will show the potatoes you somehow missed.

The skin of freshly dug potatoes will be faintly damp, tender, and easily marked. Any badly damaged specimens, the ones run through by a garden fork, should be eaten right away. Minor scratches and nicks will callus over. Discard green potatoes, which have grown at the surface. Sunlight makes them mildly toxic.

To cure freshly dug potatoes for storage, spread them in a single layer in a dry, totally dark place. (I paper over the windows of my garage.) A little clinging soil is fine; never wash them clean. Leave to dry for a week, then roll them over and cure on the other side for a week. Neighbor Del turns potatoes using a push broom. In very humid or rainy weather, I set up a fan to stir the air.

The potatoes will be ready when they feel leathery to the touch. Brush them clean at this point, if you like, but still don't wash them.

SEPTEMBER 28: Raking leaves to cover storage potatoes

Supermarket potatoes are scrubbed like children on the first day of school, but they also get gassed with fungicides to prevent rot.

Store potatoes in baskets, wooden crates, or burlap sacks in a cool, dark, damp place like a cellar or basement. Ideal conditions are 38° to 42°F with 80 to 90 percent humidity. Few of us have ideal conditions. My best advice is to work backward; that is, get your potatoes as far as possible from the arch destroyers, heat and light. The floor of a cool coat closet is a good place to stash a bushel basket of potatoes; a sunny windowsill over a radiator is not. Onions stored near potatoes provoke mutual sprouting. According to the references I have on root cellaring, the commonplace warning against storing potatoes with apples is overblown.

The best technique I've tried for storing potatoes comes from the Nearings. Place potatoes in bushel baskets to within 6 inches of the rim. On a sunny fall day when a brisk breeze pushes fallen leaves into clean windrows, scoop up armfuls and top each basket with a weightless, breathable insulating layer. (See the photo taken on September 28, page 300.) Leaves block light and maintain ideal humidity. The potatoes will last and last. If you don't have leaves, use loosely crumpled newspaper instead.

You can replant uneaten potatoes as seed for the next crop. Yields might decline in the second generation and fall off in the third as pathogens accumulate and weaken the vines. If you want to save seed, you might want to supplement with fresh seed of a different variety as a backup.

Digging and curing potatoes is an annual rite of the season and a highlight of my gardening year. Neighbor Del and I bring in hundreds of pounds to keep two households fed through the winter, with extra I send out as gifts. Do not underestimate the effect of a 5-pound box of homegrown potatoes when it arrives handsomely wrapped through the mail.

See Potato Basics, page 304, for an overview of cooking potatoes.

SWEET POTATOES

Sweet potatoes require a long, hot growing season. In the muggy South, they are an easy and productive backyard vegetable, thriving even in containers. My mother grew them in planters with flowers—some types have ornamental foliage—then found herself with a surprise at fall cleanup. "Lo and behold," she told me, "there were sweet potatoes." Another year she put a big row of sweet potatoes in her vegetable garden. Midsummer, deer came in and mowed off the vines. She replanted and the deer came back a second time. After that, my mother gave up on gardening for a while. In cool northern climates, sweet potatoes need extra coddling.

Ipomoea batatas is a tropical member of the morning glory family. To split a botanical hair, the "potato" part is not a tuber but a thickened tuberous root. The sprawly vine is propagated from sprouts called slips. If you and an elementary-school botanist ever cut a sweet potato in half, stuck it around the middle with toothpicks, and placed it cut side down in a jar of water to watch it sprout in the kitchen window, you've grown slips. (See the photo taken on May 5 on page 302.) My mom remembered from early childhood when her mother would keep a sweet potato vine in the kitchen window. Nanny grew it from the trimmings of a garden-grown sweet potato to give herself a free houseplant, a small pleasure in a time when she couldn't afford others.

In the Garden

Sweet potato slips are planted after all threat of frost is past, 2 to 3 weeks after the average last frost date. The vines become exuberant in the heat, and during the dog days of summer they run far and tangle with their neighbors. The lush vines are splendid and edible. Underground progress is far slower. It takes up to 4 months of steady heat and regular watering to get sweet potatoes to harvestable size.

MAY 5: Starting sweet potato slips

In cool climates, sweet potatoes should be planted through black-plastic mulch, the horticultural equivalent of a passive solar heat sump. The absorbed heat will speed the vines to maturity before a rogue September frost takes them out. Babying sweet potatoes in the north will be worth it if you love sweet potatoes. My no-fuss approach is to plant winter squash instead. Both crops share a common role in the kitchen—as sweet orange starchy things to roast or bake or mash—and squash has a clear advantage in the New England climate. Plus deer won't touch squash.

Slips are usually sold and shipped in a bundle of a dozen or more. Expect them to look pitiful upon arrival: pale, withered, sometimes rootless. To revive, wrap the bundle in wet paper towel and stand upright in a jar out of direct sunlight. John Coykendall at Blackberry Farm learned as a boy to drag sweet potato slips through a mud puddle to coat them with the moisture-trapping clay slurry.

Set out slips 3 inches deep and 18 inches apart in rows spaced at least 3 feet. Lift the roots before frost. Even a modest cold snap, one that might spare beans and tomatoes, will kill sweet potatoes. The day before the first sub-40°F night, hurry out with a garden fork and start digging. Fresh sweet potatoes are very brittle and tender skinned. To mature for eating and storage, spread the roots in a single layer in a very warm (80° to 85°F), humid, and airy place for 7 to 10 days.

AUGUST 14: Chen Li, with her ever-present watering can, in the background, behind sprawling, edible sweet potato vines

OCTOBER 10: Frost-tender sweet potato is a morning-glory relative.

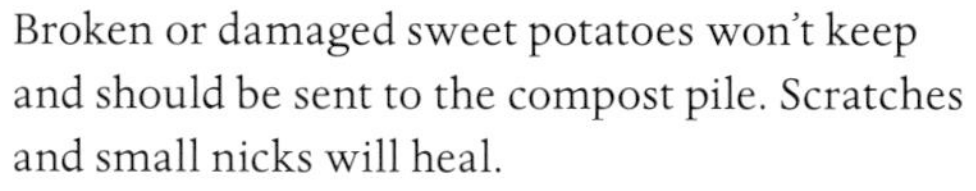

Broken or damaged sweet potatoes won't keep and should be sent to the compost pile. Scratches and small nicks will heal.

Standard varieties for home growers include the red-skinned, orange-fleshed Beauregard (90 days from transplanted slips), a good choice for northern growers, and the newer copper/orange Covington (100 to 115 days), which is more disease resistant and stores better. The red/orange Georgia Sweet (90 days) is also recommended for northern gardeners. The Japanese-type Murasaki (120 days) has purple skin and white flesh. The orange/orange Puerto Rico (100 days) is a "vineless" or "bush" sweet potato suited to small spaces and containers.

In the Kitchen

Sweet potatoes are best baked in their skins: Coat with grapeseed oil and bake for an hour, or until soft. Or peel, cube, and roast with root vegetables. Or cut lengthwise into home fries: Toss with olive oil, a generous amount of salt, smoked paprika, and a pinch of cayenne. Bake in a hot oven and turn midway through cooking. Steamed sweet potatoes can be a substitute for winter squash or pumpkin in many recipes—including pie. For the Thanksgiving table, try whipping steamed sweet potatoes with copious unsalted butter and cream. Sprinkle chopped toasted pecans over top. Or for a totally different turn, mash baked sweet potatoes with coconut cream and top with toasted shredded coconut.

POTATO BASICS

When in doubt, cook potatoes longer. There's a wide margin for error between done and overdone, but underdone potatoes are off-putting if they retain even a memory of rawness. (Sichuan cuisine's crunchy potato salad is the perhaps unique exception.) Baked, roasted, and fried potatoes should be served promptly. Boiled potatoes and potatoes for mashing, once cooked through, don't suffer too much from being left in their cooking water for a few minutes while you finish the rest of dinner. To hold them longer, for up to an hour, drain off all but ¼ inch of the cooking water, cover the pot tightly, and leave it at the back of the stove, out of the way but warm. When you're ready for them, reheat the pot on high. The potatoes will be steamy by the time the residual water boils off. Mashed potatoes will hold in a double boiler for an hour, although they stiffen and will need to be refreshed with hot milk or cream.

Organic homegrown potatoes will likely need more tidying up than the primped and presorted beauties from the grocery store. Superficial blemishes are of no concern. Cut away bruises, sprouts, scratches, or anything else that bothers you, until you reach a pristine interior. The common bacterial disease black scab causes scabby patches and little pock marks that look like insect holes. Cut them out with the tip of a paring knife. The fungal disease black scurf causes raised dark patches, like dirt that won't wash off. Scrape away with the dull edge of a knife. When all else fails, reach for the vegetable peeler or practice with a turning knife until you master the seven-sided potato of French classical cuisine.

The following roster of basic techniques will be familiar to many home cooks. I'll run through them quickly with a mind to pointing out adjustments for homegrown ingredients and recommended types—new, waxy, or starchy—and varieties.

As a rule of thumb, 2 pounds of potatoes makes 6 to 8 servings.

NEW POTATOES

ROSE GOLD, UPSTATE ABUNDANCE, OR DARK RED NORLAND • LEAFY HERBS

Carefully wash 2 pounds of small new potatoes of the same size. Avoid scrubbing off the delicate skins. Leave whole. Place in a medium saucepan fitted with a lid and add cold water to cover by 1 inch. Add enough salt to make the water quite salty, like pasta water. Cover the pot and bring to a boil over high heat. Uncover and reduce the heat to a fast simmer. Freshly dug new potatoes cook fast: 15 minutes might be enough, or even 10 if they're small. To test for doneness, stab one with a paring knife, like spearfishing. Lift it above the water, knife point down. If the potato slides off under its own weight, it is done. If unsure, cut one open and taste it. Drain when ready, using the pot lid as a strainer, and add 3 tablespoons of softened unsalted butter to the pot. Gently swirl as it melts. Add ½ cup of minced herbs, in any combination: parsley, chervil, chives, tarragon, mint, dill. Swirl a final time to distribute the herbs. Turn the dressed potatoes onto a warm serving dish and top with freshly ground black pepper. Add a sprinkle of flaky finishing salt, if you like. Serve at once.

Basil is excellent with new potatoes prepared this way. Or stir in 2 tablespoons of pesto.

SHAKEN POTATOES

NEW POTATOES OR SMALL WAXY POTATOES SUCH AS PINTO GOLD • LEAFY HERBS

Wash 2 pounds of potatoes. Leave smaller ones whole but include some larger ones, up to the size of an egg, cut in half. Peel or don't peel, as you like. Boil the potatoes until done, as described in New Potatoes (above), about 15 minutes. Reserve 1 cup of the cooking water and pour off the rest, using the lid as a strainer. Quickly, before the pot cools, add 3 tablespoons of softened unsalted butter and ⅓ cup of the reserved water. Hold the lid tightly in place and shake the pot vigorously. After 5 seconds, peek inside. The butter, potato starch, and water should have emulsified and coated the potatoes, almost like cream. If the consistency is too sticky, add 2 tablespoons of the reserved cooking water and shake three or four times. Add ½ cup minced herbs and a few grinds of black pepper. Shake once to mix, then serve immediately in a warm dish. The emulsion will break if you try to hold it. Extra-virgin olive oil will also work.

With a bit more eye-hand coordination, you can achieve the same effect by tossing the potatoes in a skillet over low heat or even tossing them in a large metal bowl.

POTATO SALAD

WAXY POTATOES • LEAFY HERBS • CELERY • ALLIUMS

Prepare and cook 2 pounds of potatoes as in New Potatoes (opposite). Drain and transfer to a large bowl. You can take the recipe in different directions.

Option One, with olive oil: Sprinkle the cooked potatoes with 3 tablespoons of champagne vinegar or red wine vinegar. Toss with any of the following: 1 cup picked parsley leaves or other herbs, some chive batons, a large handful of purslane leaves, 2 shallots or ½ onion (that have been minced and macerated in vinegar for 15 minutes), and a handful of pitted picholine or Niçoise olives. Whisk 1 tablespoon whole-grain mustard into ½ cup or more of extra-virgin olive oil and drizzle over the potato mixture. Toss to combine. Sprinkle with more vinegar to taste. Season with salt and pepper. Add crumbled bacon, if you like. Serve warm or at room temperature.

Option Two, with mayonnaise: Sprinkle the potatoes with 3 tablespoons of apple cider vinegar and toss with 1 cup homemade mayonnaise (see page 250), 1 tablespoon Dijon mustard, 1 cup diced celery, 1 cup scallion rounds (greens and whites), 2 sieved hard-boiled eggs, 1 teaspoon celery seed, and salt and pepper to taste. Add some chopped Cornichons (page 209) or sweet relish, if you like. A pinch of Old Bay seasoning will be a nostalgic garnish for some.

BAKED POTATOES

LARGE STARCHY POTATOES, SUCH AS GREEN MOUNTAIN OR KENNEBEC

Vigorously scrub 1 large potato per person under running water. Dry with a kitchen towel. Coat each potato with 1 teaspoon grapeseed oil or extra-virgin olive oil. Bake on the top rack of a 350°F oven, unwrapped, until the potato is soft when squeezed and the skin has a parchment crackle, about 1 hour. Serve immediately with unsalted butter, sour cream, chives, and trout roe or caviar.

ROASTED POTATOES WITH ROSEMARY AND GARLIC

FINGERLING POTATOES, SUCH AS LA RATTE, AND OTHER WAXY VARIETIES

Preheat the oven to 400°F. Wash and dry 2 pounds of potatoes. Trim to equal size. In a large bowl, toss with 3 or 4 cloves unpeeled, lightly crushed garlic, six 1-inch sprigs rosemary, generous salt and freshly ground black pepper, and a lot of extra-virgin olive oil. Spread in a single layer on a parchment-lined baking sheet or in a cast-iron skillet. Loosely cover with a sheet of aluminum foil. Bake for 25 minutes. Remove the foil. Continue cooking until done, about 30 minutes, turning once. Equally good at room temperature on the same day, roasted potatoes unfortunately don't reheat well after refrigeration.

GRAN'S MASHED POTATOES

GREEN MOUNTAIN POTATOES OR ANOTHER STARCHY OR ALL-PURPOSE VARIETY

Of course you already know how to mash potatoes. But the way my West grandmother made them, mashing them with their own reduced cooking stock, had a precise flavor. My dad makes them the same way, and I asked the secret. "A stick of butter," he said.

Wash and peel 2 pounds of potatoes. Cut into 1½-inch pieces, placing them in a bowl of cold water as you work. Drain and transfer to a medium saucepan. Add just enough water to barely cover the potatoes. Salt moderately and add 2 tablespoons softened unsalted butter. Bring to a boil and cook, uncovered, moderating the heat to maintain a steady boil, until the potatoes start to collapse, about 30 minutes. The cooking liquid should be reduced to a salty broth. Reserving the broth, strain the potatoes and mash in the pot with 4 tablespoons of unsalted butter, adding the reserved cooking broth as needed to achieve a somewhat loose consistency. Don't add milk or cream. The puree will hold for 1 hour in a double boiler.

SCALLOPED POTATOES

ALL-PURPOSE POTATOES, SUCH AS YUKON GEM,
OR LARGER WAXY POTATOES, SUCH AS PINTO GOLD • FRESH OR DRIED THYME

Scrub 2 pounds of potatoes. There's no need to peel them. Butter a 10-inch round baking dish. Measure 1 teaspoon fine sea salt into a small dish and 1 teaspoon chopped fresh thyme (or half as much dried thyme) into another. Working with 1 potato at a time, use a mandoline or your best knife skill to slice it into slices ⅛ inch thick. Layer the slices in the bottom of the prepared baking dish in an overlapping spiral. Season lightly with a pinch of salt, a pinch of thyme, and a turn or two of freshly ground black pepper. Repeat layer by layer until the baking dish is nearly full, lightly seasoning each layer as you go. Pour whole milk nearly to cover, 2 to 2½ cups (mixed with a large dollop of crème fraîche if you like). Loosely cover with aluminum foil and transfer to a preheated 350°F oven. Uncover after 45 minutes and use a spatula to press the top layer of potatoes down into the milk. Continue baking until the top is golden and the liquids have been absorbed, about 45 minutes, pressing with a spatula every 15 minutes.

WAXY POTATOES SUCH AS UPSTATE ABUNDANCE • SMALL CUCUMBERS • SHALLOTS • PURSLANE • BORAGE

SIDE | SERVES 8 | VEGAN, GLUTEN-FREE | SUMMER, FALL

NEW POTATOES

WITH CUCUMBERS, SHALLOTS, AND WHITE BALSAMIC

This summertime recipe is relaxed but spiffy, a shorts-and-blazer kind of thing. It brings together two ideas from Potato Basics (page 304), in that it is a potato salad and it is bound by an emulsion of cooking water and oil (see Shaken Potatoes, page 306). You literally flip the ingredients off the edge of a large bowl to make the emulsion, like how you sauté vegetables by flipping them in a skillet (*sauter* in French means "to jump"). There's an alternative technique if airborne food seems unmanageable; see the Notes, page 312.

Cucumbers and potatoes are a pair of opposites: fruit and tuber, raw and cooked, cool and warm, crisp and dense, chlorophyll and earth. A handful of summer greens adds color and texture. Purslane is my favorite. It is a self-sowing weed, although seed houses sell a cultivar named Golden Purslane. Both types have tangy succulent leaves. You could instead use small romaine leaves or frilly young mustard greens. Borage flowers could be replaced by any edible blossom, such as nasturtium, marigold petals, or even the orchidaceous bean flower.

1 shallot, minced
1 tablespoon white balsamic vinegar
Fine sea salt
1½ pounds new potatoes or small waxy potatoes, such as Upstate Abundance, scrubbed
1 or 2 cloves garlic, peeled but whole
¼ cup extra-virgin olive oil, plus more as needed
¼ teaspoon freshly ground black pepper, plus more to taste
½ pound 3- to 3½-inch cucumbers, sliced into chunky ¾-inch rounds
1 cup purslane leaves
A dozen borage blossoms

1. In a small dish, combine the shallot, vinegar, and a pinch of the salt and set aside to macerate for 15 minutes.
2. Sort through the potatoes, leaving most whole and cutting the largest few into halves or quarters of approximately the same size as the rest. Place in a medium saucepan and add cold water to cover by 1 inch. Salt the water heavily as for pasta—it should taste like your memory of seawater, to quote Samin Nosrat's advice. Bring to a boil over high heat. Reduce the heat to maintain a lively simmer and cook uncovered until a few potatoes split their skins, or a potato stabbed with the tip of a paring knife slides off, about 15 minutes. New potatoes cook quickly and you don't want them to collapse.
3. Meanwhile, grate the garlic on a Microplane grater into the bottom of a very large bowl. Add the olive oil and whisk to combine.

(CONTINUED)

4. When the potatoes are done, reserve ½ cup of the cooking water and drain them in a colander. Immediately transfer to the bowl with the garlic and oil. Sprinkle with the shallots and vinegar, ½ teaspoon salt, the pepper, and ¼ cup of the reserved cooking water. With bold confidence, grab the bowl with both hands and, with a flick of the wrist, toss the potatoes up and back into the bowl (or see Notes). The first toss or two will be watery and splashy, but within half a dozen tosses, the liquids will emulsify. Stop as soon as the sauce comes together, the consistency of heavy cream. If it turns sticky, like sour cream, or the potatoes look dry, add another 2 tablespoons of the reserved cooking water and a bit more olive oil, and toss once more to loosen.
5. Add the cucumbers and purslane. Carefully fold in with a large wooden spoon or rubber spatula. Don't overmix. Turn onto a serving platter. Drizzle with more oil. Finish with freshly ground black pepper and salt to taste and garnish with the borage blossoms. Serve at once.

Notes

In a pinch, replace white balsamic vinegar with 1 tablespoon white wine vinegar mellowed by 1 teaspoon honey.

Instead of tossing, cover the bowl with a large pot lid and shake quickly but lightly for 2 seconds. Check the emulsion and shake another second if necessary. If you shake too hard or too long, the potatoes will break apart, giving you lumpy mashed potatoes.

WAXY POTATOES SUCH AS PINTO GOLD • CELERY • ALLIUMS • HERBS | SOUP | SERVES 4 TO 6 | GLUTEN-FREE | FALL, WINTER

CLAM CHOWDER
WITH BACON AND CELERY LEAVES

New England chowder can become bland milk soup—too much dairy, not enough briny sea. Or, overly thickened with flour at roadside joints, it is halfway to wheat paste. My version is based on vegetable stock and no flour is involved. The stock takes 2 minutes to assemble and 30 minutes to simmer, enough time to prep the other ingredients. What you get for the modest effort is a clear but substantial foundation for a trio of main flavors: earthy potatoes, oceanic clams, and herbaceous celery leaves. An optional scant ½ cup of cream added at the end blanches the color and adds body without dulling the flavors.

I prefer waxy potatoes for soup, including my favorite Pinto Gold, because they hold together. Starchy varieties, such as my other favorite, Green Mountain, will break down. If using them, stir as little as possible.

Clam chowder improves if made 1 hour in advance. But don't add the clams and celery leaves until the last minute before serving. Stir them in while gently, gently reheating the pot without letting it boil.

¾ pound waxy potatoes, such as Pinto Gold, cut into in ¼-inch dice (2 cups)
4 to 5 cups Soup Stock (recipe follows)
2 pounds littleneck or Manila clams
2 ounces slab bacon, cut in ¼-inch cubes, or 2 slices thick-cut bacon, cut into ¼-inch strips
2 tablespoons unsalted butter
1 medium onion, diced
1 small celery stalk, diced
½ carrot, diced
1 bay leaf
1 sprig thyme
½ cup organic heavy cream
⅓ cup pale inner celery leaves
Freshly ground black pepper

1. In a small saucepan, cover the potatoes with 2 cups of the Soup Stock. Bring to a boil, reduce the heat, and simmer, uncovered, until nearly tender, about 10 minutes. Drain the potato cooking liquid into a 4-cup measuring pitcher or 1-quart jar. Set aside.

2. In a medium enameled Dutch oven or heavy covered pot, combine the clams and ½ cup of the Soup Stock. Bring to a full rolling boil over high heat. Cover and boil furiously for 4 minutes without lifting the lid, then check to see if any clams remain closed. If so, boil another minute. Using a slotted spoon, transfer the clams to a bowl, reserving the cooking liquids. Discard any unopened clams. Remove the meats and set aside, reserving a few of the shells for garnish. Strain the clam broth through a fine-mesh sieve lined with a damp coffee filter or cheesecloth and add to the potato cooking liquids. Add enough additional Soup Stock to make 4 cups. This soup base should be salty and delicious. Rinse and wipe the Dutch oven.

(CONTINUED)

3. Set the Dutch oven over medium heat and render the bacon. When browned, remove with a slotted spoon and drain on paper towels, leaving the fat in the pan. Add the butter, onion, celery, carrot, bay leaf, and thyme. Sweat the vegetables for 10 minutes, stirring frequently to prevent browning.
4. Add the potatoes and the 4 cups of soup base. Bring to a boil and cook until the potatoes are fully cooked, about 2 minutes. Remove from the heat, add the clams, cream, and celery leaves (saving a few to garnish the serving bowls).
5. Serve immediately in individual serving bowls, garnished with freshly ground black pepper, a few bacon bits, reserved clam shells, and a celery leaf. Leftovers will keep overnight in the refrigerator.

SOUP STOCK

1 cup dry white wine
2 small celery stalks, roughly chopped
1 medium carrot, roughly chopped
1/2 small onion
1 clove garlic, lightly crushed
4 or 5 parsley sprigs
2 small sprigs thyme
1/2 teaspoon fine sea salt
2 × 3-inch piece kombu (optional)
1/2 teaspoon Red Boat fish sauce (optional)

In a medium saucepan, combine the wine, celery, carrot, onion, garlic, parsley, thyme, salt, kombu (if using), and fish sauce (if using) with 8 cups water. Bring to a boil, uncovered. Reduce the heat to maintain a steady rolling simmer and cook until reduced by half, about 30 minutes. Strain.

CH.

21

ROOT VEGETABLES

THE EARTHBOUND COHORT OF BEETS, CARROTS, CELERIAC, PARSNIPS, RADISHES, AND TURNIPS

Amaranthaceae
THE AMARANTH FAMILY

Beta vulgaris
BEET

Apiaceae* or *Umbelliferae
THE CELERY OR CARROT FAMILY

Apium graveolens var. *rapaceum*
CELERIAC OR CELERY ROOT

Daucus carota
CARROT

Pastinaca sativa
PARSNIP

Brassicaceae

THE MUSTARD FAMILY

Armoracia rusticana
HORSERADISH

Brassica napus
RUTABAGA

Brassica oleracea var. *gongylodes*
KOHLRABI

Brassica rapa
TURNIP

Raphanus sativus
EUROPEAN RADISH,
INCLUDING SPRING AND FALL/WINTER VARIETIES

Raphanus sativus var. *longipinnatus*
ASIAN RADISHES SUCH AS DAIKON

The catchall kitchen category "root vegetables" is haphazard, botanically speaking. Among the vegetable families with rooty kin, umbellifers include walrus-tusked parsnips and colorful carrots—which can take the shape of cigars, dreidels, or a piano virtuoso's long fingers. Another umbellifer is celeriac (suh-LAIR-ee-ak), or celery root, an ill-favored strain of the same plant that gives us elegant stalk celery. Bloody beets belong to the amaranth family and are chard's twin—same genetics, different personalities. Turnips are brassicas, as are prim radishes; both can be eaten raw or cooked, although entrenched habit expects radishes to be served crispy cold and turnips buttery hot. Unloved rutabaga is a hybrid between a turnip and wild cabbage. Turnip-shaped kohlrabi resembles a root vegetable but, in fact, is a swollen stem. Close enough.

Root vegetables store their energy as sugar, rather than starch, like potatoes. For better or worse, a rumor of good health precedes them. Vegetarians, well-preserved retirees, and clean-living bloggers are forever repeating that roasted root vegetables are "good for you." In their favor, cooking concentrates their natural sugars, giving them caramelized edges and candied flavors.

Root vegetables are crunchy-fresh when raw. Daikon is a cylinder of brittle water, and kohlrabi is a sort of Teutonic jicama—crisp and bland. Both become turnip-y when cooked. Snappy carrots are a kiddie snack; grated carrots can be tossed with tasty bits (raisins, toasted spices, grated lemon zest, crumbled feta) for a chewy salad. Celeriac's lumpy surface encloses pale beauty, an interior of cream marbled with milk. Grate raw for céleri rémoulade, the mayonnaise-y French bistro specialty, or cook and mash with potatoes—especially intriguing alongside stews and braised meats in the long dark months. See Bashed Roots, page 334.

Root vegetables often prop up wintry meals because they keep so well in storage, a characteristic of biennials. To be precise about botanical nomenclature, carrots and parsnips are true roots but the bulbous part of a beet, turnip, or radish is a root-ish swollen stem, like kohlrabi. To make a still narrower clarification, celeriac's swollen stem is a hypocotyl. In most plants, the hypocotyl is the embryonic stem attached to the cotyledon within the seed. Celeriac is one of a few plants to mature the hypocotyl. In the garden, the swollen-stem vegetables tend to ride higher in the soil, like a fishing bobber, while the true roots plunge deep as if to avoid the light.

Root vegetables are cool-weather crops and often evoke an autumnal mood for me. But spring is the season for the quick red radish, which also comes in pink, purple, and white. It jumps up and is harvested within weeks. Two-toned French Breakfast radish was introduced to America as early as 1870, according to Seed Savers Exchange. The made-up name has the ring of salesman's hyperbole—"so mild a Frenchman would eat them at breakfast"—the way Vidalia onions are claimed to be "as sweet as apples."

Radishes can be cooked with their tops, like turnips. The large Asian types, best known by the Japanese name daikon (from big + root), are simmered in broth or salted and fermented in earthenware pots, as for the Korean kimchi called kkakdugi. Daikon is mild, with just enough mustard-y bite in its leaves to discourage garden pests. Pretty Watermelon radishes, another Asian type, are usually sliced translucently thin and served raw. The European winter radish Black Spanish can be roasted. Horseradish, cultivated since antiquity for its sinus-burning bite, has been considered at times both condiment and medicine. The root has nothing to do with the equine tribe; the name emerged in Shakespeare's era and meant rough or harsh, as in a hoarse voice. It was the Falstaffian radish, vulgar but fun.

JULY 3: Bull's Blood beets, on left, with Early Blood Turnip, two heirloom beet varieties

In the Garden

Root vegetables are duller than tomatoes and less profligate than squash but easy to grow and fun to pull. The radish is the garden's fastest nonleafy crop, on pace with the minor salad greens. Small turnips aren't far behind. On the other hand, if you get tired of seeing your arugula and lettuce bolt before you're half done with them, the larger root vegetables are less twitchy. Carrots, beets, parsnips, celeriac, and winter radishes are biennials programmed to bloom in their second year.

As a group, the root vegetables are cold-weather crops. Sweet spring turnips turn cottony in the summer and in hot weather radishes run to ruin too soon, like a hard-drinking poet. Most roots can be sown as soon as the ground thaws. Seedlings tough out frosts. For fall harvests, sow in July and August. Mature roots improve with the onset of chilly weather. Vegetable-cooking guru Deborah Madison and influential chef Dan Barber both have enthused about the extraordinary sweetness of overwintered carrots, a special kind of crop popularized by Maine's four-season organic farmer Eliot Coleman.

Spring-planted varieties of carrots, turnips, and radishes are small and quick to maturity. Full-sized, full-flavored winter varieties require longer growing periods.

Root vegetables have broadly similar care requirements although spacing varies by species depending on size at maturity. All prefer well-drained, loamy or sandy soil for obvious reasons. Heavy clay or compacted ground can stunt them. Gravelly soil causes deformities. Consistent water is essential to prevent forked roots, cracks, and hollows. Except for celeriac, root vegetables are sown directly in the garden rather than started indoors because they dislike having their roots jostled.

ROOTY VEGETABLES

The word *fruit* has a specific botanical meaning, but there never has been a precise definition for *vegetable*. A vegetable is the part you eat. The word in its modern sense—plant-based foodstuffs that produces disgust in picky children—emerged some five hundred years ago from the now-archaic *wort,* which in turn derived from the Old English *wyrt,* for plant, and an even older Proto-Indo-European word for root. Wort defined the opposite of *weed*. Hence beneficial St. John's Wort versus noxious bindweed. The vegetable was, in this conceptual framework, a plant that is good for you, an idea still current among nutritionists.

Generations ago, cabbage was colewort: cole (from Latin *caulis,* stalk) + wort = the stalk vegetable. Today a vegetable can be, among other things, a root (carrot), leaf (spinach), stem or stalk (celery), flower bud (broccoli), immature seed (sweet corn), or fruit (tomato, cucumber, squash).

Common usage rarely produces orderly taxonomies. And so we exclude other edible plant parts from the category *vegetables*. Wheat, rice, and other dried seeds are *grains*. Are dried beans vegetables? If grouped with dried peas and lentils, they are all *pulses.* Potatoes inhabit a category of one, more substantial than a mere vegetable (manly meat and potatoes) but less wholesome than vegetables in the eyes of carb-phobic dieters, for whom it is a horrifying *starch.* Or perhaps potatoes are a category of two, depending on what you do with sweet potatoes.

The subterranean crops—turnips, carrots, parsnips, beets, and some look-alikes—are loosely grouped as "root vegetables." The term is botanically meaningless but intuitively right. It also distantly echoes the Indo-European sense of vegetable-as-root and, furthermore, brings back the Chaucerian certainty that worts are good for you.

Grow root vegetables in rows or beds. Carefully spaced rows will give you a more uniform crop. Carrots can be spaced quite closely. I start pulling turnips small, so for the sake of efficiency I'll sometimes prepare a 4-foot-square bed and broadcast the seed with a heavy hand, thinning as the crop grows in. You eat the thinnings. My Hudson Valley friends Andy and Barbra Rothschild and their four children eat radishes even before the roots swell. Andy brings in great mounds of thinnings and Barbra sautés them whole.

Jump-up radishes and small turnips can be interplanted between larger crops—cabbages or broccoli, for instance—or sown opportunistically wherever you have an opening. I keep radishes near lettuces and resow both until hot weather. Slow-growing parsnips and celeriac need the entire growing season and perhaps are best relegated to an out-of-the-way spot where they won't be disturbed.

Raised beds are ideal for root vegetables. Only the smallest varieties will grow in pots and window boxes.

Beets are a good choice for novice gardeners. Each "seed," a BB-sized pellet, is a dried fruit containing several fused seeds. Thin successively as seedlings crowd each other. Beets can be pulled quite small—their virtues are present from infancy—or left to swell and bulge like a blue-ribbon steer, less symmetrical when fully grown but equally palatable. I like the refined flavor of Early Blood Turnip (68 days), an 1820s heirloom sold by Monticello's Thomas Jefferson Center for Historic Plants. Detroit Dark Red (60 days) is another favorite heirloom. Italian Chioggia (55 days) has a concentrically striped interior, yielding bullseye rounds when sliced. Bull's Blood is grown for its maroon leaves as well as its roots (35 days for baby leaf; 58 for roots). A few types of beets are cylindrical, like a roll of silver dollars, and cleave into consistent slices rather than the loose-change variability you get with globe beets. Golden beets can be grown interchangeably with red.

Cherry Belle (25 days) is the round radish of one's imagination. Pink Lady's Slipper (25 days) provides a complementary hue. White Icicle (30 days) is daikon white. The large Asian radishes come in purple, green, and Watermelon (55 days). The European winter radish Spanish Black (65 days) also takes longer to mature. Large radishes can be stored like carrots, as described in the following In the Kitchen section.

Hakurei Japanese turnips are ready in 38 days—they're the variety to sauté with their greens. The traditional two-tone turnip is Purple Top White Globe (50 days). Large Gilfeather (85 days), a green-shouldered knob, was discovered in Wardsboro, Vermont, before 1900 and endures on Slow Food's Ark of Taste. The ole timey rutabaga, out of fashion now, is husky and yellowed, like old cheese. Its turnip-cabbage parentage is a strange cross, to be sure, but both are brassicas. Folksy synonyms include swede and neep. Planted in late spring or early summer, neeps are harvested after the second hard frost.

Carrot seeds can take 3 weeks to sprout and must be kept moist the whole time. Cover the row or bed with damp burlap, such as ratty potato sacks, the ones nibbled by mice. Remove the protection when seedlings emerge. Ferny carrot tops are pretty—once they adorned Victorian mantlepiece arrangements—and the varied root colors provide visual interest in the kitchen. Those grown in deep loose soils will look best. Choose early types such as Mokum (48 days) for fresh spring eating. Blunt Nantes (68 days) is a classy traditional carrot for the main-season harvest. The stubby Red Cored Chantenay (70 days), a favorite of mine, is suited to long storage. The lovely Tonda di Parigi (65 days) is top-shaped. Bagged grocery store carrots, the ones as long as rulers, require bottomless fluffy soil. Double-dig the bed and sow the hybrid Sugarsnax 54 (68 days) if you're intent on growing prodigies.

Parsnips are planted in spring and harvested late, like Brussels sprouts. Neighbor Del reserves his highest praise for parsnips. "The pork of vegetables," he calls them. Not everyone so highly esteems their out-front flavor, which is vibrant but unvaried, like color field painting. Sow parsnips as you would carrots, either in rows or beds.

Celeriac must be started indoors with meticulous care. The seed has a diva's sensitivity to temperature change. I buy starts at the farmers' market around last frost. Celeriac and celery can be planted together. (Read about celery in chapter 18, page 234.)

As a rule of thumb, a 10-foot row of root vegetables, or a block 3 feet square, yields up to 10 pounds of edible roots, depending on whether you pull them at baby size or allow them to bulk up. One-eighth ounce of beet seeds, a typical packet, will sow 20 feet.

ROOT VEGETABLES—DIRECT SOWN

TYPE	DIRECT SOW*	SEED SPACING	DEPTH	THIN TO	ROW SPACING	EMERGENCE	TIME TO MATURITY	SUCCESSION PLANTING
Beets	2 to 4 weeks before last frost	1 inch	1/2 inch	3 inches	12–18 inches	5–21 days	48–60 days	Every 3 weeks
Carrots	Around last frost	3/4 inch	1/2 inch	2 inches	12 inches	Up to 3 weeks	48–75 days	Every 3 weeks until August
Parsnips	Midspring	1 inch	1/4 inch	3 inches	18–24 inches	3 weeks	120 days	
Radishes, spring	Beginning in early spring	3/4 inch	1/2 inch	2 inches	6–12 inches	4–6 days	21–26 days	Every 10 days until hot weather, again in late summer
Radishes, winter	As soon as the ground can be worked	2 seeds every 5 inches	1/2 inch	1 per 5 inches	12–18 inches	4–16 days	55–70 days	
Turnips	Beginning in early spring	1 inch (small), 2 inches (large)	1/2 inch	2 inches (small), 4 inches (large)	12–18 inches	7–14 days	35–55 days	Every 3 weeks through late summer

CELERIAC—START INDOORS

START DATE	DEPTH	EMERGENCE	TRANSPLANT DATE	SPACING	ROW SPACING	TIME TO MATURITY
10–12 weeks before last frost	1/4 inch	10–21 days	After last frost	6 inches	18 inches	90–115 days

JULY 5: Tonda di Parigi carrots

OCTOBER 10: Purple Top White Globe turnips

In the Kitchen

Being earthborn, root vegetables will need extra cleanup before they look presentable. Storage carrots, parsnips, and brittle daikon radishes should be coaxed up with a garden fork. Hose them off outside and trim tattered or yellowed leaves. Small radishes and turnips come in cleaner, like those enviable people who always appear neat and unwrinkled. On hot days I'll dunk radishes in a basin of cold water for 15 minutes to refresh the tops.

Trim root vegetables as soon as you bring them in. (See What to Do with the Tops, at right.) Wash roots a second time under running water. Store in a loosely closed plastic bags for a week or more. For longer storage, fall-harvested carrots, parsnips, turnips, beets, and large radishes should be kept in a cool, humid environment. A slatted box covered with a burlap sack would work.

If you'd like to experiment with overwintering carrots and parsnips, simply leave them in the ground and cover with 6 to 8 inches of mulch before the first hard freeze. Cold will increase their sugar content by up to 300 percent. Stake the row or bed so that you'll know where to dig after the tops die back.

The root vegetables are so useful in the kitchen. Raw and roasted options are mentioned at the head of the chapter. Caramelize sliced root vegetables in a sauté pan with unsalted butter, ghee, or bacon fat, then splash with water/wine/stock, cover, and steam until tender. Vinegar and strong spices—cumin with carrots, red chile flakes with turnips—balance the sweetness. Or sprinkle carrots with sugar to glaze.

Roots make splendid purees. A memorable dish from the time I lived in France was carrots cooked in milk and mashed with crème fraîche. Root-vegetable soup can bring one flavor into exquisite focus—the connoisseur's cream of parsnip—or harmonize an array of flavors, as in life-giving borscht. Root vegetables bolster many winter stews. For beef stew, I add diced carrots at the outset for seasoning and chopped carrots at the end to brighten the final dish. Carrots are essential for stock.

WHAT TO DO WITH THE TOPS

Beet greens, turnip tops, radish greens, and even carrot tops have flavors well worth exploring. When you bring root vegetables in from the garden, trim the tops and set aside the smaller, more tender leaves for cooking. Store separately in the refrigerator, wrapped in a kitchen towel.

Beet greens can be cooked interchangeably with chard, although they are somewhat stemmier and tougher. Small turnip tops cook quickly, while larger tops can be braised like hearty greens. Radish greens are not too fuzzy when small. They cook up like turnip greens—see Radish Tops and Tails with Zesty Umami Butter (page 332). Radish greens can be used also to make Julia Child's simple, perfect recipe for watercress soup.

I usually save a handful of carrot tops to make carrot-top gremolata (see Braised Beef with Carrots, page 345) or to bulk up Universal Green Sauce (page 248). Snipped into fringy bits as you would fennel frond, they add a strong chlorophyll/resin flavor to tossed salads. A little bit goes a long way. One is not likely to use all the carrot tops produced in the normal course of things. They are too abundant and too potent. Think of them as a bonus herb rather than a second vegetable, the way beet tops are a second vegetable. Compost the excess.

A final tip from parsnip advocate David Tanis, a friend who compares their flavor to "a carrot crossed with a squash, with chestnut undertones." To improve their texture, David blanches parsnips in salted water for 5 minutes, then roasts them or finishes them with butter in a hot skillet.

BEET BASICS

Whatever you do, wear an apron when working with red beets, or at least change out of your good clothes. The juice makes dramatic spots. They usually rinse out but not before panic sets in.

Raw beets are firm and earthy-sweet. Paler varieties are milder. Cooked beets have a dense, fine-grained texture and can be off-puttingly sweet. In any form, beets go a long way. I think of them as an accent—deeply colored, deeply flavored—rather than a cornerstone vegetable, such as broccoli or squash. Thinly sliced raw beets brighten a salad. A small serving of baked or pickled beets jolts the palate (and stains the teeth) with a distinctive intensity not to everyone's liking. In any case, a bunch of 4 to 6 medium-sized beets will easily yield 6 servings and supply a small pile of greens as a by-product.

RAW BEET SALAD

Trim several medium beets and set the greens aside. Peel and shred the roots on a box grater. Dress the crimson mulch with a strong vinaigrette fortified by Dijon mustard and toss with a handful of mixed chopped herbs (dill, chervil, fennel, tarragon, parsley, cilantro, chives). The result is a flavorful if somewhat dense salad. Crumbled feta adds a spikey dimension. Sprinkle it over top rather than tossing with the beets to avoid the queasy look of pink cheese.

DRESSED BEETS

Beets must be cooked and peeled before dressing. There are two approaches: baking and boiling. To bake, leave the root intact and trim stems to 1 inch long. Rub with olive oil to coat and salt liberally. Crowd into a baking dish, add ½ cup water, and cover tightly with foil. Bake at 300°F until tender when pierced with a skewer, about 1 hour for medium beets, 1½ hours for large. Once cool enough to handle, rub off the skins and wipe clean with paper towels. Alternatively, boil whole beets until tender, about 45 minutes for medium and 1 hour for large. As with baking, leave the root and 1 inch of stem intact to reduce bleeding. Drain, and when cool enough to handle, slip off the skins. Rinse and pat dry with paper towels.

Slice the prepared beets into any shape for serving. The simplest dressing is melted unsalted butter and chopped herbs, especially anise-scented tarragon or chervil. Chives and mint also pair well. Walnut oil is a terrific substitute for butter. A grating of lemon zest and a squeeze of lemon juice will balance the flavors. A dollop of sour cream alongside will add a rich, velvety dimension.

WARM BEET SALAD

Cook and peel 4 to 6 medium beets as described in Dressed Beets. While they are cooking, macerate a minced shallot in 2 tablespoons of red wine vinegar. Whisk ⅓ cup extra-virgin olive oil or walnut oil into the shallots. Add dribbles of balsamic or sherry vinegar for extra flavor. Pour the vinaigrette over peeled, sliced beets and toss. Salt somewhat aggressively, and season with freshly ground black or white pepper. Taste and add more vinegar as needed. Add a lot of chopped herbs and toss another time or two. Turn out onto a bed of dressed bitter greens, such as escarole or radicchio. Sprinkle with crumbles of feta, chèvre, or blue cheese, and top with a scattering of toasted walnuts, pistachios, hazelnuts, or pumpkin seeds. In winter, I might tuck in sectioned citrus, especially blood oranges.

To use red and golden beets in the same dish, dress them separately and combine at the last minute to avoid staining.

Other forceful flavors to consider include whole-grain mustard whisked into the vinaigrette, fresh horseradish grated over top, a sprinkle of whole spices (caraway, fennel seed, cumin), or a chiffonade of celery leaf.

BEET GREENS

Beet greens can be cooked like chard, although they are somewhat stemmier and tougher. As you prep beets, set aside the greens. Trim stems to the base of the leaf. Blanch the leaves in boiling salted water until tender, 5 to 7 minutes. Drain. Once cool, squeeze dry. Coarsely chop and reheat in a sauté pan with unsalted butter and perhaps a bit of minced shallot. Squeeze a lemon wedge over top before serving.

CARROTS • CILANTRO • GREEN CORIANDER | SIDE | SERVES 6 TO 8 | VEGETARIAN, GLUTEN-FREE | SUMMER, FALL

CARROT RIBBONS
WITH TOASTED SEEDS AND SPICES

A freshly dug carrot can be shaved thin enough to see through—"like gold to airy thinness beat" in John Donne's immortal phrase. Lay the carrot on a cutting board and steady it by the stem. Draw a sharp vegetable peeler down its length to lift an orange ribbon. Repeat. After several strokes, turn the carrot's planed face down and shave the other side. Continue, back and forth, until only the flat core remains. The shavings will be orange and loopy, supple but firm enough to tease into haystacks.

Raw carrots are suited to any sort of vinaigrette, from sweet to tangy to spicy. Confidently add other ingredients for texture and taste. Raisins and diced pineapple push carrot salad into church-supper territory, homey and familiar. This livelier variation has preserved lemon and a birdfeeder mix of toasted seeds. If you have bolted cilantro in the garden, scatter green coriander berries over the top.

The vibe here is Moroccan, but not insistently so. Pair the salad with Mediterranean summer cooking, such as rosemary chicken and zucchini from the grill. Or, in a carefree mood, pack it in a mason jar for a picnic with egg-salad sandwiches. Ribbons of extra-sweet storage carrots would brighten the midwinter table before a main course of pork stew or lamb tagine.

1½ pounds carrots, in ribbons
2 tablespoons minced preserved lemon
⅓ cup extra-virgin olive oil
2 tablespoons fresh lemon juice
2 tablespoons sherry vinegar
1 tablespoon honey
2 cloves garlic, minced
1 teaspoon cumin seeds, toasted and ground
1 tablespoon sweet paprika
¼ teaspoon cayenne pepper
1 teaspoon fine sea salt
½ cup sunflower seeds, toasted
½ cup pumpkin seeds, toasted
¼ cup sesame seeds, toasted
1 tablespoon coriander seeds, toasted
2 tablespoons green coriander berries (optional)
Cilantro leaves, for garnish

1. In a large bowl, toss together the carrots and preserved lemon.
2. In a small bowl, whisk together the olive oil, lemon juice, vinegar, honey, garlic, cumin, paprika, cayenne, and salt. Pour over the carrots and toss to coat. Refrigerate for 30 minutes or up to 2 hours.
3. Immediately before serving, sprinkle with the toasted seeds, toasted coriander, and green coriander berries (if using). Toss to combine. Scatter a handful of cilantro leaves over top.

RADISHES • PARSLEY

ALSO FOR: SMALL TURNIPS, SUCH AS HAKUREI

SIDE

SERVES 4

PESCATARIAN, VEGETARIAN OPTION, GLUTEN-FREE

SUMMER, FALL

RADISH TOPS AND TAILS
WITH ZESTY UMAMI BUTTER

Radishes come on fast, 3 weeks to maturity, swelling a little larger every day until the center splits. They don't hold in the garden. Braising radishes in butter is how to use them up before they get pithy. There is no sense in scrapping the tops when you're simultaneously laboring to grow arugula or mustard greens, two closely related brassicas. The fine, short hairs that cover radish greens soften with cooking. The roots also transform, losing their bite to become turnip-y sweet. Roots and greens, tops and tails, tangle in the pan. Small turnips can be cooked the same way.

Butter-braised radishes could come to the table with just a squeeze of lemon, but here I finish them with a compound butter of lemon zest, parsley, and umami-rich anchovies. (You could replace anchovies with a splash of fish sauce.) To make a vegetarian version, fortify the compound butter with a dab of grated garlic, miso, and/or finely chopped nori.

12 to 15 large radishes or 10 to 12 Hakurei turnips
6 tablespoons unsalted butter, at room temperature
1 salt-cured anchovy (rinsed, boned, and chopped)
1 tablespoon minced fresh parsley
Grated zest of ½ lemon
1 teaspoon fresh lemon juice
¼ teaspoon fine sea salt
⅛ teaspoon freshly ground black pepper
½ cup water or broth
Grilled bread (optional), for serving

1. Trim the radish tops to ½ inch and remove the roots. Cut into halves or quarters so that they're all the same size. Discard any greens that are large, tough, or yellow. Rinse the rest, drain, and keep separate from the roots.
2. In a small bowl, use a fork to mash together 4 tablespoons of the butter with the anchovy, parsley, lemon zest, lemon juice, ⅛ teaspoon of the salt, and the pepper.
3. In a large sauté pan, melt the remaining 2 tablespoons of butter over medium heat. When it foams, add the radishes and remaining salt. Sauté until slightly colored at the edges, about 5 minutes.
4. Add ¼ cup of the broth, cover, and steam for 5 minutes, shaking occasionally. Add the greens and more broth, if the pan is dry. Cover and steam until the roots are tender and the greens silky, 5 to 7 minutes.
5. Remove from the heat and add half the compound butter. Swirl to coat the radishes. Taste and add the rest if you like, or serve it alongside with grilled bread (if using).

CELERY ROOT • POTATOES • PARSLEY | SIDE | SERVES 6 TO 8 | VEGETARIAN | FOUR SEASONS

ALSO FOR: TURNIPS, PARSNIPS, CARROTS

BASHED ROOTS

The Scots have a colorful vocabulary for root vegetables, which is useful because the dishes being described look utterly dull. Neeps and tatties are a duo of rutabagas and potatoes, which are cooked separately and served with haggis. Bashed swedes are mashed rutabagas, and they go with game. The word *neeps* derives from the Latin *napus,* and *swedes* refers to the putative homeland of the *rotabagge,* or baggy root. *Bashed* means what it usually does but is so much more expressive than mashed.

This recipe, in plainer terms, makes a baked puree, which sounds dismal, I know, but the dish itself is delicious and so old-fashioned it feels new. Celery root (or other root vegetables) are peeled and cooked in milk, then mashed with potatoes that have been cooked separately. The quantity of potatoes is flexible. Their role, besides stretching the dish with an economical filler, is to bind and lighten the pureed root vegetables. The mixture puffs slightly as it bakes. It can be prepared in advance and baked off at the last minute—an advantage for complex holiday meals. On that count, bashed turnips, celery root, or rutabagas would sit well on the Thanksgiving table. In the meantime, celery root is an ideal accompaniment for beef.

2 tablespoons unsalted butter, melted, plus more for the pan
2 cups whole milk
2 pounds celery root, peeled and cut in 1-inch pieces, or other root vegetables, similarly peeled and cut
1½ pounds all-purpose potatoes, such as Keuka Gold or Kennebec
1 tablespoon finely minced parsley
½ teaspoon fine sea salt
1 tablespoon sugar
¼ teaspoon white peppercorns, ground
1 ounce Gruyère, shredded (about ¼ cup)

1. Preheat the oven to 400°F. Lightly butter a 10 × 7-inch oval gratin pan or a similar baking dish.
2. In a medium saucepan, pour the milk over the celery root. Bring to a boil and simmer until the celery root is soft, about 20 minutes. Drain.
3. Meanwhile, cook the potatoes in salted boiling water in a medium saucepan over high heat until soft, about 20 minutes. Drain.
4. Puree the celery root and potatoes together in a food mill or ricer. Whip in the parsley, the 2 tablespoons unsalted butter, the salt, sugar, and pepper. Turn into the prepared gratin pan and spread evenly. Scatter the cheese over the top. Bake until lightly browned, about 15 minutes. Serve immediately.

ALSO FOR:
HAKUREI TURNIPS,
RADISHES, PARSNIPS,
CAULIFLOWER

ROASTED CARROTS

WITH CHIMICHURRI AND MEYER LEMON YOGURT SAUCE

This recipe is a cheat. It comes to the table looking like a cheffy restaurant dish—two sauces! But one of the many benefits of roasted root vegetables is quick prep, and the two sauces take less than 10 minutes combined. I put them together while the carrots are in the oven, around the time I open a bottle of wine. (They can be made up to a day in advance as well.) The inspiration was twofold. First, I was looking for another way to use up carrot tops. Second, I'd been thinking about how to make roasted carrots a bit more forgiving, less likely to turn bitter and hard if roasted a minute too long. The answer is to blanch them before they go into the oven, and the result justifies the extra step.

There's enough flavor and color in this dish for the entire meal, so the other elements could be as simple as a green salad and chicken kebobs or crispy chicken cutlets. Or go big with stewed chickpeas and kofta, or other dishes brought into harmony with the two sauces.

2 pounds carrots, cut into 3-inch batons
2 tablespoons grapeseed or extra-virgin olive oil
1 teaspoon cumin seed
1/2 teaspoon white pepper, ground
1/2 teaspoon coriander seed, ground
1/2 teaspoon fine sea salt
1/4 cup picked carrot greens
1/2 cup Chimichurri (recipe follows)
1/2 cup Meyer Lemon Yogurt Sauce (recipe follows)
Meyer lemon zest, for sprinkling

1. Place the oven rack at the top position and preheat the oven to 400°F. Line a baking sheet with parchment or foil.
2. Blanch the carrots in a large pot of salted boiling water for 2 minutes. Drain and spread on a platter lined with paper towels to dry.
3. Toss the carrots, oil, cumin, pepper, coriander, and salt in a large mixing bowl. Transfer to the prepared baking sheet and spread in a single layer. Set aside the mixing bowl. Roast the carrots for 10 minutes. Turn the carrots with a spatula. Continue roasting for 8 to 10 minutes, or until the edges begin to caramelize.
4. Transfer the carrots to the mixing bowl and toss with the carrot greens and 2 tablespoons of Chimichurri. Turn out onto a serving platter. Splash with a few tablespoons of the yogurt sauce. Sprinkle zest over the top. Serve the remaining Chimichurri and yogurt sauce alongside.

(CONTINUED)

To adapt for Hakurei turnips, radishes, parsnips, cauliflower: Trim vegetables into bite-sized pieces and blanch for 2 minutes. Proceed with the recipe, substituting parsley leaves for carrots greens. Likewise use additional parsley in place of carrot tops for the Chimichurri.

CHIMICHURRI

½ cup lightly packed coarsely chopped carrot tops
½ cup lightly packed coarsely chopped parsley leaves
4 cloves garlic, peeled and roughly chopped
½ teaspoon fine sea salt
1 teaspoon paprika
¼ teaspoon red chile flakes
¼ cup extra-virgin olive oil
1 tablespoon freshly squeezed Meyer lemon juice

Finely chop the carrot tops and parsley leaves together. In a mortar, crush the garlic and salt. Add the chopped herbs, paprika, and chile flakes and pound a few times to bruise. Stream in the olive oil. Stir in the lemon juice.

MEYER LEMON YOGURT SAUCE

½ cup whole milk Greek yogurt
1 tablespoon organic heavy cream or buttermilk, plus more as needed
Zest from ½ Meyer lemon, removed with a Microplane or zester
¼ teaspoon fine sea salt
2 tablespoons extra-virgin olive oil

In a small bowl, whisk together the yogurt, cream, zest, and salt. Stream in the olive oil, whisking all the time. Add more cream, as needed, to achieve a pourable consistency. Serve in a small pitcher. Store in an airtight container in the refrigerator for up to 3 days.

PURPLE TOP TURNIPS • THYME | SIDE | SERVES 6 TO 8 | VEGETARIAN | SPRING, FALL, WINTER

ALSO FOR: KOHLRABI

TURNIPS OR KOHLRABI
BAKED IN CREAM WITH THYME

This dish is parallel to scalloped potatoes, although I bake potatoes in milk rather than cream, lest they become too rich. Turnips have a sharper flavor, which cuts through cream. They turn silky as they bake. This is a dish for people who didn't know they liked turnips. It is even better with kohlrabi, which, as mentioned earlier, is not a root but a swollen stem. The closest description is the sweet and nutty flavor of broccoli stems. You'll never wonder again what to do with kohlrabi. Always peel them deeply—the outer layer is fibrous. I go over them twice with a sharp vegetable peeler.

This dish is a universal side player in winter but is especially fine with braised beef and dark stews. Serve a little salad of bitter escarole or peppery chlorophyll-rich cress, along with chunks of baguette for sopping.

2 tablespoons unsalted butter, plus more for the pan
1¼ teaspoons fine sea salt
1 cup fresh bread crumbs, lightly toasted
Several thyme sprigs
2 pounds medium turnips or kohlrabi, peeled and sliced into ⅛-inch slices
Freshly ground black pepper, to taste
1½ cups organic heavy cream

1. Preheat the oven to 375°F. Place a foil-lined sheet pan on the bottom rack to catch spills. Lightly butter a 9½-inch glass pie pan or similar baking dish. Measure the salt into a small dish.
2. Melt 1 tablespoon of the butter in a small dish. Add the bread crumbs and a few leaves stripped from the thyme sprigs. Stir to combine. Set aside.
3. Layer the turnips in the baking dish. Lightly season each layer with salt, pepper, and whole thyme leaves. Use a bit more salt, pepper, and thyme on the top layer.
4. Stir together the cream and remaining salt in a small dish. Pour into the baking dish, aiming for a gap between slices to avoid disturbing the top layer. The cream will not cover the turnips. Cut the remaining 1 tablespoon of butter into bits and scatter over the top. Transfer to the oven, checking that the baking dish is centered over the sheet pan on the bottom rack.
5. After 15 minutes, use a spatula to press the top layer into the cream. Repeat at 15-minute intervals. Total cooking time will be about 1 hour, on until the turnips are very soft when tested with a fork. Allow to cool for 5 minutes. Scatter bread crumbs over the top immediately before serving.

LARGE BEETS • CABBAGE • ALLIUMS • CELERY • CARROTS | MAIN | SERVES 8 | OMNIVORE, GLUTEN-FREE | FALL, WINTER

TANGY BEET SOUP
WITH CABBAGE AND CELERY LEAF

Fall arrives quickly in the Berkshires. The last week of August can be hot and muggy. Goldenrod in full bloom draws bees like fruit flies to split tomatoes. Then a cool night arrives in early September, only a week after the last sweaty day. The asters open their purple eyes to watch for the first flush of fall color—a slash of yellow on a young sugar maple or the five-fingered blush of Virginia creeper, autumn caught red-handed. From one week to the next, from goldenrod to asters, I'm suddenly done with sliced tomatoes and ready for brothy pots, such as this hearty beet soup.

Beet soup travels widely under the name borscht, although a red-stained soup of fall vegetables is not uniquely Russian. There is no fixed standard. The right way to make borscht is how your mother or grandfather made it. Often the beets and other vegetables are grated and simmered in beef stock. Not always. My personal mental reference for borscht is the vegetarian version at B&H Dairy, home of kosher comfort food in New York's East Village, where I ate in my hungry twenties.

This beet soup starts with chicken stock rather than beef or pork stock, because it takes much less time to boil a tender chicken. Large beets work for this recipe.

The final broth should be earthy but high-strung. Tangy vinegar, sour kvass, or sauerkraut juice brings the root-y sweetness into pitched balance. (If you don't have homemade, check the refrigerated case at your grocery store or wherever you'd buy probiotic foods.) Remember that kvass and kraut juice are salty brines, so salt lightly in the early stages. At the last minute, some might balance the zing with a spoonful of honey.

Serve beet soup in large bowls with a bit of the boiled meat for substance, a dollop of sour cream for comfort, and chopped celery leaf for raw intensity. On the table, where the salt cellar usually sits, I put out a small pitcher of sauerkraut juice.

STOCK

1 whole chicken (about 3½ pounds)
1 medium onion, quartered
1 large carrot, halved
2 celery stalks, broken in half
1 teaspoon fine sea salt

SOUP

2 tablespoons unsalted butter
1 medium onion, diced
2 celery stalks, diced
2½ pounds large beets, peeled and cubed, tops reserved
2 pounds cabbage, thinly sliced
2 cups kvass or sauerkraut juice, plus more for serving
2 tablespoons apple cider vinegar, plus more to taste
1 teaspoon fine sea salt, plus more to taste
Sour cream, for serving
Freshly ground black pepper
Celery leaf chiffonade, for serving

(CONTINUED)

1. **Make the stock:** In a stockpot, combine the chicken, onion, carrot, celery, salt, and 3 quarts water. Bring to a boil, uncovered, over high heat. Reduce the heat to maintain a low simmer and carefully skim the surface. Simmer until the thigh joint is loose, 45 to 50 minutes. Remove from the heat.
2. Transfer the chicken to a large bowl or sheet pan. Discard the vegetables. Strain the stock. When the chicken is cool enough to handle, pull the meat into large pieces, discarding the skin and bones.
3. **Make the soup:** In a large pot, melt the butter over medium heat. When it foams, add the onion and celery and cook gently, stirring occasionally, until the onion colors, about 10 minutes.
4. Add the beets and cabbage and stir until coated. Add 2 quarts of the chicken stock. Bring to a boil. Adjust the heat to maintain a steady simmer and cook, uncovered, until the beets are nearly tender, about 20 minutes.
5. Meanwhile, sort through the reserved beet tops. Discard large or yellowed leaves and trim the stems short.
6. After the soup has cooked 20 minutes, add the kvass, vinegar, and salt. Taste and adjust as needed with more vinegar and/or salt. Stir in the reserved beet greens and simmer 10 minutes longer.
7. Serve in large bowls. Place a few pieces of pulled chicken in the bottom of each. Top up with soup. Finish with a spoonful of sour cream, generous grindings of black pepper, and a heavy sprinkle of celery leaves—they are part of the flavor balance, not merely a garnish. A small pitcher of kvass or sauerkraut juice on the table allows everyone to add more salty, sour, fermented funk to taste.

ROOT VEGETABLES • GARLIC • HERBS | MAIN | SERVES 6 TO 8 | VEGAN OPTION | SPRING, FALL, WINTER

SHEET-PAN ROOT VEGETABLES
WITH SAUSAGE, ROSEMARY, AND GRAPE SALAD

One year while I was working on this book, Del and I invited two neighboring families to join us in raising pigs for the freezer. A pair of Gloucester Old Spot piglets arrived from my farming friends Barbra and Andy Rothchild and established their new home in a paddock behind my barn. They rooted up every living thing including Japanese knotweed, which I had begun to believe was immortal. Many admirers came to marvel at how fast they grew. We fed the pigs organic local grain supplemented with pulled weeds and other garden trimmings, as well as kitchen scraps and vegetable leftovers. The compost pile went hungry, but the pigs flourished. In the end, our four households divvied up 120 pounds of homemade pork sausage.

This recipe features sausage and has an autumnal ingredients list of root vegetables, but it comes off as fresh and vegetable-forward because it is dressed with a little grape salad and scattered with lots of chopped fresh herbs. The quantity of sausage is meant to be skimpy. The meat is a flavorful garnish rather than the main event—and it could be left out altogether. The sheet pan goes straight to the table, just don't forget a trivet. Serve with something green, such as a fall salad of baby mustard greens, sautéed spinach, or broccoli tossed in a hot skillet, see page 146.

- 1 pound medium turnips, peeled and cut into quarters
- 1 pound carrots, peeled and cut in 2½-inch sections
- 1 pound mixed radishes (whole) and daikon (cut in 1-inch chunks)
- 2 cloves garlic, peeled and halved
- 1 cup seedless grapes
- 3 tablespoons extra-virgin olive oil, plus more for the pan
- 1½ teaspoons minced fresh rosemary, plus more for garnish
- ½ teaspoon minced fresh thyme
- 1 teaspoon fine sea salt
- Freshly ground black pepper
- 4 sweet Italian sausages, cut in half

GRAPE SALAD

- 1 shallot, minced (about 2 tablespoons)
- 1 tablespoon red wine or champagne vinegar
- Salt and freshly ground black pepper
- 1 cup seedless grapes, halved
- 2 tablespoons extra-virgin olive oil
- Pinch of minced rosemary
- Pinch of minced thyme
- ¼ teaspoon fine sea salt
- ¼ cup picked parsley leaves, lightly packed

(CONTINUED)

1. Place a rack at the top position in the oven and preheat the oven to 400°F. Lightly oil a sheet pan.
2. Blanch the turnips in a large pot of salted boiling water for 2 minutes. Remove with a slotted spoon and spread on a platter lined with paper towels to dry. Blanch the carrots in the same water for 2 minutes, drain, and spread to dry.
3. Combine the turnips, carrots, radishes, garlic, grapes, olive oil, rosemary, thyme, salt, and several turns of freshly ground black pepper in a large mixing bowl. Mix to coat. Turn onto the prepared sheet pan. Tuck in the sausage. Roast for 25 to 30 minutes, or until the sausages are browned and the vegetables lightly caramelized.
4. In the meantime, make the grape salad. Combine the shallot and vinegar in a small dish with a pinch of salt and some freshly ground black pepper. Set aside to macerate for 15 minutes. In a small mixing bowl, toss the grapes, olive oil, rosemary, thyme, and salt. Add the macerated shallots with their liquid. Toss. Immediately before serving, add the parsley and toss to coat.
5. To serve, allow the sheet pan to cool for 5 minutes. Scatter the grape salad over the top. Sprinkle with the remaining rosemary.

CARROTS • SMALL ONIONS • PEAS • ALLIUMS • HERBS

ALSO FOR: TURNIPS, PARSNIPS

MAIN

SERVES 8

OMNIVORE, GLUTEN-FREE

FOUR SEASONS

BRAISED BEEF
WITH CARROTS AND CARROT-TOP GREMOLATA

When I was very young, Mrs. Braswell, the neighbor who kept me after school, would sometimes help my mother by cooking supper. Her pot roast began with chuck, a flavorful shoulder cut sold cheap because it is tough. After hours of cooking, Mrs. Braswell's roast would shred into long fibers, like the Mexican dish ropa vieja. My mother's beef stew came to much the same end. Before work she'd fill a Crock-Pot with cubed chuck, carrots, celery, potatoes, and onion, half-cover everything with water, and throw in a bouillon cube. By the time I got home from school, the house smelled of dinner. It was spoon food: soft meat and vegetables saturated with long-cooked flavor, delectable pap.

The following recipe lands in a similarly nostalgic comfort zone but with double vegetables. You braise the beef with the usual carrots, celery, onion, and garlic, then swap out the mushy bits for additional lightly cooked vegetables to serve. I like to use whole small carrots, such as round Tonda di Parigi in the photo on page 347.

Small onions or shallots could be among the additional vegetables—even small green scallions if you have them. Fall-planted peas could join the mix, or blanched green beans from the freezer. Turnips could replace the carrots. Add small, boiled potatoes if you like.

Gremolata is a sprinkled condiment discussed in chapter 18. Here I use carrot tops instead of the typical parsley for the fun of utilizing another aspect of a homegrown ingredient. You could also finish the dish with a grating of fresh horseradish—a special friend to beef. Horseradish is a low-maintenance perennial that's ready to dig in braising season. Serve the stew in wide bowls with mashed potatoes or celeriac puree.

Like most stews, this one makes excellent leftovers. It also freezes well.

BRAISED BEEF

1 tablespoon grapeseed oil
3 pounds chuck roast, cut into 3-inch cubes (see Notes)
1 cup vermouth
2 tablespoons red wine vinegar
2 tablespoons tomato paste
1 teaspoon Red Boat fish sauce
1 large carrot, halved
1 celery stalk, broken in two
1 medium onion, halved and stuck with 1 whole clove
1 small head garlic, halved horizontally
Small herb bundle: 1 bay leaf, parsley sprigs, thyme sprigs, celery leaves
1 teaspoon fine sea salt
1/2 teaspoon black peppercorns, ground

ADDITIONAL VEGETABLES

2 tablespoons unsalted butter
3/4 pound small carrots, trimmed and scraped
Fine sea salt
1/2 pound small white onions, peeled
1 cup green peas

GREMOLATA

1/3 cup lightly packed chopped carrot tops
2 cloves garlic, peeled
Grated zest of 1 lemon

BEURRE MANIÉ

1 heaping teaspoon all-purpose flour
1/2 tablespoon unsalted butter

(CONTINUED)

1. Preheat the oven to 300°F.
2. **Braise the beef:** In a large Dutch oven, heat the grapeseed oil over high heat until it shimmers. Working in two or three batches, sear the beef until deeply browned, 5 minutes per side. Once all the beef is browned, pour off all but 1 tablespoon of the fat.
3. Add the vermouth to the Dutch oven and cook to reduce by half. Turn off the heat and stir in the vinegar, tomato paste, and fish sauce. Return the browned meat to the pot along with any juices. Tuck in the split carrot, celery, onion halves, garlic, bouquet garni, salt, and pepper.
4. Add 1 quart water, or enough to cover by two-thirds. Cover and bring to a boil over high heat. Transfer to the oven and cook until the beef is fork-tender, about 2½ hours.
5. Remove from the oven and allow to cool, uncovered, for a few minutes. Using a slotted spoon, fish out the spent vegetables. (They make a delicious mushy snack on toasted country bread.) At this point you can refrigerate overnight or for up to 3 days. When ready to cook, skim off the congealed fat. (It will be delicious for frying eggs or roasting potatoes.)
6. If you haven't refrigerated the beef, spoon off the fat or transfer the cooking liquids to a fat separator and separate, returning the broth to the pot.
7. **Cook the additional vegetables:** In a large skillet, heat 1 tablespoon of the butter over medium heat. Add the carrots and sauté until lightly colored, about 7 minutes. Add a large pinch of salt and ¾ cup water. Cover and simmer until tender, 10 to 15 minutes. Remove the carrots and set aside.
8. In the same pan, heat the remaining 1 tablespoon of butter over medium heat. Add the small onions and sauté until lightly colored, about 5 minutes. Add a pinch of salt and ½ cup water. Cover and simmer until tender, 10 to 15 minutes, adding more water if the pan dries out. Set aside with the carrots.
9. Fill a medium bowl with ice and water. Bring a pot of salted water to a boil. Blanch the peas in the boiling water for 1 minute. Shock in the ice bath and drain.
10. **Make the gremolata:** Mince together the carrot tops, garlic, and lemon zest.
11. About 30 minutes before you're ready to serve, gently reheat the braised beef. Add the parcooked carrots and onions and simmer for 20 minutes to heat through.
12. **Meanwhile, make the beurre manié:** In a small bowl, blend the flour and butter.
13. Increase the heat under the pot and bring to a boil. Add half the beurre manié and swirl to melt and combine. Boil for 1 minute. If you'd like the sauce thicker, add the remaining beurre manié and cook 1 minute longer. Add the peas and gently stir to combine. Remove from the heat.
14. Sprinkle with a tablespoon of the gremolata. Pass the remaining gremolata at the table.

Notes

Replace 1 pound of chuck with beef shank for beefier flavor.

For the sake of time management, make the recipe over 2 days. Braise on the first, finish on the second. The flavors will deepen overnight in the refrigerator and the fats will congeal, making it easier to degrease.

CH.

22

SQUASH AND PUMPKIN FAMILY

GOURDLIKE CUCURBITS

Cucurbitaceae
THE GOURD FAMILY

Cucurbita argyrosperma, aka *C. mixta*
CUSHAW

C. pepo
SUMMER SQUASH, ACORN,
AND OTHER THIN-SKINNED WINTER SQUASH,
SOME PUMPKINS

C. maxima
HUBBARD AND OTHER WINTER SQUASH,
SOME PUMPKINS

C. moschata
BUTTERNUT AND OTHER WINTER SQUASH,
SOME PUMPKINS

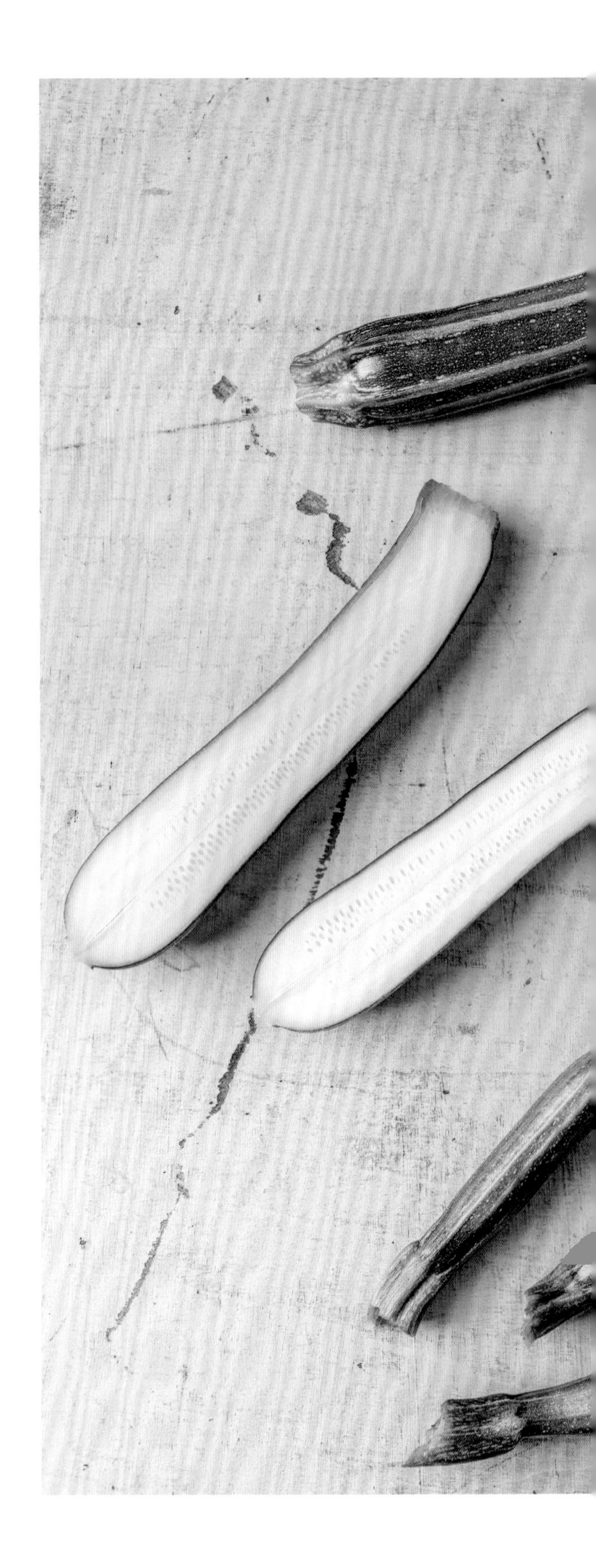

SEPTEMBER 15:
Pantheon, a cocozelle-type hybrid zucchini

This chapter covers a vegetable family that makes mischief of taxonomies. You say zucchini, I say jack-o'-lantern, both are of the species *Cucurbita pepo.* Picked immature, squash is vegetal, moist, yielding, tender, pale-fleshed, fast-cooking—the culinary virtues of zucchini. Left on the vine until the end of the growing season, it is thick-skinned, solid, hollow but weighty, deeply colored, starchy, sweet, dense, and long-storing fruit—pumpkin. Or, as some people say, winter squash. There's no botanical difference between pumpkin and winter squash, only habitual distinctions of language and use. Ornamental Connecticut Field pumpkin and oven-ready delicata squash differ, botanically speaking, as a Tennessee preacher differs from a Kentucky minister.

Yet other types of pumpkin belong to an altogether different species. France's bright-red Rouge Vif d'Étampes—the Cinderella pumpkin—and Japanese kabocha-type pumpkins, such as red kuri, are *C. maxima,* which also gives us the giant American heirloom blue Hubbard squash. Cheese-shaped pumpkins, such as stackable Long Island Cheese and sexy Musquée de Provence, are Rubenesque varietals of *C. moschata,* whose tan exemplar is the sleek, solid-necked butternut squash. A fourth species, *C. argyrosperma,* formerly *C. mixta,* includes the giant striped crookneck known to its scattered advocates as cushaw, but few Americans born in the twenty-first century will have heard of it.

A chart on page 355 breaks down popular squash varieties by type and species.

The details matter little, except on one count. Squash readily cross-pollinates, frustrating seed savers who want to isolate preferred genetics. Plant breeders use squash's promiscuity to develop new hybrids. A stellar novelty called Robin's Koginut was midwifed by Michael Mazourek of Cornell University, who also developed Honeynut and its utterly delicious successor, Honeypatch (formerly 898 Squash). Horticultural evolution hasn't tapped out. Future heirlooms continue to emerge through skillful cross-breeding and artificial selection.

OCTOBER 10: Howden pumpkins for jack-'o-lanterns, Robin's Koginut to cook

This chapter covers summer squash and winter squash/pumpkin. They are among the most rewarding vegetables to grow and cook—among the ranks of herbs, greens, and tomatoes as essential backyard vegetables.

SUMMER SQUASH

Homegrown zucchini are much tastier than the dull spongy things sold at grocery stores. The name means little *zucca,* an Italian pumpkin, and varieties are distinguished by color, which ranges from netted celadon to black jade. A few sport zebra stripes, such as the Italian cocozelle types. Cocozella di Napoli was the nineteenth-century antecedent to our modern zucchini, which first appeared in Milan in 1901, according to squash expert Amy Goldman. The word *zucchini* came to the US in a 1921 seed catalogue. The Englishman's vegetable marrow is also a kind of *C. pepo,* but if he wants zucchini, he'll sow

AUGUST 13: Yellow crookneck summer squash is smooth-skinned and dense when 5 inches long—prime for eating. It turns warty and seedy when mature.

what he calls courgettes, using the French diminutive for *courges,* squash. In Spanish they are *calabacitas*—little gourds. The clear theme is to pick zucchini small. Straight from the garden, they have a mild flavor and tender flesh suited to any cooking method. They can also be shaved thin and eaten raw, as with Zucchini Carpaccio (page 364) served with a tonnato-style pink salmon sauce.

Yellow crookneck squash, a direct Native American inheritance, has a delicate but pronounced squashy flavor, as unmistakable in its way as fresh corn. UFO-shaped pattypan is blander but comes in yellow, white, and celadon green, like silicone gadgets produced in three colorways.

All types of summer squash mix it up in the pan. They absorb the flavors of new onion, fresh garlic, thyme, and basil. They pair sweetly with savory tomatoes, peppers, and eggplants. Ratatouille, the Provençal vegetable stew made with all of the previously mentioned, expresses the layered sensuality of the summer garden better than any other dish. (See page 446 for the recipe.)

A squash plant is called a vine, but most summer squash maintain a bushy growth habit. They require ample space nonetheless because their unwieldy leaves are like a bouquet of parasols. A single zucchini plant will spread 3 to 5 feet—a serious space hog in a raised bed. At least you don't need many. Two hills of zucchini is *plenty* for most families. You inevitably have more summer squash than you can use.

WINTER SQUASH

If summer squash is vigorous, winter squash is rampant. I've had cushaw vines exit the garden, climb a peach tree, and set fruit among the branches. One year I left a 10-foot buffer between hills of Koginut squash and a row of sweet potatoes and, even so, keeping them apart was like separating quarrelsome dogs. Most winter squash will outgrow a raised bed or small garden. (Look for one of the bush varieties, such as Gold Nugget, Sweet Dumpling, or Burpee's Butterbush.) With sufficient room to run, two hills of butternut or another small-to-medium winter squash will bolster the gardener's self-esteem with their galloping development. Winter squash blooms with self-sacrificing abandon and fruits recklessly. Ruined at last by its productivity, it daubs the swollen fruit with fleshy color. The voluptuous shapes turn openly lewd after frost exposes them to view.

Cooked winter squash is rich and orange, an endowment of sweet summer sunlight. As it roasts slowly to pudding-like smoothness, the youthful squashy flavor turns soulful. Throughout the fall, I bake Honeypatch several times a week as a side dish—saving the leftovers for breakfast. The secret is to pour a tablespoon of thick cream into the seed cavity as soon as the squash come out of the oven. It soaks in, disappearing. No one needs to know. A roast

JULY 12: Young cushaw vines at Monticello

JULY 13: Zucchini

stuffed pumpkin, served whole and carved at the table like a ceremonial goose, can be the showpiece of a festive menu. In northern gardens, winter squashes are the climate-appropriate surrogate for tender sweet potatoes.

The finest culinary varieties of winter squash/pumpkin hardly need more seasoning than butter, but they marry with the ravishing flavors of curry, tomatoes, spice, chiles, maple syrup, brown sugar or wine. My favorite recipe in this book is Chicken and Pumpkin Braised in Amontillado (page 376), which puts sherry's shimmery taste up against the sweet depths of winter squash.

Winter squash stores as well as any vegetable. I once bragged to John Coykendall, the head gardener at Blackberry Farm in Tennessee, about a cushaw I kept for 18 months. He said he once grew a particularly fine pumpkin and entered it in the county fair. It won first place. A year later, he still hadn't cut it, so he entered the pumpkin again. It won second.

In the Garden

Squash seeds are large and pleasing in the hand, finely sculpted and minutely textured, like wooden tokens. Long before Columbus, squash seeds were shared widely by ancient gardeners, who carried them in pouches and pockets, a universal currency. Wild squash was likely domesticated in southern Mexico, but it spread throughout the Americas as far as climate allowed. Squash is the ground-shading sibling in the Three Sisters, the indigenous agricultural triad: Its sprawling vines and broad leaves protect the soil beneath corn and beans, smothering weeds and holding in moisture.

Squash is a warm-season crop. It is put out late, after all danger of frost has passed. Some winter squash and pumpkins need 110 days to mature. In northern climates, sow as late as possible to escape spring frost and as early as possible to escape fall frost. Squash seeds are always sown directly because seedlings are disturbed by transplanting. Or so goes the standard advice.

I have a different take. In my garden, juicy-stemmed squash seedlings are enjoyed by cutworms, slugs, grasshoppers, and voles. To give the plants a fighting chance, I often start seeds in small pots and transplant after the second true leaves emerge. They tolerate the change of scenery if I get them in the ground at the right moment—well-rooted but not yet root-bound. (See page 67 for a photo.) Biodegradable pots further reduce the risk of transplant shock—although gardeners should avoid peat pots, because peat bogs are now recognized as an environmentally important nonrenewable resource. Collar squash seedlings with a strip of newspaper to stop cutworms. A dusting of diatomaceous earth puts off the grazers. Voracious voles are a true nuisance, but they will investigate a snap trap baited with cucumber, gullibility being their one decent trait.

Squashes like well-drained soil, full sun, and consistent water—the usual growing conditions. With their vigorous greenery, they need extra feeding. I mulch each hill with compost and refresh the mulch midseason. That said, I've found that winter squash will manage on unimproved soil with surprising self-reliance. Erratic watering can cause squash to drop blossoms and grow misshapen fruit. Even in good conditions, the first few fruits may fail to develop, likely due to pollination problems. Let bees and flies do their work.

On hot days, squash leaves will droop in a show of misery, like teenagers in a mope. Chances are the vines will recover their turgid composure after the sun goes down, when water uptake from the soil catches up with transpiration, or loss of water vapor through the leaves. Do not be tempted to throw water on a sulky squash plant at midday. Stick to a regular regimen of infrequent deep watering. Usually all will be fine. Only occasionally is wilting a sign of trouble—read more following.

Expected pests include squash bugs, striped and spotted cucumber beetles, and squash vine borers. The first is a loathsome thing. The adult looks like an elongated stinkbug. Females lay bronze or brass-colored oval eggs, half the size of sesame seeds, on the undersides of leaves, often clutched where the leaf rib forks. Nymphs emerge hemoglobin-red but soon turn ghostly blue-gray. They swarm and skitter restlessly like nightmare goblins.

Pill-shaped cucumber beetles hide under leaves and congregate inside blossoms. They can be picked off on cool mornings, before their metabolism speeds up, or when they are lost in reproductive oblivion. The eggs are orange. In addition to chewing leaves and flowers, cucumber beetles transmit bacterial wilt. If vines don't perk up after watering in the evening or early morning, they might have the dreaded wilt. There's nothing to be done about it, although squash is generally more tolerant than cucumbers, for which it is a terminal diagnosis.

Wilted leaves in the morning could also be the effect of squash vine borers. The maggoty larvae emerge from the soil, bore into the vine, and tunnel their way through, disrupting water flow. If you look closely at the stem just above ground level, you might see bore holes or piles of frass, chewed-over debris. Try slicing into the stem lengthwise to reveal the villain and pluck it out. Mound soil over the cut and keep watered. The vine might recover. *C. pepo* and *C. maxima* are vulnerable.

As usual, prevention is the best approach. Pests overwinter in the soil, so outbreaks can be reduced by rotating crops. Every day in the early summer, check vines for pests and take measures to prevent a population boom. To control squash bugs, squash bugs. (The verb comes from the Latin *quassare,* to shatter, while the noun derives from the Narragansett word *askutasquash.*) Or knock pests into a pail of soapy water.

Sunshine and good airflow are the best defense against diseases such as powdery mildew. Don't crowd plants.

Summer squash vines will eventually peter out from exhaustion. Succession plantings at 3-week intervals provide backup. Later plantings will be less in step with pest life cycles.

Choosing summer squash varieties is like shopping for penny loafers: There are many choices within a narrow range. The striped, ribbed Italian heirlooms such as cocozelle (53 days from seeding) and Costata Romanesca (60 days) have the edge for flavor. Black Beauty zucchini (50 days), developed in Connecticut in 1957, is long, dark, and handsome. Newer hybrids such as Raven (48 days) promise higher yields and disease resistance. Eight Ball (50 days) is round. Some zucchini varieties are yellow—so peculiar. Yellow summer squash is of a different feather, and I'm devoted to the flavor of heirloom yellow crookneck (58 days). The newer hybrid Gentry matures 2 weeks earlier.

Trialing winter squash is a project for a lifetime. I'll vouch for the superlative culinary value of Robin's Koginut (110 days) and Honeypatch (110 days), both offered by Row 7 Seeds. An excellent organization in Oregon, the Culinary Breeding Network, has a program called Eat Winter Vegetables, which publishes online results from taste trials. The section on winter squash is especially commendable as it also tells when to eat each variety. Some sweeten as they age and don't reach peak eating until midwinter. Others have excellent flavor but a short shelf life, so should be eaten first. Still others can be held back until the very end of the storage cycle with no loss of quality. See eatwintervegetables.com for reviews. I wish similar initiatives would take root and thrive in every region of the country.

The iconic carving pumpkin for jack-o'-lanterns is Howden (115 days), introduced in the 1970s by John Howden of Sheffield, Massachusetts, where his son continues to raise sweet corn. Winter Luxury (100 days) is a noted pie pumpkin.

Given their mature size, I sow squash in widely spaced hills. Allow 5 seeds per hill and select the strongest 2, if growing summer squash, or the strongest 3, for winter squash and pumpkin, which sprawl away from their roots and thus avoid crowding each other.

To grow the Three Sisters, start by planting corn in a 12-foot-square block. Sow one hill of winter squash or pumpkin at the center of the block. Two weeks later, thin corn seedlings to 12 inches apart and sow 2 or 3 bean seeds per stalk, thinning to the best 1 or 2. The squash will soon enough cover the ground beneath the corn and then some.

Count on 7 to 10 seeds per gram for *C. pepo.* Other species have larger seeds. In any case, a single seed packet of any variety will supply the average home gardener. Seeds remain viable for 3 to 5 years.

SAVING SQUASH SEED

Every variety of squash will cross-pollinate freely within its own species. This is particularly troublesome for *C. pepo*. That zucchini crosses with yellow crookneck, another summer squash, is easy enough to accept. But it will just as readily cross with acorn winter squash or Winter Luxury pumpkin. Seeds from such crosses could look like either parent or a montage of the two. The next generation won't "sow true," or reliably exhibit parental traits.

In their avidity for hybridizing, sometimes squash go one step further and breed across species. This messes with the very definition of a species as a group of organisms able to share genetic material, or interbreed. Squash go for it, rules be damned. The only certainty, according to garden writer Lee Reich, is that *C. pepo* will not cross with *C. maxima*.

The takeaway for the seed saver—someone who strives to preserve and perpetuate a specific genetic strain—is to grow only one type of each species per year. If maintaining species purity is crucial, limit yourself to one variety of *C. pepo*—a summer squash for fresh eating—and one kabocha-type of *C. maxima* for storage to avoid the risk of unwanted hybrids.

TYPES OF WINTER SQUASH

	SUMMER SQUASH	WINTER SQUASH	PUMPKIN
C. argyrosperma aka ***C. mixta*** fruit stalk smooth, hard and angled		Cushaw	
C. maxima fruit stalk soft, round, and prickly; corky when fruit ripens		Buttercup, Hubbard, Candy Roaster, kabocha, red kuri	Atlantic Giant, Rouge Vif d'Étampes
C. moschata fruit stalk smooth, hard, flared		Butternut, Honeynut, Robin's Koginut	Long Island Cheese, Musquée de Provence, Black Futsu
C. pepo fruit stalk pentagonal; prickly stems and leaves	Zucchini, yellow crookneck, pattypan	Delicata, acorn, spaghetti	Connecticut Field, New England Pie, Howden, Winter Luxury

Note: Stalk descriptors from garden writer Lee Reich.

SUMMER SQUASH

DIRECT SOW	HILL SPACING	SEED SPACING	DEPTH	EMERGENCE	THIN TO	TIME TO MATURITY
1 week after last frost	6 feet	5 per hill	1 inch	5–10 days	2 per hill	48–54 days

Note: To start indoors, sow 2 seeds per cell three weeks before average last frost. Pinch the weaker. Transplant after all danger of frost has passed.

WINTER SQUASH, PUMPKINS

DIRECT SOW	HILL SPACING	SEED SPACING	DEPTH	EMERGENCE	THIN TO	TIME TO MATURITY
1 week after last frost	6–12 feet	5 per hill	1 inch	5–10 days	2 per hill	90–120 days

Note: Start indoors as for summer squash. Small-fruited winter squash can be spaced 6 feet apart. Allow 12 feet between winter squash and pumpkins. Cushaw know no limits.

DECEMBER 3: Clockwise from center: green-striped cushaw, blocky Robin's Koginut, Honeypatch, and Butternut

IN THE KITCHEN

Summer squash are prime eating when 5 to 8 inches long. At baby size, they are bland. Left too long on the vine, they get seedy. Check squash plants daily and harvest with a knife or clippers, taking 1 inch of stem. The plants are brittle. Don't step on vines or break leaves when pushing them out of the way.

Somehow there's always a zucchini that escapes notice and makes its startling appearance a week later, anaconda-thick. Large fruits are still edible. Scoop out the seeds with the edge of a tablespoon. Cut the flesh into ¾-inch chunks and cook as usual. Yellow crookneck squash loses some of its appeal with size, but provided the specimen isn't truly enormous, you can scoop out the seeds and chop the flesh for soup or stew. Or else slice the solid neck into rounds and compost the bloated body.

Summer squash blossoms are edible. Pick in the morning once they are fully opened. There are male and female blossoms—the thicker-stemmed female blossoms attach to a tiny fruit. Picking females will reduce your harvest, which could be a benefit if you're overwhelmed by squash. The flavor of squash blossoms approaches zero but they are gorgeous stirred into pasta or risotto, tossed in salad, stuffed and fried, or used as an edible garnish.

Summer squash grow fast and clean. I bring straight-neck varieties into the kitchen stacked like cordwood and loop crooknecks over my fingers like candy canes. All are best eaten at once but will keep at room temperature for a day or two and up to a week in the refrigerator, losing quality each day.

The uses of zucchini are truly limitless, which comes close to matching the supply. See Zucchini Basics (page 358) for an overview. Yellow crookneck squash can be stewed in its own juices with unsalted butter, onions, and thyme or dill. The flavor carries well in soups and gratins. Goat cheese complements it. Summer squash out of the freezer is charmless mush.

AUGUST 24: My kitchen collaborator Dash in the garden picking squash blossoms (left), and the open-face squash blossom omelets he made us for lunch (above)

Winter squash are harvested late, after the fruit has matured on the vine. Many varieties signal ripeness by turning orange. Others will be ready when the stem dries; on *C. maxima*, the stem turns corky. Don't rush the harvest. Vines will survive a light frost, and mature fruit will be unfazed. Once fully ripe, bring in the keepers to cure. Slightly underripe ones, those with green patches, will mature off the vine. (See the photo on page 350.) Cure in a warm place such as a sunny room for 2 weeks. Wipe down and store in a cool spot—I keep them in the front hall with apples. Pumpkins should not touch in storage, but smaller squash will share a basket.

Storage quality depends on the variety. I plan to use Honeypatch by Christmas. Robin's Koginut fades by March. Kabochas last longer. The most durable storage varieties, such as the cushaw, will keep until summer.

SEPTEMBER 15: Goldenrod indicates the end of summer, New England aster the start of fall.

ZUCCHINI BASICS

Zucchini always have a place on my summer table, no matter what else is being served. They adapt to any cooking method. In a sauté pan, they quickly turn pale gold and can be tossed with pasta and grated Parmigiano-Reggiano for an instant meal. If I light the grill, I'll fit zucchini around the edges of the fire. If I have the oven going, I'll put in a sheet pan of zucchini to bake. The leftovers will find their way into a lunch frittata, or I'll chop them up with extra-virgin olive oil and a little grated garlic to go on grilled toast smeared with fresh ricotta or goat cheese, a tartine snack. Zucchini even shows up for weekend breakfast, when I griddle them to serve alongside bacon and eggs as if they were diner potatoes. Stuffed baked zucchini makes an easy supper everyone enjoys, including teenagers. Sliced into spears, firm young zucchini become a delicious refrigerator pickle. And, when all else fails, zucchini bread.

This overview of basic techniques skims the surface by necessity. It is intended as a reminder of things to do with zucchini, a list of prompts.

Two pounds of zucchini will serve six to eight people as a side.

SAUTÉED ZUCCHINI

This technique—sautéing over fierce heat such that the pan remains in near-constant motion—gilds paper-thin slices of zucchini without caramelizing them. It approximates one of those contorni Roman restaurants drop on the table with such offhand elegance. The flavor is the quintessence of just-picked zucchini.

Choose prime young zucchini, 6 to 8 inches long, and slice as thinly as possible, no thicker than a dime. Sauté in a thin layer in a very hot skillet with generous extra-virgin olive oil, salt, pepper, and 2 cloves lightly crushed peeled garlic. Don't overcrowd the pan and cook at the highest heat until just cooked through, 5 to 7 minutes. You must keep the zucchini moving constantly to prevent browning. Cooking zucchini this way is how I learned to flip vegetables in a sauté pan without flinging them across the stove. If slices start sticking, it never hurts to add more oil. Once the zucchini are barely cooked through, turn them out into a serving bowl and scatter with fresh herbs (basil, mint) if you like. Finish with freshly ground black pepper or Aleppo.

ZUCCHINI WITH PASTA

Cook ½ pound spaghetti according to package direction. Meanwhile, cook 1 pound zucchini as directed in Sautéed Zucchini (opposite). Reserving a scant ¼ cup of the pasta cooking water, drain the pasta and toss with the zucchini, copious grated pecorino or Parmesan, and enough of the reserved water to make a sauce. For a richer, creamier version, toss the pasta and cheese with a raw egg plus an extra yolk before adding the cooked zucchini for a final toss to combine.

BAKED ZUCCHINI

Slice 5 or 6 prime zucchini, 6 to 8 inches long, lengthwise. Slash each cut face in several places. Layer in the bottom of a large bowl with thyme sprigs and 2 cloves crushed and peeled garlic. Drizzle with 2 tablespoons extra-virgin olive oil and season generously with salt and pepper. Roll the zucchini halves in the oil to coat. Set aside for 30 minutes if you have time, rolling every 10 minutes. Transfer to a sheet pan lined with parchment, placing the zucchini cut-side down and saving the seasoned oil. Cook in a 325°F oven, uncovered, for 20 to 25 minutes. Turn cut-side up and bake until the zucchini is soft and shrunken, its flavor concentrated, another 20 to 25 minutes. Transfer to a serving platter and drizzle with the reserved seasoned oil or fresh extra-virgin olive oil. To serve, sprinkle with coarse finishing salt, freshly ground black pepper, and chopped basil, parsley, or thyme.

GRILLED OR BROILED ZUCCHINI

Marinate the zucchini as for Baked Zucchini (above) for at least 30 minutes to draw water. Pat dry and coat very lightly with grapeseed oil. Grill over medium coals, turning often, until very tender and deeply browned on all sides, 10 to 12 minutes total. (Alternatively, brown under a hot broiler, turning once, for 6 to 8 minutes per side.) Serve as for Baked Zucchini.

DEEP-FRIED ZUCCHINI

Zucchini is one of the best vegetables to deep-fry. I like a very light coating rather than a heavy batter.

Slice small zucchini into ¼-inch-thick coins or 2½-inch batons. Toss with salt and place in a colander to drain for 30 minutes. Rinse and dry thoroughly. Toss with a handful of all-purpose flour lightly seasoned with fine sea salt, freshly ground black pepper, and a little cayenne. Working in small batches, deep-fry in a saucepan with 3 inches of hot vegetable oil until golden, 2 to 3 minutes. Drain on crumpled paper towels. Sprinkle with salt while hot. Serve with lemon wedges.

STUFFED ZUCCHINI

Stuffed zucchini is a crowd pleaser, especially when made with the big Mediterranean flavors of Italian sausage, feta, and pine nuts. Garden helpers Chen and Deeva tore into it with the awesome appetites of college students engaged in outdoor work. Because it is a filling dish, a meal-in-one, stuffed zucchini needs only a tossed salad on the side and Lemon-Herb Pound Cake (page 263) for dessert. The recipe is flexible and forgiving.

Use 2 pounds large zucchini, 8 to 10 inches. Blanch in boiling water for a scant 5 minutes, then drain until cool. Cut lengthwise and gently scoop out the interior with a tablespoon, leaving an intact shell to stuff. Chop the flesh and set aside in a large bowl. Brown a pound of sausage in a large skillet. Drain, discarding the fat, and add to the bowl with the chopped zucchini. In the same skillet, soften a diced onion in some extra-virgin olive oil, along with chopped garlic, and season with salt and pepper. Add it to the bowl with the zucchini and sausage. Add ½ cup pine nuts, ¼ pound crumbled feta, ¼ cup grated Parmesan, ½ cup bread crumbs, and quite a lot of chopped marjoram and chopped parsley to the bowl and toss to combine. Stuff the zucchini shells with this mixture and top with ½ cup bread crumbs tossed with 1 tablespoon olive oil. Bake in a 400°F oven until browned and crunchy, about 25 minutes. Finish with chopped parsley and a drizzle of your greenest, most peppery olive oil.

ZUCCHINI CASSEROLE

My father, who is a lazy cook, achieves superb results with this easy mixed-squash casserole when he has an abundance of summer vegetables. It is full of flavor and an eyeful at the table.

Slice 2 pounds of zucchini or mixed summer squash into ¼-inch-thick rounds or lengthwise into planks. Combine in a large bowl with ¾ pound sliced Roma tomatoes, ½ pound cubed fresh mozzarella cheese, and ¼ cup extra-virgin olive oil. Add several chopped cloves of garlic, several tablespoons of chopped parsley and chopped chives, a dozen large basil leaves, ½ teaspoon dried savory or thyme, and 1 teaspoon fine sea salt. Toss everything to combine. Spread the mixture into a lightly oiled 13-inch casserole or other baking dish. Cover with foil and bake in a 350°F oven for 30 minutes. Uncover and use a spatula to press the top layers into the juices. Continue baking uncovered until the juices are reduced and the squash very tender, about 30 minutes longer. If you like, throw a handful of grated hard cheese over the top to brown when you remove the foil.

PRIME ZUCCHINI • HERBS | STARTER | SERVES 4 TO 6 | PESCATARIAN, GLUTEN-FREE | SUMMER

ALSO FOR: OTHER SUMMER SQUASH

ZUCCHINI CARPACCIO

One year in September I took a day trip to the Morton Arboretum outside of Chicago to see the hedge collection. My morning flight landed at O'Hare at hungry o'clock, and among the terminal's dismal lunch offerings, stir-fried vegetables from an Asian-themed fast-food franchise seemed the least-bad option. Even setting the bar that low, I was amazed, after a summer of eating from the garden, by the terrible quality of the vegetables. The zucchini and broccoli were flavorless apart from a bitter edge—embalmed. No wonder some people hate vegetables.

When you bring zucchini straight in from the garden, the flavor is fresh as grass and mildly sweet. Tiny beads of sap rise from its cuts, countertop dew. The smell is indefinite but clean, like the air over a lake. Raw zucchini wants a flavorful sauce. This recipe runs parallel to vitello tonnato in the Italian repertoire. To make color an essential ingredient, I've replaced tuna with salmon—*salmonato*.

- 7½ ounces canned salmon, drained and boned
- 1 cup mayonnaise
- 2 teaspoons anchovy paste
- ⅛ teaspoon Red Boat fish sauce
- 3 tablespoons salt-cured capers, rinsed and drained
- 3 tablespoons fresh lime juice
- Fine sea salt and freshly ground black pepper
- 2 pounds (6-inch-long) zucchini
- Chopped fennel fronds or cilantro leaves, for garnish
- Niçoise olives (optional)

1. In a blender or using an immersion blender, process the salmon, mayonnaise, anchovy paste, fish sauce, capers, and lime juice until smooth. Taste and adjust the seasoning with salt and pepper.
2. Thinly slice the zucchini on a mandoline and arrange on a large serving platter. Pour a portion of the sauce in the middle and jiggle the platter to center evenly. Garnish with herbs and olives (if using). Serve the remaining sauce on the side.

Note: The carpaccio can be varied in many ways; for example, served simply with extra-virgin olive oil, lemon juice, pecorino, pine nuts, and chopped mint.

YELLOW SUMMER SQUASH • SWEET ONION • THYME | SIDE | SERVES 6 TO 8 | VEGETARIAN, GLUTEN-FREE | SUMMER

BUTTER-BRAISED CROOKNECK SQUASH

Like corn, yellow squash has a specific flavor that is difficult to describe by comparison to something else. It tastes like itself.

In this simple dish, versions of which I have eaten every summer of my life, sliced yellow squash is heated very gently in butter. It sweats and braises without browning; a too-hot pan will caramelize the juices and spoil the flavor. Sweet onion, thyme, and black pepper are the modest enhancements. A handful of chopped dill could replace thyme—an idea shared by cookbook author Tom Hudgens, mentioned elsewhere, who learned it from David Tanis, who learned it from the late Winona Holloway at Deep Springs College—a venerable legacy, indeed.

The best-tasting variety of yellow summer squash is yellow crookneck. Straight-neck yellow squash was developed for commercial growers: The club shape packs and ships better. Robert Frost said that what gets lost in translating poetry is the poetry. What was lost in straightening crookneck squash's elegant swan-neck was the flavor. Yellow zucchini is something else entirely; it tastes like zucchini minus the chlorophyll.

As for the rest of the menu, I'd serve this side dish as part of a Southern vegetable plate—maybe with a basket of fried chicken on the side. Squash also goes with many fish dishes that pair with sweet corn, such as wood-grilled mackerel, pan-seared bluefish, or scallops browned in bacon fat.

- 4 tablespoons unsalted butter
- 1 medium sweet onion, cut into ½-inch-thick slices
- 1 teaspoon thyme leaves
- 1 teaspoon fine sea salt
- ½ teaspoon freshly ground black pepper
- 2 pounds yellow summer squash, cut into ¼-inch-thick rounds

1. In a large skillet, heat the butter over medium heat until it foams. Add the onion, thyme, ½ teaspoon of the salt, and ¼ teaspoon of the pepper. Sweat until softened but not browned, 3 to 4 minutes.
2. Add the squash and the remaining ½ teaspoon of salt. Turn to coat. Reduce the heat to low, cover, and cook until the squash is very soft, 20 to 25 minutes. Turn at least once during cooking.
3. When ready to serve, sprinkle with the remaining ¼ teaspoon of pepper and carefully turn a last time. If the pan juices are too watery, remove the cover and turn the heat to high for 1 to 2 minutes. Don't step away from the pan! Shake gently and remove from the heat as soon as the juices reduce.

YELLOW SQUASH • ALLIUMS • HERBS | STARTER OR MAIN | SERVES 6 TO 8 | VEGETARIAN OPTION, GLUTEN-FREE | SUMMER

SUMMER SQUASH SOUP

This recipe builds on Butter-Braised Crookneck Squash (page 367). It turns a downhome side dish into a nice lunch by simmering yellow squash with stock, onion, and a few sliced potatoes for substance. The flavor is very pure and clear. You can tip heavy cream into each serving bowl for a richer taste and a visual flourish. Serve with a hunk of baguette, butter, and rillettes or jambon cru. To elevate the soup for dinnertime, puree a portion until thick and velvety. I'd still garnish the bowl with a pour of cream.

4 tablespoons unsalted butter
1 medium onion, diced
1 clove garlic, lightly crushed and peeled
2 or 3 sprigs thyme
1 teaspoon fine sea salt
2 pounds yellow squash, cut into ½-inch-thick slices
½ pound waxy potatoes, such as Pinto Gold, cut into ¼-inch-thick slices
1 quart chicken or vegetable stock

1. In a Dutch oven or large saucepan, melt the butter over medium heat until it foams. Add the onion, garlic, thyme, and ½ teaspoon of the salt. Sweat until the onion is translucent, about 5 minutes.
2. Add the squash and the remaining ½ teaspoon of salt. Turn to coat and stir frequently until the cut edges soften, about 5 minutes.
3. Add the potatoes. Add the stock and up to 2 cups water to nearly submerge the vegetables. Bring to a boil. Reduce to a simmer and cook, uncovered, until the vegetables are tender, about 25 minutes. Discard the garlic and thyme stems.
4. You can serve the soup clear, if you like, with or without a small pitcher of heavy cream alongside. Or to thicken the soup, ladle about half (solids and liquids) into a blender. Pulse until smooth. Return the puree to the pot and swirl to combine while reheating briefly over medium heat.

SUMMER SQUASH
BAKED IN GOAT CHEESE CUSTARD

The zucchini casserole that appears earlier in this chapter has a generally Mediterranean vibe and takes its name from the original French source, an oval baking pan, *une casserole*. This recipe is closer to an all-American casserole, the mish-mashy delight of weeknight meals and potluck suppers, meant to be cooked in a super-sized rectangular glass baking dish. It is based on my West grandmother's macaroni and cheese, which she made not with béchamel but savory egg custard. Goat cheese complements the flavor of squash, and Parmesan adds umami. The quantities here are enough for a crowd. Baked squash casserole can be served hot, at room temperature, or reheated the next day. Accompany it with almost anything, but my choice would be pork chops or pork tenderloin off the grill plus a plate of sliced Green Zebra tomatoes and a few salad greens.

8 tablespoons unsalted butter
2 medium yellow or sweet onions, cut into ½-inch-thick slices
2 cloves garlic, minced
1 teaspoon fresh thyme leaves
½ teaspoon dried savory
1 bay leaf
¼ teaspoon red chile flakes
¼ teaspoon freshly ground black pepper, plus more to taste
1½ teaspoons fine sea salt
4 pounds yellow crookneck squash, sliced in ¼-inch slices
1 cup dried bread crumbs
4 large eggs
1 cup organic heavy cream
2 ounces extra-sharp white cheddar cheese, grated (about ¾ cup)
4 ounces fresh goat cheese, crumbled (about ½ cup)
1 ounce Parmesan cheese, grated (about ⅓ cup)

1. Preheat the oven to 350°F. Coat a 9 × 13-inch baking dish with 1 tablespoon of the butter.
2. In a rondeau or very large pan, melt 5 tablespoons of the butter over medium heat until it foams. Add the onions, garlic, thyme, savory, bay leaf, chile flakes, black pepper, and ½ teaspoon of the salt. Sauté until translucent, about 5 minutes.
3. Add the squash and the remaining 1 teaspoon of salt. Increase the heat to high and sauté until softened, about 10 minutes, stirring often to prevent browning. Transfer the squash to a colander set in the sink to drain for 10 minutes.
4. Meanwhile, melt the remaining 2 tablespoons of butter. Stir in the bread crumbs to coat. Set aside.
5. In a very large bowl, whisk together the eggs, cream, cheddar, goat cheese, and Parmesan. Add the drained squash and black pepper to taste. Stir to combine.
6. Turn the mixture into the prepared baking dish and top with the bread crumbs. Transfer to the oven and bake, uncovered, until deeply browned and bubbling, 45 to 50 minutes.

SMALL WINTER SQUASH | SIDE | SERVES 8 | VEGETARIAN, GLUTEN-FREE | FALL, WINTER

BAKED HONEYPATCH SQUASH
WITH POURED CREAM

I include this "recipe" as a testimonial for Honeypatch winter squash and its predecessor, Honeynut. Everyone I serve it to marvels: What is the secret? The secret is the garden. The varietal does all the work. The other secret is cream—some secret. The technique works with familiar Butternut as well; increase the cooking time.

4 Honeypatch or Honeynut squash
½ cup organic heavy cream
Fine sea salt and freshly ground black pepper

1. Preheat the oven to 400°F. Line a sheet pan with parchment.
2. Halve the squash lengthwise and scrape out the seeds with the edge of a teaspoon. Place cut-side down on the lined sheet pan and bake until the necks are soft and the juice around the edges is foamy and browned, 35 to 40 minutes.
3. Remove the sheet pan from the oven and let rest for 3 minutes—not too long. Gently turn each piece of squash face up. Pour cream into the seed cups to fill, about 2 tablespoon each. Sprinkle with salt and freshly ground black pepper. Serve immediately before the cream is absorbed or wait a few minutes to hide the evidence.

WINTER SQUASH OR PUMPKIN • ALLIUMS • CANNED TOMATOES | MAIN OR SIDE | SERVES 8 | VEGETARIAN | FALL, WINTER

WINTER SQUASH STEW
WITH WALNUT OIL, CINNAMON, AND HONEY

One ice-gripped evening at home in New England, the phone rang and chef Wes asked me what I was doing. I said was I sizing up a Koginut squash, unsure of my next move, and he pronounced the generous words "warming North African spice." This is a dish to bring the solar flavors and hot colors of squash and tomato to the winter table.

The warming influence of spice in this recipe is subtle—and no doubt some cooks will want to double down and perhaps throw in turmeric, cloves, cardamom, and so on. But balance is the game here, triangulating among three final ingredients added as flavor garnishes: a slick of walnut oil, a bare sprinkle of ground cinnamon, and a drizzle of honey. (Single-origin Peni Miris cinnamon from Diaspora Co. spice importers is a favorite—bone dry, tense, and savory.)

I serve this meat-free stew at a party I give every fall to celebrate the chestnut harvest, making it in a rondeau the size of a washtub and serving it over rice, perhaps with the backup of long-cooked greens or garbanzo beans. The next day, I have leftovers alongside roasted fish splashed with chermoula. A condiment of salt-cured lemon would not be out of place.

3 tablespoons grapeseed or extra-virgin olive oil
1 large onion, diced
2 cloves garlic, minced
1 teaspoon dried savory or thyme
1 teaspoon coriander seeds, toasted and ground
1/2 teaspoon cumin seeds, toasted and ground
1/2 teaspoon cayenne pepper
1/2 teaspoon ground cinnamon, plus more for sprinkling
1/4 teaspoon freshly ground black pepper
1 teaspoon fine sea salt, plus more to taste
3-pound winter squash or pumpkin, peeled and cut into 1-inch chunks
2 cups canned tomatoes with juice
2 tablespoons tomato paste
2 to 3 cups vegetable stock or water
Sourwood or other raw honey
Walnut oil
Grated lemon zest or chopped preserved lemon
Fresh cilantro and mint leaves, roughly chopped

1. In a large Dutch oven, heat the grapeseed oil over medium heat until it shimmers. Add the onion, garlic, savory, coriander, cumin, cayenne, cinnamon, black pepper, and salt. Sauté until translucent, 5 to 7 minutes. Add the winter squash, stir to coat, and cook until the edges soften, 5 to 7 minutes.
2. Add the tomatoes, tomato paste, and enough stock to nearly cover the squash. Bring to a boil. Reduce the heat to maintain a steady simmer, partially cover, and cook until the squash is very soft, 45 minutes to 1 hour. Don't overstir, but shake the pot occasionally to prevent sticking. Add more stock as needed.
3. Once the squash is tender, uncover and adjust the stew's consistency, either adding more liquid or boiling off excess as needed. Taste and adjust the salt.
4. Ladle into individual serving bowls and finish by drizzling with honey and walnut oil. Sprinkle with some cinnamon. Garnish with lemon zest. Scatter with the chopped herbs.

SMALL WINTER SQUASH OR PUMPKIN SUCH AS KOGINUT, BLACK FUTSU, OR WINTER SWEET KABOCHA | MAIN | SERVES 6 TO 8 | OMNIVORE, GLUTEN-FREE | FALL, WINTER

CHICKEN AND PUMPKIN
BRAISED IN AMONTILLADO

Copper-colored Amontillado is a mouth-filling dry sherry. It was renowned during the age of sail, when fortified wines withstood long sea journeys and oxidative flavors were welcomed by the drinking public. (Edgar Allan Poe's early readers could well appreciate the fatal allure of a cask of Amontillado.) The wine tastes of toasted almonds and walnuts, bitter marmalade, and briny green olives.

The overall effect of this rustic braise belongs to the Spanish countryside. A truer version might use rabbit in place of chicken. A salad of bitter chicories and a dessert of quince paste and Manchego complete the Iberian theme. Put out rough hunks of bread to sop the alluring pan juices.

If you're lucky enough to get a chicken with giblets—as scarce these days as Amontillado—cook the liver in butter with a bit of minced shallot while the chicken and pumpkin are in the oven, then mash it with a spoonful of the braising liquid. Spread on grilled toast that has been rubbed with garlic to serve.

1 whole chicken (3½ pounds), cut into 10 pieces and salted in advance
Fine sea salt
1 small winter squash with edible skin, cut into ¾-inch segments or chunks
3 tablespoons extra-virgin olive oil
1 large yellow onion, sliced
4 cloves garlic, lightly crushed and peeled
1 cup Amontillado sherry
1 tablespoon pimentón or paprika
2 tablespoons Marcona almonds, toasted and roughly chopped

LIVER TOASTS (OPTIONAL)
1 tablespoon unsalted butter
1 small shallot, minced
1 chicken liver, trimmed
1 to 2 tablespoons braising liquid from the chicken
⅛ teaspoon ground white pepper
Fine sea salt
2 slices rustic country bread
1 clove garlic, peeled but whole

1. Preheat the oven to 375°F.
2. In a large bowl, toss the winter squash pieces with 1 tablespoon of the olive oil and ½ teaspoon salt. Set aside.
3. In a cazuela or large ovenproof skillet, heat the remaining 2 tablespoons of olive oil over medium heat until it shimmers. Add the onion, garlic, and ½ teaspoon salt. Stir occasionally until the onion sweats, 3 to 5 minutes. Push aside and add the chicken pieces, skin-side down. Brown lightly, 6 to 8 minutes. No need to overdo it.

(CONTINUED)

4. Turn skin-side up. Add the sherry. Tuck in the pumpkin pieces. Sprinkle with pimentón. Cover loosely with foil.
5. Bake for 20 minutes. Remove the foil and turn the pumpkin pieces or spoon over juices to keep them from drying out. Bake until the chicken is done, 25 to 30 minutes longer. Set aside, loosely covered, until ready to serve.
6. **If making the liver toasts:** In a small pan, melt the butter over medium heat. Add the shallots and cook for 1 minute to soften. Add the liver and brown on both sides until barely cooked through, 2 to 3 minutes per side. Remove from the heat.
7. Spoon out 1 or 2 tablespoons braising liquid from the chicken and add to the pan. Add the white pepper and a big pinch of salt. Mash into a paste. It should be highly seasoned.
8. Toast the bread and rub with the whole garlic while still hot. Cut into toast points and spread a dab of chicken liver on each.
9. To serve, reheat the chicken and pumpkin under a hot broiler for 2 minutes, or until the edges darken. Garnish with the chopped almonds. Offer the liver toasts alongside.

KOGINUT, BLACK FUTSU, OR SIMILAR • ALLIUMS • HERBS | MAIN | SERVES 10 TO 12 | VEGETARIAN OPTION | FALL, WINTER

THANKSGIVING KOGINUT
FOR CLAUDIA

With their sculpted forms and foundry colors, winter squash and pumpkins, of all garden produce, come closest to grandeur. Melons can be as large but who puts them out as decoration? The giant squash alone are magnificent. They adorn the threshold and decorate tabletops, a tribute to Lares and Penates, the Roman household gods of home and pantry. Their culinary moment arrives at Thanksgiving. Neglected too long, they collapse in on themselves like abandoned hay barns.

Tender-skinned squash, such as Koginut and Black Futsu, can be roasted and brought to the table whole, as dramatic as any bronzed fowl. Stuffing them with cornbread, chestnuts, herbs, and sausage (or mushrooms for vegetarians) gilds the pumpkin and creates an air of occasion. You carve the thing with flashing steel and serve it in steaming wedges.

The proudest compliment I received as I worked on this book came from Claudia Martin, daughter of neighbors Del and Christine, who asked her mother to make stuffed Koginut for Thanksgiving the year after I'd taken it to their family holiday. Here's the recipe, for Claudia.

- 2 (3-pound) Koginut or other blocky hollow tender-skinned squash, seeded (see Notes)
- 2 teaspoons vegetable oil
- 1/2 pound bulk breakfast sausage
- 8 tablespoons (1 stick/4 ounces) unsalted butter
- 2 medium onions, diced
- 2 celery stalks, diced
- 2 cloves garlic, minced
- 1 tablespoon chopped fresh thyme
- 1 teaspoon dried savory
- 2 teaspoons fine sea salt
- 12 ounces day-old cornbread, cut into 1-inch cubes (about 6 cups)
- 3 tablespoons minced fresh parsley
- 1 tablespoon chopped fresh sage leaves
- 1/4 teaspoon cayenne pepper
- 1/2 teaspoon freshly ground black pepper
- 1/3 cup grated Parmigiano-Reggiano cheese
- 1/2 cup strong stock, plus more as needed
- Cranberry sauce, for serving

1. Preheat the oven to 350°F. Line a sheet pan with parchment.
2. Coat each squash with 1 teaspoon of oil.
3. Crumble the sausage in a large skillet and cook over high heat until browned, 8 to 10 minutes. Drain in a colander, reserving the fat to fry eggs over the long weekend. Transfer the sausage to a large bowl. Wipe out the skillet.

(CONTINUED)

4. Using the same skillet, melt 4 tablespoons of the butter over medium heat until it foams. Add the onions, celery, garlic, thyme, savory, and 1 teaspoon of the salt. Sauté until softened but not browned, about 10 minutes. Turn into the bowl with the sausage.
5. Melt the remaining 4 tablespoons butter on the stovetop in a small saucepan.
6. Add the cornbread to the sautéed vegetables, along with ½ teaspoon of the salt, the sausage, parsley, sage, cayenne, black pepper, and Parmigiano. Drizzle with the melted butter and sprinkle with the stock. Fold to combine. The consistency should be moist but not sticky. If dry, sprinkle with more stock. Taste and add the remaining ½ teaspoon of salt, as needed.
7. Spoon the mixture into the squashes. With each spoonful, bump the squash against the counter to level the stuffing or, at most, gently settle it in place with the spoon, but do not press or pack tightly. Put on the lids and pin in place with a skewer or toothpick. Insert a meat thermometer into the center of one squash. Place on a the lined sheet pan.
8. Bake, uncovered, for 45 minutes. Check to ensure the squash aren't browning too quickly. Continue baking until the internal temperature reaches 185°F and the squash feels fully tender when poked with the handle of a wooden spoon, 20 to 25 minutes longer.
9. To serve, remove the skewers and carry to the table on a cutting board or platter decorated with autumn leaves. Set aside the top and slice the squash as if it were a cake, following the natural seams. Sauce with cranberries.

Notes: Clean the squash for stuffing as you would hollow out a pumpkin for a jack-'o-lanterns. Carefully cut around the stem and remove it like a lid, then scrape out the seeds and fibers with the edge of a spoon.

For a vegetarian option, substitute ½ pound of sliced fresh mushrooms for the sausage. Cook the mushrooms in a dry skillet over medium heat for 5 minutes, shaking or turning frequently, to purge water. Add 2 tablespoons unsalted butter, ¼ teaspoon fine sea salt, and a few thyme branches. Continue cooking until the mushrooms are shriveled and juicy, about 8 minutes, turning frequently. Transfer to a large mixing bowl. Proceed with the recipe.

CH.

23

TENDER GREENS

THE SALAD BOWL COHORT

Asteracea

THE COMPOSITE FAMILY, ALSO KNOWN AS THE ASTER, DAISY, OR SUNFLOWER FAMILY

Chicorium spp.
VARIOUS ENDIVES, INCLUDING FRISÉE, ESCAROLE, RADICCHIO, ETC.

Lactuca sativa
LETTUCE

L. sativa var. *capitata*
ICEBERG

L. sativa var. *crispa*
LOOSELEAF

L. sativa var. *longfolia*
ROMAINE

Taraxacum officinale
DANDELION

Brassicaceae

THE CABBAGE FAMILY

Brassica rapa var. *japonica*
MIZUNA

Diplotaxis erucoides and D. tenuifolia
WILD ARUGULA, ROCKET, SYLVETTA

Eruca vesicaria
ARUGULA

Lapidium sativum
GARDEN CRESS

Nasturtium officinale
WATERCRESS

MAY 15: Clockwise from top: Red Romaine, Salad Bowl, and Deer Tongue

Amaranthaceae

THE AMARANTH FAMILY, ALTHOUGH SOME GARDEN SPECIES ARE PERHAPS BETTER KNOWN BY ***Chenopodiaceae,*** THE GOOSEFOOT SUB-FAMILY

Spinacia oleracea
SPINACH

Caprifoliaceae

THE HONEYSUCKLE FAMILY

Valerianella locusta
MÂCHE OR CORN SALAD

Portulacaceae

THE PURSLANE FAMILY, MORE RECENTLY ASSIGNED TO ***Montiaceae***

Claytonia perfoliate
MINER'S LETTUCE

Portulaca oleracea
PURSLANE, VERDOLAGAS

In garden terms, tender sometimes means the opposite of cold hardy. Here it means the opposite of tough, as in tender mercies. Arugula is as hardy as kale but less hearty. Most tender greens do best in the shoulder seasons, spring and fall. They suffer in oppressive heat, as every tender spirit does.

Lettuce, the stalwart of the salad bowl, likes to grow no matter how modest the space and it gets by with inexpert care. Lettuce also illustrates the concept of garden biodiversity. Seed catalogues since the eighteenth century have celebrated the many shapes, textures, and flavors of lettuce. Very few of its many guises are ever seen in the supermarket.

"Lettuce has an ancient lineage," writes Peter Hatch, the former director of gardens and grounds at Monticello and an authority on the early American garden and its antecedents. More to the point, he continues, *salad* has an ancient lineage. Egyptians, Greeks, and Romans ate lettuce "dressed raw with oil and vinegar," Hatch writes. The habit remained current into Jefferson's day—he set aside a part of his garden for "raw sallads" to distinguish those crops from stewing greens, perhaps a surprise to anyone who thinks Southern vegetables were always cooked to mush. The salad cult continues until now at such restaurant chains as Sweetgreens.

As mentioned in an earlier chapter, Jefferson instructed his gardeners to sow a thimbleful of lettuce seed every Monday from February through September. (A thimble holds 1 gram of lettuce seed, enough for 25 feet.) The gardens at Monticello also gave space to what Jefferson called "small-salading," diminutive greens such as cress, mâche, arugula, and nasturtium, which is a flowering annual vine grown for its peppery round leaves and spicy seedpods pickled for homegrown "capers."

I take a similarly ecumenical approach to the salad bowl. Lettuces and the minor greens vie for space in my salad patch with a few salad-y herbs, such as chervil and parsley, and salad-friendly radishes, which mature at the fast pace of small greens.

Lettuce can be sorted into several broad categories. To begin alphabetically, Batavian or summer crisp is heat tolerant. Tender Bibb or butterhead, a longtime favorite for its delicate texture and mild flavor, is also known as Boston after a popular variety that came to represent the generic type—the Kleenex of butterhead lettuce. (Limestone is another synonym.) Then there is the blandly textural iceberg, first introduced to the public in 1894 as crisphead. Shaggy loose-leaf lettuce is nonheading; my grandfather seeded it in tobacco beds and my grandmother "killed it" with a dressing of hot bacon fat and cider vinegar, a seasonal delicacy. Additional types include lollo, as in Lollo Rosso, a frizzy redhead. Oakleaf has two colors, red and green. Red leaf is as the name describes. Lettuce geeks dispute how to categorize some varieties. The pointy New England heirloom known as Deer tongue, or Matchless, is described variously as bibb, loose-leaf, or oakleaf.

Heat-tolerant romaine, or cos, is another complete category. Both names derive from putative origins, either in sunny Rome (*laitue romaine* in French) or on the balmy Greek isle of Cos. The chef's favorite Little Gem, or Sugar Cos, is usually considered a compact romaine, although some seed houses class it with the Bibb lettuces.

The newest lettuce category, introduced by Dutch plant breeder Rijk Zwaan, is sometimes described as "one-cut" and at other times by the registered trade name Salanova. Its showy rosette of same-sized leaves looks fabulous on display in a farmers' market.

Finally, the unusual celtuce, also called sword lettuce or stem lettuce, is grown frequently in Taiwan but rarely elsewhere. It is a cooking vegetable, a reminder that head lettuce and romaine can also be braised or grilled. Celtuce is esteemed for its thickened stalk more than its lanceolate leaves.

A chart on page 388 breaks down the types.

The wider cohort of tender greens receives similar care in the garden, with modest adjustments

MAY 5: Red romaine in the cold frame, with Green Salad Bowl in the foreground and Lollo Bionda

MAY 5: Spinach that overwintered in a greenhouse behind a roadside farm stand in Vermont. Notice how larger outside leaves have been picked, leaving the central rosette intact.

for each species' mature size and temperature tolerance. Mâche is 2 inches wide and survives snow cover—sow in the fall and sow thickly. The cut-leaf endive known as frisée, perhaps my favorite green to grow, needs space and mild spring or early fall temperatures: It can grow to a foot wide. As is often the case, varietal names hold clues. The 1855 French heirloom Brune d'Hiver, "winter brunette," is cold hardy, while Parris Island romaine, developed in 1952 on a muggy South Carolina sea island, tolerates heat and humidity.

Botanically, the salad patch encompasses several families. Lettuce is a composite, kin to asters, as anyone can plainly see when it blooms. Arugula and cress are brassicas; they sparkle with mustard-like edible blossoms. Spinach is a goosefoot, and mâche a distant relative to honeysuckle, of all things. I also make room for oddities, such as salad burnet, an Elizabethan potherb that smells like cucumber.

In addition, almost any of the hearty greens from chapter 17 can be gathered small for salad. Sow more thickly than recommended and pick at the baby stage. Multicolored chard, ruffled Siberian kale, frilly mustard, tatsoi, and mizuna are well suited to early harvest. Spinach can go either way, tender or hearty. Some smooth-leaf spinach varieties reach baby size in 20 days, while the heavily savoyed varieties such as Bloomsdale take 7 weeks to mature as a clump of thick, chard-like leaves that won't melt away in a hot pan; they make a chewy mouthful when raw.

In the kitchen as well as the garden, tender greens encourage a mix-and-match approach. "Mixed salad," one says, or *mesclun* for an air of sophistication.

Tender greens adapt well to containers and small gardens. They emerge quickly and mature before impatience sets in. Many are suited to repeat harvests: Pick the larger outside leaves and let the smaller interior ones grow. Or if you gather the full head—the ruffled wrapper leaves around a paler, folded center—you'll have time to replant a second crop. A single packet of seeds will give you enough to experiment. Sow them liberally, wherever there's a patch of bare soil.

The habit of starting dinner with a trip outside to fill your salad bowl is a touchstone pleasure of cooking from the garden.

In the Garden

Lettuce can be gathered as a loose-leaf mix of baby greens or left to mature into full heads. For the former, you sow seeds practically on top of each other, or else broadcast thickly over a prepared bed, then harvest with scissors, a seedling haircut. For the latter, plant in rows and thin successively as the heads fill in. The same options apply to arugula, mizuna, cress, and so on. I used to buy the tiniest greens I could find at the farmers' market but now, as a gardener, I let my salad greens grow up a bit. I enjoy watching the plants fill out, even letting a few pop off for the sake of pollinators—and for the sake of life's unfolding. Cleopatra looked back at the "salad days" of her youth as a time when she was "green" in judgment and "cold" in blood, adolescent in thought and body, an unformed girl. Arugula blossoms, lettuce puffballs, and riotous bolted spinach show plants in their maturity, as they achieve their telos, or Aristotelian purpose. Remember: Plants belongs to nature, not the gardener.

Tender greens do best in the cool days of spring and fall. Lettuce may fail to germinate above 75°F.

JULY 3: Très Fine Maraîchère Olesh curly endive, better known as frisée

The salad-bowl greens are best grown out quickly in rich soil. They are light feeders. You can incorporate them into various rotations and/or tuck them in among other crops. Lettuce is well suited to tight spaces, containers, and raised beds.

All the tender greens have delicate, moist leaves and require regular water. Heat-stress and drought will cause them to bolt in a desperate attempt to set seed before withering. Extend the season by sowing late successions in partial shade, such as beneath a bean tripod or in tall corn's shadow. Romaine and Batavian types are the most heat-tolerant lettuces.

The tender greens are also relatively untroubled by disease and pests, although slugs and snails will find them, and herbivores of all sizes will take a bite, from grasshoppers to groundhogs to moose. Know that arugula will attract flea beetles. Organic growers sow arugula under floating row cover for protection, or else plant late in the season to desynchronize from pest life cycles. Or you learn to tolerate perforated leaves. Of the three approaches, the last is most dependable.

Because tender greens grow fast, they also pass quickly, life in a single generative burst. The key to keeping your salad bowl full is succession plantings. Sow less than you think you need at any one time, but sow more often. Instead of putting out a 10-foot row all at once—20 heads that will mature at the same time—plan 4 weekly sowings.

By the same token, I'll sprinkle seeds throughout the garden wherever I find bare ground to fill. If a row of peas comes up short, I'll finish it with 3 feet of arugula—enough for a few salads. Or I'll stick nasturtium seeds at either end of my tomato trellis. In early spring, I'll scatter cress on a 4-foot square patch where later I'll plant summer squash. Or I'll sow mâche in the tidy bed left behind after harvesting garlic. Tender greens are also suited to interplanting between larger crops. Spinach sown between cabbage will be up and gone in a flash.

For baby greens, sow thickly. You could plant in rows of 4 to 6 seeds per inch, sow a 2- to 4-inch band, or broadcast over a bed, estimating 60 seeds per square foot. Cover seeds with ⅛ inch soil and gently pat down. Harvest when leaves are 3 to 4 inches tall by cutting with scissors 1 inch above the growing point. The roots will send up a second flush of leaves. Quality declines with the third flush.

A final reminder that tender greens usually don't hold well in the garden or the kitchen. Plan for the changeable abundance of variety rather than the all-at-once overload of monocropping.

See the following chart for recommended lettuce varieties.

RECOMMENDED LETTUCE VARIETIES BY TYPE

TYPE	GARDEN NOTES	KITCHEN NOTES	TRAITS	VARIETIES
Batavian or summer crisp	Large and sturdy, forming tightly packed heads; slow growing	Ruffled leaves; crisp; not bitter	Heat tolerant; germinates at up to 80°F	• Muir, the most heat tolerant • Sierra, a red-tinged 1992 French variety
Bibb	Forms small bunched heads with ruffled outer leaves. Often grown for mini-heads.	Tender, sweet, delicate	Matures early; easily torn; leaves bruise	• Buttercrunch, green • Brune d'Hiver, bronze-tipped heirloom • Deer tongue, aka Matchless, a pointy American heirloom
Butterhead	Like Bibb, but larger heads; ruffled outer leaves cover a folded heart	Tender, sweet, buttery; green or red	Delicate	• Boston, the group's heirloom namesake • Nancy, an improved Boston type • Pirat, burgundy tipped • Tennis Ball, parent of Boston, a low-care heirloom favorite
Iceberg	Forms a large "closed" or cabbage-like head with wavy wrapper leaves; slow growing	Crisp! Mild; pale interior	Can be fussy to grow	• Crispino, the standard
Lollo	Forms large loose bunches rather than heads	Deep color; frizzy leaves	Can be bitter when large	• Dark Red Lollo Rossa, red • Ilema, green
Looseleaf	Nonheading	Green or red; crinkled, pebbled, or ruffled; juicy	First to mature; vitamin- and mineral-rich; less heat tolerant	• Black Seeded Simpson, a popular heirloom
Oak leaf	Good for cutting at baby stage; forms loose heads	Attractive red or green, deeply lobed leaves	Beautiful in the garden	• Italienischer, green, shapely, robust • Red oak leaf
One Cut	Forms a dense rosette of wavy leaves around a central core	Head separates into tidy, uniform leaves	The Next Big Thing in commercial lettuces	• Salanova varieties
Romaine	Tall, upright, head-forming; baby leaf, mini-head, and full size varieties; single harvest or multiple harvests	Crisp, sturdy, sweet; green, red, and speckled	More heat tolerant than most lettuce	• Little Gem, chef's favorite for romaine crunch, cute size • Parris Island, the green standard • Spotted Trout, speckled German heirloom. My favorite. • Truchas, deep red leaves

TENDER GREENS AND HEARTY GREENS, *BABY SIZE*—DIRECT SOWN

DIRECT SOW	DEPTH	SPACING	ROW SPACING	TIME TO HARVEST	SUCCESSION PLANTING	HARVEST SIZE
Early spring	1/8–1/2 inch	4–5 per inch, or thickly in 2- to 4-inch bands, or 60 per square foot in beds	2 inches	20–30 days, depending on type	Every 2 weeks until hot weather; again in late summer	3–4 inches

Note: Cover with 1/8 inch soil for smaller seeds (lettuce, arugula), 1/4 inch for medium (kale, mustard, mizuna), and 1/2 inch for larger (spinach). Sow chard and beets at 2 seeds per inch. Harvest greens by cutting with scissors 1 inch above growing point. Leave roots undisturbed for a second and third flush.

TENDER GREENS, *FULL SIZE*—DIRECT SOWN

TYPE	DIRECT SOW	SEED SPACING	DEPTH	THIN TO	ROW SPACING	EMERGENCE	TIME TO MATURITY	MATURE SIZE
Arugula	early	1 inch	1/4 inch	6 inches	12 inches	4–8 days	40 days (20 days for baby)	12 inches tall
Cress, garden	early	2 inches	1/4 inch	4 inches	6 inches	7–14 days	45 days	8 inches (pick small)
Endive (frisée)	early	4 inches	1/8 inch	12 inches	12 inches	7–14 days	50–65 days	10 inches tall, 12 inches wide
Mâche	early or fall	1 inch	1/4 inch	4 inches	12 inches	10–14 days	55 days	3 inches tall
Lettuce	4 weeks before last frost	3 seeds every 10 inches	1/8 inch	1 seedling per 10–12 inches	12–18 inches	5–10 days	50–60 days, depending on type (25–30 days for baby)	8–12 inches tall, 12 inches wide depending on type
Spinach	6 weeks before average last frost	1 inch	1/2 inch	6 inches	12 inches	5–10 days	35–40 days (25 days for baby)	8 inches tall

Note: Eat the thinnings.

In the Kitchen

Don't harvest lettuce or other tender greens until you're ready for them. In warm weather, gather as early in the day as possible. The leaves will be turgid with moisture drawn up through the roots overnight. Well-watered soil means crisp salad.

As already mentioned, there are three ways to harvest tender greens: Cut a swath of baby leaves with scissors (the haircut method), selectively pick larger outer leaves (repeat harvest), or slice the head/bunch/rosette from the roots at ground level with a sharp knife (single harvest). For the last, trim the coarse outer wrapper leaves while still in the garden and toss on the compost pile.

Immediately plunge harvested greens in cold water to "remove the field heat" as professional growers say. Ice water would not be excessive. I asked Jan at Mill River Farm, which grows acres of salad greens, the ideal water temperature for washing greens and she said, "33°F." Swish the greens gently to flush out bugs and coarse organic debris. Trim or sort out ragged or yellowed leaves.

At this point your greens will look like what you buy at a farmers' market—tidy but not table-ready. To store, roll heads or loose leaves in a kitchen towel like a burrito. Wrap in waterproof beeswax sheets (or tuck in a plastic bag). In prime condition, tender greens will keep in the refrigerator for 3 to 5 days.

When ready to use, I'll wash salad greens a third time. I'm always surprised how much they need it. Pull apart heads or separate leaves from the basal rosette. Plunge one handful at a time into a large bowl of water. Briskly stir the water into a whirlpool and watch as the leaves swim through it. Specks of debris and bewildered aphids will rise to the top, sand and grit will swirl at the bottom, and small worms will tumble through the current like fish in a waterspout. Change the water after each handful of leaves.

Tear larger leaves into manageable pieces while respecting the original shape. There is no truth to the kitchen myth that cut edges brown more quickly than torn edges. Cuts and tears both oxidize if held too long before serving.

Before dressing salad, always thoroughly dry the greens. For twenty years I used an antique salad spinner that looked like a loosely woven wire basket with long handles. You fill the basket, step outside, and furiously whirl the thing 360 degrees like Pete Townshend on his guitar. It's an arm stretcher. Then while getting equipped for the arrival of garden helpers Chen and Deeva, I bought an OXO salad spinner, a grudging exception in a years-long campaign to de-plasticize my house. The thing is a marvel, essential equipment for anyone making salad regularly.

Scatter spun greens on a clean kitchen towel to air-dry before dressing. Or else roll them in a kitchen towel like a jelly roll and store in the refrigerator until ready to use.

In the recipes ahead, I give examples of tossed salads and composed salads. Tossed salads are mostly what I make at home. You put everything in a big bowl and dress at once. (See Tossed Salad, page 398.) A composed salad (see page 401) is for company—you dress the elements separately and arrange them on a serving platter. A composed salad can also become an attractive light meal on its own.

Tender greens should be dressed at the last moment before serving. To speak of minor details, I like salad at room temperature, not direct from the refrigerator. My kitchen collaborator Dash likes to serve salad on chilled plates, an old-fashioned elegance.

Don't forget that tender greens also can be cooked. Braised lettuce is a delicate antiquarian dish as evocative of May as buttered asparagus. Sturdier lettuce such as romaine can be grilled, and the bitter chicories—endive and radicchio—love fierce heat. Scorch them mercilessly, advises David Tanis, ideally over a wood fire. Serve grilled lettuces and chicories with a strong sauce of chopped herbs and anchovies. Consult the recipe for Grilled Romaine on page 404.

Or add a handful of young peppery greens such as arugula or cress into your favorite pasta.

Either throw them in the pot with your pasta to blanch for a few seconds before you drain off the cooking water, or simply toss them with the pasta and sauce before serving. The greens will wilt in the heat. Do the same with shredded radicchio.

Finally, why not add overlarge or surplus tender greens to a pot of braised mixed greens (see Young Greens Braised with Spring Onions, page 225). They won't take much cooking time, on par with chard, so add them toward the end.

APRIL 14: In the Berkshires, forsythia blooms 4 to 5 weeks before average last frost—an indicator to plant lettuces and other salad bowl greens.

TENDER GREENS • HERBS • ALLIUMS • MOST ANYTHING ELSE

STARTER, MAIN, SIDE

YIELD VARIES

VEGETARIAN, VEGAN, AND GLUTEN-FREE OPTIONS

SPRING, SUMMER, FALL

SALAD FROM THE GARDEN

You already know how to make a salad, and this isn't a recipe. It's a strategy, or a point of view. When you have a garden, the particulars of your salad bowl change every year, every season, every week. A garden is flux. Salad from the garden requires a flexible approach.

The word *salad* suggests to me a jubilee of ingredients, most of them raw, each chosen to add its specific color, texture, or flavor, and all of them united by a dressing that balances succulence (oil) with vivacity (vinegar, lemon juice) and, sometimes, fatty opulence (dairy). There is a nearly infinite range of expression within those broad parameters, and salad can be very simple. Consider the classic insalata tricolore of chopped arugula, radicchio, and endive dressed with oil and vinegar.

But if a salad becomes too simple, it ceases to be a salad. I'll put a bowl of naked greens on the table and pinch from it during dinner, dropping a few leaves on my plate and pushing them around in whatever juices remain, a plate dressing. (I think it's fine to use your fingers because other foods that are clean to the touch may be picked up—oysters on the half shell, toasted almonds, a hard-boiled egg, an apple, bread, poached asparagus.) Likewise, a handful of plain arugula riding on a platter of sliced roasted pork or a tangle of frisée alongside a roasted chicken are welcome additions to the table. I wouldn't call any of them a salad.

A salad is dressed before serving and, more to the point, it is salted. The words salad, *ensalada, salata, salade, insalata, Salat, Sałatka,* and *салат* all derive from the Latin *sal*. The primacy of salt, I think, tells us something essential: A salad is not meager fare but a fine dish worthy of the once-rare mineral commodity used to pay Roman soldiers, hence *salary*.

Salad is a delicacy, an ornament to the meal, both less and more than essential nourishment. Salad splinters a crowd. There are food-as-fuel types for whom frou-frou salad is not "real food." The health-obsessed who take it for "fiber." The lunch-table aesthete who craves sensory intrigue and luxury without an obligation to eat much—the lobster salad set. And the Francophile who serves it after the main and before cheese, an elegant move.

The cook-gardener's salad is a sampling of what's growing that day. It brings the garden to the table more directly than any other dish, like how an oyster brings you the sea.

Salad from the garden comes together quickly as an add-on to any meal. The results are both studied and ad hoc. Salad expresses a fixed idea—freshness, seasonality—but is freewheeling in its particulars. It is a daily practice, always new. Salad is the whole that is greater than its parts, the artist's assemblage, a meaningful arrangement of otherwise minor objects. Salad is a plan launched into action, seeds becoming cuisine.

Variety is key. Too much of one thing and a salad is tiresome, like a dullard who weighs down the dinner conversation. But too many disparate things thrown together create a dismal effect, like a pilfered refrigerator. A successful salad is a reunion rather than a jumble. It somehow evokes the spirit of a vegetable garden, the joyful air of increase, in the same way that a wildflower bouquet captures the flush of a meadow at its living peak. A salad amounts to a sketch of nature,

a portrait of the season. As modest as it may be, salad expresses the maker's creativity: what you do with what you've got in a moment of controlled spontaneity. Salad can be punk, thrashing all the rules.

A tip: Make the vinaigrette first, before you prep the other ingredients, because it needs to rest at two points.

Another tip: Make salad in a very large mixing bowl. It will toss easily if the ingredients fill it only halfway. Also, dress the salad in the kitchen to avoid flinging drippy greens all over the tablecloth and to hide the fact that you toss it with your hands, by far the best tools for the task. Transfer the dressed salad into a handsome serving bowl and fluff it into a mound. At the table, make a show of "tossing the salad" with a pair of salad tongs, but only as a ritual gesture, a tiny spectacle to draw attention to the table's common focus.

KEY ELEMENTS

Vinaigrette—1/2 tablespoon per person, plus 1 tablespoon for the bowl. (See recipe, page 397.)

Mild greens—a handful per person, plus an extra handful for the bowl

- Include 2 or 3 types of lettuce: green, red, romaine, oak leaf, ruffled, frilly, and so on, ideally all of a similar size

Flavorful greens—a small handful per person—spicy, bitter, tangy, and chlorophyll-rich dark greens, one or several, such as:

- *Spicy:* arugula, cress, radish greens, baby mustard, mizuna, and/or nasturtium leaves
- *Bitter:* frisée, dandelions, radicchio, endive
- *Tangy:* sorrel, purslane, wood sorrel (oxalis)
- *Dark:* spinach, mâche, baby chard, baby lacinato kale, baby beet greens

Herbs—a large pinch or more per person, including any combination of:

- Picked whole leaves: Parsley, chervil, cilantro, tarragon, lovage, dill, mint, basil, shiso, etc.
- Chopped carrot tops
- Whole celery leaves

Alliums—a pinch per person, such as chive batons or scallion rounds

Crisp or firm ingredient—a handful for the entire salad

- Asparagus tips, blanched in salted water and shocked
- Peas or snap peas, blanched and shocked
- Celery chevrons
- Apple slices (Cameo, Cortland, Empire, and Ginger Gold are nonbrowning)

Colorful ingredients, one or two—a handful for the entire salad

- Carrot ribbons
- Radish coins
- Small beets, roasted or boiled, peeled, and sliced

Fine sea salt and freshly ground black pepper

Flowers—a sprinkling over the top

- *Whole flowers:* nasturtium, borage, arugula, violet, or pansy blossoms
- *Flower petals:* dandelion, marigold, or rose petals, pulled from the blossom
- *Herb blossoms:* chive blossoms (separated into bells), thyme, savory, marjoram

OPTIONAL ELEMENTS

Protein—a handful or two for the entire salad

- Cooked chickpeas or white beans, at room temperature
- Grilled or poached salmon or chicken, sliced
- Grilled steak, sliced

Juicy or soft ingredient—a handful for the entire salad

- Tomato chunks or cherry tomatoes, halved
- Grilled asparagus or other grilled vegetables
- Pears, cherries, figs, other fresh fruit pieces
- Dried fruit
- Cold roasted fingerling potatoes, sliced, or cold boiled new potatoes, quartered

Cheese crumbles—a small handful for the entire salad

- Fresh goat cheese
- Feta
- Blue cheese

Crunchy, crackly, savory ingredients—a sprinkling

- Flaky finishing salt
- Croutons
- Toasted nuts
- Toasted seeds
- Za'atar, dukkah, other spice blends

1. Make the vinaigrette (recipe follows). Set aside.
2. Add the following to a very large bowl:

   Mild greens + flavorful greens + herbs + alliums

   Using your clean hands, toss until combined. Scoop from the bottom, lifting and turning. Adjust the mixture, adding more greens, herbs, and alliums until the elements are balanced to your liking. Be bold with the herbs.
3. Now add the following to the bowl:

   Crisp or firm ingredient/s + protein (if using)

   Toss to combine.
4. Drizzle half the vinaigrette over top and toss to coat.
5. Add the following ingredients:

   Colorful ingredient/s + juicy or soft ingredient/s (if using)

   Drizzle with half of the remaining vinaigrette. (You'll have used three-quarters of the original quantity.) Gently toss two or three times, only enough to distribute.
6. Add the cheese crumbles (if using) and toss a time or two.
7. Taste the salad, trying different elements alone and in combination. Sprinkle with the remaining vinaigrette, if needed. Season with fine sea salt and freshly ground black pepper to taste. Add the following elements:

   Flowers + crunchy, crackly, savory ingredients (if using)

   Give it one final toss to distribute.
8. Turn the salad into a serving bowl and fluff it into a mound. Salad is best eaten right away, before salt wilts the greens. Although limp salad at the end of the night is still tasty in its way, salad doesn't keep well until the following day.

VINAIGRETTE

MAKES ½ CUP

This recipe makes enough to lightly dress a large salad—the amount I'd serve for 8 people, or maybe 10, depending on what else is on the table. If your salad includes a lot of protein or absorbent ingredients like roasted potatoes, the dressing won't stretch as far. I learned in France that vinaigrette is elevated by a spoonful of meat juices, so I look around while cooking for a source—the dripping from a roast chicken or the beautiful juices that gather in a platter of sliced pork. Whisk them in at the end.

Experiment to know which olive oil and vinegar you like best. I keep a bottle of mild, fine Flor de Aceite extra-virgin olive oil from Spain's Nuñez de Prado. I usually combine two vinegars, a base red wine vinegar from Badia a Coltibuono, plus a few drops of expensive sherry vinegar for flavor. Vary the acid to suit the salad. A simple salad of young spring greens might be best with mild champagne vinegar, apple cider vinegar, or lemon juice.

Scale up this recipe as needed or cut it in half. Vinaigrette can be made a day ahead.

1 shallot, diced
2 tablespoons red wine vinegar
¼ teaspoon fine sea salt
1 teaspoon sherry vinegar
¼ teaspoon Dijon mustard
⅛ teaspoon fish sauce (optional)
Freshly ground black pepper
⅓ cup best extra-virgin olive oil
A small clove garlic, lightly crushed
Pinch of fresh thyme, savory, or tarragon
Pinch of dried thyme, savory, or oregano
1 tablespoon meat juices, if handy

1. In a small dish, cover the shallot with the red wine vinegar. Add ⅛ teaspoon of the salt. Macerate for 15 minutes.
2. Pour the shallot vinegar into a small bowl, reserving the shallots. To the shallot vinegar, add the remaining ⅛ teaspoon of salt, the sherry vinegar, mustard, the fish sauce, if using, and several turns of freshly ground black pepper. Whisk until blended. Stream in the olive oil, whisking all the while.
3. Stir in the reserved shallots, garlic, fresh and dried herbs, and optional meat juices, if using. Taste and adjust the seasonings and balance as needed. Set aside for at least 15 minutes for the flavors to blend.

ARUGULA • CHERVIL | STARTER, SIDE | SERVES 8 | VEGETARIAN, GLUTEN-FREE | START OF SUMMER

TOSSED SALAD

ARUGULA WITH CHERVIL, CHERRIES, AND BLUE CHEESE

When my kitchen collaborator Dash came for a late-June visit, the arugula had gone looney with the gorgeous weather. It looked like a weed patch, bolted waist high and jittery with frenzied bees. We snapped off each stem at the base, shook out the bees, and kept at it until we had a rangy bouquet. Underfoot, the chervil was also blooming. I left a few to go to seed, mother plants for the fall crop, but cut the rest to the ground mercilessly. Hot weather reaches the Berkshires by July 4; the chervil was doomed. Better to clear space for cilantro.

Back in the kitchen we had Hudson Valley cherries and blue cheese. The associative logic of our salad, which turned into the recipe below, went something like this: anise-scented chervil gets along with cherries, which get along with blue cheese, which gets along with arugula, as everything does. A perfect recipe has three ingredients, but we allowed for four, because of the chervil. In its abundance, chervil might be counted a ferny salad green rather than an herb. The simple vinaigrette of sherry vinegar includes some lemon juice to flatter the cherries, fruit on fruit. Then a scattering of salt, freshly ground black pepper—and *basta*. On to the more complicated parts of the meal.

Or not. A well-made salad can have the satisfying *completeness* of a more elaborate dish. Serve this one alongside pink slices of duck breast and a stack of grilled bread rubbed with new garlic for a complete start-of-summer supper.

1 shallot, diced
2 tablespoons sherry vinegar
1 tablespoon fresh lemon juice
1 tablespoon honey
1/2 teaspoon fine sea salt
1/2 cup extra-virgin olive oil
3 quarts lightly packed arugula leaves and blossoms
2 cups lightly packed chervil sprigs
1 pound sweet cherries, pitted
4 ounces Stilton or other blue cheese, crumbled
Freshly ground black pepper

1. In a small bowl, combine the shallot, vinegar, lemon juice, honey, and salt and whisk until the honey dissolves. Macerate for 15 minutes.
2. Whisk in the olive oil. If the pitted cherries have released any juices, whisk them in.
3. In a very large bowl, combine the arugula, chervil, and cherries. Toss to combine. Dress with half the vinaigrette and toss until coated. Add the blue cheese, several turns of freshly ground black pepper, and the remaining vinaigrette, as needed. Toss a time or two to combine. Turn into a serving bowl and fluff into a heap.

ARUGULA, CRESS, OR MIXED GREEN • PEA PODS • HERBS | STARTER, LIGHT MAIN | SERVES 4 AS A MAIN | VEGETARIAN, GLUTEN-FREE | SPRING

COMPOSED SALAD

PEPPERY GREENS WITH GRAPEFRUIT AND SMOKED TROUT

The composed salad comes down to us from a more composed era. I think of yesteryear's elegant luncheon menus and the casual sexism of the term "ladies who lunch." Julia Child taught America how to make *salade composée* in 1961. The most famous of the lot, Salade Niçoise, required what must have seemed to the tossed-salad homemaker like a ton of work. Child separately dressed Bibb lettuce, tomatoes, blanched green beans, and potatoes before "interspersing them with a design of tuna chunks, olives, eggs, and anchovies." The result qualified as a main-course summer salad in her view. And who would dispute Julia Child? To this day, composed salads appear on a thousand menus, sometimes bulked up with manly red meat and rebranded as steak salad.

This recipe is an inland cousin of Niçoise salad. The tuna is replaced by smoked trout from the Hudson Valley. Instead of summery green beans, I use spring snow peas, a flattened sugar pea with edible hulls. Green, yellow, and purple snow peas can be interplanted. Snow peas mature in about 60 days, as peppery greens reach their prime. Garden arugula, *Eruca sativa,* the most common spicy green, is lance shaped, while the weedy wild type, *Diplotaxis erucoides,* is deeply cut, almost frilly. Other flavorful greens to grow for this salad include cress. (Several types are offered by Southern Exposure Seed Exchange.) If you have access to wild watercress—lucky you.

Each element of this salad could and should change according to the month. Replace sugar peas with multicolored filet beans, which also come in green, yellow, and purple. (See the photo on page 119.) As soon as new potatoes come in, add them to the arrangement, either roasted or boiled. Summer-blooming nasturtium produces beautiful round leaves that are almost too spicy on their own but could be balanced with juicy, sweet, heat-tolerant Little Gem. And so on.

1 small red onion
3 grapefruits
Vinaigrette (page 397)
½ pound snow peas, blanched and shocked in an ice-water bath
8 ounces smoked trout, in large pieces
8 cups lightly packed mixed small lettuces, arugula, cress, and other tender greens, washed and dried
1 cup lightly packed plucked fennel fronds or other leafy herb, washed and dried
Fine sea salt and freshly ground black pepper
¼ cup Niçoise olives, pitted
2 tablespoons sunflower seeds, toasted

(CONTINUED)

1. Thinly slice the red onion, cover with iced water, and set aside for 15 minutes.
2. Peel and supreme the grapefruits. Squeeze the scraps into a small bowl. Measure out ¼ cup of the juice and whisk into the Vinaigrette.
3. Slice the cooked snow peas on the bias. In a large bowl, dress with 2 tablespoons of the vinaigrette. Toss to coat and transfer to a serving platter, heaping them on one side.
4. Drain the onion and add to the same bowl. Dress with 1 tablespoon of the vinaigrette and toss to coat. Place the onion in a pile beside the pea pods.
5. Arrange the smoked trout and grapefruit sections on the same half of the platter, leaving the other half for the salad greens.
6. Add the greens and herbs to the bowl. Drizzle with 2 tablespoons of the vinaigrette and toss until coated. Taste and season with salt and pepper, as needed. Transfer to the serving platter.
7. Tuck the olives around the edges. Sprinkle with the sunflower seeds. Drizzle the remaining vinaigrette over the smoked trout, grapefruit sections, and greens.

HENRY VIII'S SALAD

Traditional English cookery once included a dish, largely forgotten now, that corresponds to salade composée. It was called salmagundi. The word entered the language from who-knows-where in 1674, when a Stuart king sat on the throne, and attached to a likely older preparation of sliced roasted meat or shredded chicken, cooked and raw vegetables, hard-boiled eggs, and anchovies garnished with dressed salad greens. Henry VIII might have eaten salmagundi, and yet in many respects, it also sounds like the grab-and-go chef salad of the twenty-first century. The distinction is care: Each element of salmagundi was prepared separately and arranged artfully on a serving platter. Salmagundi was composed. The chaotic chef salad is a gallimaufry.

BIBB LETTUCE, LITTLE GEM, OR RADICCHIO • ALLIUMS • HERBS | SIDE | ONE HEAD MAKES TWO SERVINGS | VEGETARIAN | SPRING, FALL

BRAISED AND GRILLED LETTUCE

Remember: Vegetables mature in generations. You can no more hold them back than you can keep a litter of puppies from becoming dogs, and a home garden is frankly inconvenient on this count. Farmers plan successive plantings for weekly market sales all season long. Grocery store produce buyers manage complex supply chains to guarantee an uninterrupted year-round supply. But the backyard garden has nature's pulse: anticipation followed by excess, surplus followed by lack. Too little, too much. Too much, too little.

There are workarounds. Cabbage holds well and summer squash bear repeatedly. Potatoes, the roots, and onions will keep. You can pickle beets, preserve tomatoes, dry chiles, and freeze corn.

Then there is lettuce. It doesn't hold, doesn't last, doesn't pickle, doesn't freeze. When the garden gives you more than you need for salad, cook it. Chefs like to grill Little Gem, and in a zero-waste mood, I've also grilled bolted romaine, which balances bitterness with char. Serve with an anchovy-strong vinaigrette for an on-trend rustic vibe.

Braised lettuce, on the other hand, sounds fussy and antiquarian, like something from the forgotten entre-guerre bourgeois cooking of *La bonne cuisine de Madame E. Saint-Ange*. Too bad. It's a wonderful springtime dish for a season when I'm starved for fresh chlorophyll but also want something warm and comforting because the weather hasn't hotted up yet.

Braised lettuce has melting leaves and a tender heart. The taste is buttery and rich. Bits of sautéed allium cling like confetti after the tickertape parade. The effect of the finished dish can be cozy or rather elegant. For the former, serve alongside slices of roast lamb with flageolet beans. For the opposite, play mild against mild. Serve braised lettuce on a warm plate with tender fish fillets strewn with chervil and tarragon.

If you have fresh peas, add a handful to the lettuce in the final 5 minutes of cooking time.

(CONTINUED)

BRAISED LETTUCE

Choose small, compact heads, such as Bibb or Little Gem. Remove floppy outer leaves, cut the heads in half, and rinse well. For 4 halves, melt 2 tablespoons unsalted butter in a lidded sauté pan over low heat. When it foams, add a minced shallot and a few rounds of chopped spring onions and/or tiny green garlic. Season with salt and pepper. Soften for a minute or two. Lay the lettuce into the pan cut-side down. Reduce the heat to low, cover tightly, and stew for 15 minutes, shaking occasionally. If the pan threatens to run dry, add a tablespoon of water at a time. Carefully turn the lettuce cut-side up and continue cooking until the core is tender, about 10 minutes. If you have peas, add them in the final 5 minutes. Transfer to a warm serving platter. Whisk 2 tablespoons butter into the scant pan juices. Spoon over the lettuce.

GRILLED ROMAINE

Little Gem is ideal for grilling, although the crisp hearts of larger romaine will work as well. Trim and discard outer leaves. Cut the heads in half and rinse carefully, with special attention to the base of each leaf where grit gathers. Dry in a salad spinner, then pat with a kitchen towel. Rub all over with grapeseed oil or exra-virgin olive oil. Grill over lively heat for 2 or 3 minutes per side, just long enough to mark and wilt the lettuce. Remove to a serving platter. Season with fine sea salt and freshly ground black pepper. Splash with a strong sauce, such as mustardy vinaigrette or Universal Green Sauce (page 248) made with extra anchovies. Strew with fresh herbs, sprinkle with Aleppo pepper, and serve with croutons or grilled bread drizzled with olive oil. Serve alongside Greek gigante beans or other white beans. Radicchio, endive, and other small chicories can be prepared the same way.

CH.

24

TOMATOES AND KIN

FRUITING NIGHTSHADES

Solanaceae
THE NIGHTSHADE FAMILY

Capsicum annuum
BELL PEPPER, JALAPEÑO, CAYENNE, NEW MEXICO, BIRD CHILE, MANY OTHERS

C. chinense
HABANERO

Physalis ixocarpa
TOMATILLO

Solanum lycopersicum, also *Lycopersicon esculentum*
TOMATO

S. pimpinellifolium
CURRANT TOMATO

S. melogena
EGGPLANT

AUGUST 23: Brandywine, Black Krim, Sungold, and Sweet 100

The summer garden belongs to the tomato. And vice-versa, for that matter. Once you've grown delicious Cherokee Purple, Green Zebra, or German Striped, there is no question of buying generic off-season tomatoes the rest of the year. It is better to miss the gems of summer than to waste one's hopes on worthless imposters. Plus, homegrown tomatoes have a second life in the pantry, where their essential flavor remains undiminished and their lycopene color persists in hues of brick and hemoglobin. Half of the tomatoes neighbor Del and I grow are San Marzano, the standard Italian paste tomato, along with a few Amish Paste, an American heirloom suited to our local climate. Canned tomatoes—whole, crushed, sauce, juice, or paste—may not replace the just-picked flavors of August but they have an unmatched carmine intensity and umami force. The tomato's second self is in no way inferior to its first.

Peppers, chiles, and eggplants appear in this chapter because they are close tomato relatives, even if the family resemblance is not immediately apparent. All of them are fruiting nightshades—such an ominous name—and share similar growing requirements. Tomatoes provide the case study. Once you have a feel for them, you know what to do with the others, which are sturdier and less needy. A second set of tomato relatives, tomatillos and ground cherries, are easier still. They exhibit the weedy profligacy of wild plants.

The nightshades in this chapter worship the sun, a reminder of their ancient origin in the equatorial sunbelt. Tomatoes, peppers, and tomatillos are New World species. Eggplants likely derived from a wild plant in Africa or South Asia before being domesticated in Southeast Asia. In each case, what we eat is the plant's fruiting body. (Tomatoes, peppers, and eggplants are technically berries.) The 1893 case of *Nix v. Hedden* ruled the tomato a vegetable because, in the unanimous decision of a U.S. Supreme Court, whose gastronomic credentials are not recorded, it is normally eaten with the main course, not dessert. Potatoes, a nightshade cousin from the high Andes, are genetically close enough to hybridize with tomatoes in a laboratory setting, but they have substantially different growing requirements.

AUGUST 13: Neighbor Del with New Girl, an improved version of Early Girl

Still other nightshade relatives are inedible, including poisonous belladonna, hallucinogenic jimsonweed, and addictive tobacco, all of which produce toxic alkaloids as biodefense against pests. Ironically, the molecules that repel pests also attract humans, who wish to experience the strange effects brought on by a (hopefully) nonlethal dose of the compounds.

Alkaloids produced by the greenery of tomatoes, peppers, and eggplants taste bitter. Tomato leaves provoke vomiting if eaten in quantity, although there is no harm in the culinary trick of swishing a few green leaves through a finished tomato sauce to freshen it with a scent of live vines. Another chemical defense is capsaicin, the fire in chiles.

TOMATOES

Wild tomatoes originated in the coastal deserts of western South America, in what is now Peru and Bolivia. The Aztec name *tomatl* means "plump fruit." Theirs were smallish and yellow, perhaps visually like the modern hybrid cherry tomato Sungold. The tomatl reached Spain at some point before 1522, according to Waverly Root, and soon made its way throughout southern Europe. New names played on the color. In Italy, *pomo d'oro,* "golden apple," became the modern *pomodoro,* even though any Italian child will tell you tomato sauce is red. The color is a novelty from the 1570s. Modern tomatoes are capable of blushing red, yellow, orange, and pink, as well as remaining green when ripe or veering into white, striped, mottled, purple, or near-black. Color is a distinguishing trait—Cherokee Green was selected from Cherokee Purple—but not a reliable guide to flavor. Tiepolo-pink Brandywine is no less rich than Cadillac-red Mortgage Lifter, and Green Zebra looks jejune but tastes ripe in the dark. Be that as it may, we eat with our eyes first.

Europeans from the start judged the tomato beautiful but spent several hundred years debating its edibility. Southern Europeans, and especially Italians, were braver on this front. (In Germany, witches and warlocks ate tomatoes to transform themselves into werewolves, dark sorcery preserved in the Linnean name *Solanum lysopersicum,* or wolf-peach nightshade.) The 1760 Vilmorin seed catalogue listed tomatoes among ornamentals, granting it food status only in 1778. Thomas Jefferson is widely credited with popularizing the tomato as a foodstuff in the United States, courageously planting them no later than 1781. During his presidency, he found time away from the grave matters of state to record which types of produce were sold in Washington's vegetables markets. Tomatoes appeared from July 16 through November 17—a curiously long season. The late offerings must have come from storage. My uncle David told me that one year he kept his final picking of Mountain Fresh tomatoes on the cement floor of his basement storage room until . . . June. Were he and Aunt Barbara not sixty-year members of Clover Hill Presbyterian Church, I would suspect a tall tale. Still, I know from my own informal experiments that keeping tomatoes into the New Year is no stretch. Bring in green tomatoes before the last frost and store in a cool spot, arranging them on newspaper so that they don't touch. Slowly, slowly they turn to a pallid shade of their normal ripeness.

Before 1830, tomatoes were fluted and bottom-heavy like a pouch gathered at the top. Surviving examples of the form include heirloom Costoluto Genovese ("ribbed from Genoa") and Cuore di Bue, a sauce tomato named for its oxheart shape. In the nearly two centuries since, the tomato's malleable genetics have been manipulated to create many shapes, sizes, and colors, such that today the seed houses of the world offer more than four gajillion named varieties. The smallest I grow is the heirloom Cades Cove Currant, an example of the second cultivated tomato species, *S. pimpinellifolium.* It's not much to look at, but the varietal history overlaps with my family story. A cove, in this instance, is a flat-bottomed mountain valley ringed by ridges. Cades Cove was an early settlement in the southern Appalachians of eastern Tennessee, where my mother's ancestor John Burchfield established a homestead in 1823. Viny in habit, perhaps like ancestral wild tomatoes, Cades Cove Currant grows rampantly and self-seeds. Each fruit weighs a tenth of an ounce.

The largest tomato I grow is the heirloom Brandywine, a pink-skinned beauty from 1885 that is instantly recognizable in the garden by its leaves, which resemble potato leaves. Brandywine is a stingy bearer, but fruits can weigh 1 pound apiece and each is a masterpiece of flavor. The heirloom Mortgage Lifter dependably bears even larger fruits, although growing monster tomatoes, like driving monster trucks, is a passion that eludes me. Many varieties are already productive enough. In a good year, my paste tomatoes average nearly 10 pounds per vine. I never wished for more.

By the late twentieth century, the pursuit of uniform tomato cosmetics—round, red, and impervious to long-distance shipping—nearly ruined a good thing. Breeders forgot to account for taste. The industrial food system's drive toward unnatural consistency led to CGN-89564-2, known to consumers as Flavr Savr, the first genetically modified crop to receive FDA approval for human consumption. Consumers decided it wasn't fit to eat and the experiment failed. Since then, the farmers' market movement has brought tomatoes back from the brink of success. If anything, public taste for picturesque appearances has turned "heirloom" into another marketing cliché, but at least it encompasses a broader view of beauty.

In their natural habitat, tomatoes are cold-sensitive perennials. Plant after all danger of frost has passed. In the early fall, they will stop blooming as nighttime temperatures fall below 50°F, and, at some point, shorter days will cause the plants to stall out, even if frost doesn't kill them first. When I lived in Southern California, tomato season ran from June through Halloween. Uncle David in Tennessee picks tomatoes from Independence Day until the first of October. In my New England garden, I will have cherry tomatoes in late July, followed 2 weeks later by slicers and beefsteaks. Canning tomatoes peak at Labor Day and soon thereafter everything succumbs to the autumnal chill.

Every gardener will have to fine-tune the general method for growing tomatoes to suit the specifics of his or her garden—climate, locale, and weather. Remember always that tomatoes are a sun fruit. To thrive they need a full 8 hours a day. Large tomatoes won't ripen with much less. In marginal locations, such as a partially shaded garden plot or in a cool, foggy region, stick with cherry tomatoes and small-fruited types.

Tomatoes also suffer heat stress and, at temperatures over 95°F, will fail to set fruit because of poor pollination.

Look to the varietal name for clues. Anna Russian, San Francisco Fog, and Glacier signal cool-climate adaptions. Early Girl is quick to the table. Beefsteak, Big Boy, Big Zac, and others of their magnum ilk will require maximum sunshine and a long growing season.

TOMATILLOS

The name means "little tomatoes," but tomatillos belong to an altogether different genus, *Physalis*. They are known in Mexican cooking as *tomates verdes,* and they give salsa verde its color and tang. The upright, branching plant is robust. Each fruit grows inside a papery husk, the calyx, which is shaped like a Japanese lantern. Inside, the tomatillo's smooth skin is tacky, like surf wax. (The specific epithet, *ixocarpa,* means sticky fruit.) If you leave green tomatillos on the bush to ripen, they turn yellow or purplish and develop a slightly musky, fruity flavor that is pleasant if less tangy in raw salsas. I'll sometimes cut tomatillos into wedges as a simple salad to serve alongside tacos, or slice them into fine matchsticks to garnish pork stew, as if they were radishes.
A smaller, sweeter *Physalis* goes by the name ground cherry, husk cherry, or Cape gooseberry. The bursting golden fruit is fun to eat out of hand.

PEPPERS

Peppers and hot chiles are a world unto themselves, and I give them short shrift here. Grow them if you crave their provocative flavors and lipstick colors.

Peppers and chiles demand heat and sun. They grow less enthusiastically in New England than in my father's Tennessee garden, where his jalapeño plants regularly grow chest-high and fill 5-gallon buckets with blazing hot, fire-engine red chiles. I ferment my share into hot sauce, which mellows with time. My Hudson Valley garden buddy Andy Rothschild also grows peppers by the bushel in his hoop houses. It seems too obvious to be true, but chiles grown in warm conditions taste spicier—healthy and vigorous plants accumulate more capsaicin, filling the seeds and seed membranes with fury.

Ripe peppers and chiles turn yellow, orange, and red. Green ones are not distinct varieties but unripe fruit. You pick at the stage you prefer.

Green peppers and chiles have a vegetal, grassy, slightly bitter flavor. In Louisiana, green bell peppers belong to the holy trinity, the mix of diced onion, celery, and peppers that lays the foundation for Creole and Cajun dishes. Fire-roasted green chile is the sine qua non of New Mexican cooking. Minced green jalapeño or serrano adds the pique to Mexico's pico de gallo.

At maturity, peppers and chiles turn sweet, fruity, rich, velvety. The words *pimentón* (smoked paprika), *pimiento* (cherry peppers), and *pimento* (as in pimento cheese) come from the same Latin root as pigment. They are ripe with color. Think of staining paprika and kimchi's gochugaru, or the spicy Martian landscape of chile powders and chile flakes: cayenne, guajillo, Aleppo, piment d'Espelette, ancho, Marash, Urfa biber, and Kashmiri.

AUGUST 7: Jimmy Nardello Italian frying peppers

EGGPLANTS

The first eggplants to reach Europe lived up to the name—small, oval, white. The modern Italian types are large and pendulous, often purple-black but sometimes white striped with violet. Mild Japanese eggplants are longer and thinner—emoji shaped. Little round Thai eggplants are green-and-white variegated. China and India have their own regional subtypes. (The preferred name in Britain, aubergine, comes from Sanskrit via Moorish Spain, where the Arabic word was Hispanicized as *berenjena.*) A related species, *Solanum gilo,* is known as African or Brazilian eggplant. As with most crops, the smaller-fruited ones yield earlier. I've found that both slim Japanese types and heavyset Italian types grow in New England but without

AUGUST 23: Galine, a hybrid Italian eggplant for Northern growers

"EGGPLANT" BY KATERI KOSEK

My local newspaper, the monthly *Monterey News*, established in 1970, is *very* local. Monterey has a thousand full-time residents. Contributions from citizen-reporters include nature notes and obituaries, as well as the occasional piece of creative writing, such as this poem from Kateri Kosek. "Eggplant" appeared in the November 2020 issue, well after the end of garden season, and exemplifies Wordsworth's insight that poetry is emotion recollected in tranquility. The speaker of the poem recalls the muggy tedium of garden chores, when there is too much to do. Any gardener knows the overwhelmed feeling. On the verge of exhausted surrender, she finds a surprise behind the weeds, a moment of wonder nearly erotic in intensity.

Later, you'll give up altogether,
after it's clear the tomatoes
need staking,
after even the sturdy upright
tomatillo, soaked and heavy
after a long rain, splits in two
and self-destructs,
but now, sifting through grasses,
you part some leaves
and see them—
purple and curved and fully
there, growing all this time
behind a curtain of weeds,
growing
in spite of you, surprising
but inevitable
like two people
tendrilling towards each other
through months of cool dark
shade,
so that when the door
flings open
they are ready,
sun-starved,
shining.

the impressive vigor and yield of eggplants grown in hotter climates.

In the Garden

Tomato plants demand more attention than other crops. Or perhaps I willingly give them more, the way you can't help but favor one child in a group. The essentials for growing blue-ribbon tomatoes include a setup for germinating seeds indoors, favorable weather for transplanting, healthy soil, judicious amendments, staking or trellising, pest control, endless sunshine, unwavering care, steadfast hope, whispered sweet talk, and the luck of the Irish. What have you gotten yourself into? It is worth noting that author William Alexander had a national bestseller with his gardening memoir *The $64 Tomato,* named after his estimated per-tomato cost at the end of a Brandywine summer. The reality is not quite so dire, at least for anyone willing to forgo platonic perfection.

Tomato seeds are often started indoors, in cells or pots, 6 to 8 weeks before last frost and transplanted about 1 week after last frost.

Seeds started in trays should be pricked out when the first set of true leaves emerges. I'll oversow in trays if I'm concerned about seed viability, then prick out the strongest seedlings. For more on pricking out, see chapter 6, Sowing and Transplanting (page 56).

Germinate seeds in a warm spot with full sun or use heated grow mats and artificial lights. If seedlings outgrow their cells before you can get them in the ground, transplant to a larger pot to hold. Don't let them get rootbound.

Put out tomato starts only after the ground has warmed and all threat of cold has passed. Harden off seedlings for a week before transplanting. Again, refer back to chapter 6 (page 56) for details.

Plants sold by commercial greenhouses might be available earlier than you need them; keep them in a warm, sunny spot until the weather is right. Large transplants don't necessarily confer advantage. If anything, leggy seedlings are more likely to be rootbound, the consequence of being held too long. If early growth stalls in a too-small pot, the plant risks transplant shock. An ideal transplant will be compact and robust, like a compressed spring ready to release.

Unlike most vegetables, tomatoes should be planted deeply. Pinch off the cotyledons and first leaf, perhaps even the second, for a long, bare stem. Also pinch any precocious flower buds to force the plant's energy into growth. Dig a hole deep enough to bury two-thirds of the stem, putting the plant in the ground "up to its neck." The buried stem will sprout roots. Protect transplants from cutworms with a paper sleeve or by placing a plastic cup, bottom removed, or a tin can, opened at both ends, over it.

Black plastic mulch will benefit tomatoes in cool climates and reduce disease caused by splashing water.

Most tomato varieties will need a sturdy support. Unlike cucumbers, which grasp a trellis with tendrils, or pole beans, which wrap themselves around any upright structure, tomatoes will need to be tied up. Attach the leader, or main stem, as it grows, as well as selected side-shoots. (See right for notes on pruning to maintain form.) Many attachment gizmos are sold; I use cheap biodegradable sisal baling twine.

An alternative to staking is the tomato cage, which provides exoskeleton support. Premade ones sold for a few dollars at a garden center are too flimsy to be of any use. I've had tomato vines flatten them to the ground, like an elephant sitting on a folding chair. Make your own with hog wire, pre-cut hog panels, or concrete reinforcing remesh. (Small-mesh chicken wire won't work.) Bend a 7- to 10-foot length into a loop and secure at the ends with wire twists. You'll want to anchor each cage with a heavy stake, such as a metal T-post.

With their big vines and long growing season, tomatoes are the very definition of heavy feeders. Give them deep, well-drained soil with balanced fertility. I mulch tomato transplants with compost and side-dress the vines with more compost once they've set fruit. Excessive nitrogen overstimulates the vines, making them jungly and neglectful of flowering. Water tomatoes regularly to avoid blossom-end rot. (See Secondary Nutrients and Trace Elements, page 36.)

JULY 21: Sungold

Optional pruning helps train the plant to a stake or trellis. Encourage upright growth by pinching suckers, the side shoots that emerge at each leaf junctions along the main leader. Tie the leader every foot or so. Leave well-placed side shoots to grow, tying them to the support. Pruned vines will produce fewer but larger fruit. There's no need to get carried away. The leafy canopy is a chlorophyll-powered solar array that supports growth and ripening. Pruned suckers root quickly in water and can be used for a succession planting. If vines overrun their trellises at midseason, pinch out the tops to force side growth.

Pests include the tomato hornworm, an apple-green caterpillar with diagonal cream stripes. The first one you find might be a shock: They can grow 4 inches long.

Various blights and pathogens will likely do more damage over the course of the summer, spreading upward from the lowest leaves. The cumulative injuries lead to demise the way Hemingway's man went bankrupt—gradually, then suddenly. Some favorite heirlooms will be most vulnerable. In the northeast, few tomato vines grown outdoors remain unscathed. (Tomatoes grown in a high tunnel or hoop house, where foliage is protected from rain splashes and roots can be targeted with drip irrigation, are less prone to waterborne disease.)

Choose varietals with proven resistance to *Fusarium,* nematodes, tomato mosaic virus, *Verticillium,* or other diseases of your region. Late blight is the worst of all, an all-but-hopeless situation. Cultural controls include crop rotation and generous spacing for airflow and sunlight. (Remember that transplants will look absurdly far apart at first.) Mulch will reduce splashing. Avoid spraying with water when irrigating. If there are active pathogens afoot, take care not to spread them. Keep away when vines are wet with dew or rain and disinfect clippers with rubbing alcohol after each plant.

For detailed guidance on diagnosing disease and choosing preferred organic treatments, refer to online resources such as the aptly named Got Pests? at Maine.gov. The site covers gardening problems from weeds to blight to cutworms to deer with clear photos and concise written descriptions. It links to fact sheets from university research departments and agricultural extensions around the country. (Advice is practical but broadminded: A web page on garden weeds includes advice on organic controls as well as chemical herbicides, plus guidance on edible weeds.) Other online guides include state agricultural extension websites and the Clemson Cooperative Extension Home & Garden Information Center. Finally, Johnny's Selected Seeds and other seed companies publish libraries of free online information.

When choosing tomato varieties to try, be aware that so-called indeterminate types keep growing and blooming all season, while determinate types, sometimes called bush tomatoes, stop at a genetically determined height. Most popular tomato varieties are indeterminate. Certain paste tomatoes are determinate, so that the crop matures simultaneously for convenient processing. Smaller determinate types, which don't necessarily need staking, are suited to containers, although many small-fruited indeterminate, including Sweet 100, a productive red cherry tomato, will also thrive if staked or trellised.

The best advice is to adopt an experimental mindset. Trial multiple varieties, rotating a selection of 3 or 6 or 9 per year, until you hit upon the one best suited to your climate and taste. Planting varieties with staggered maturity dates will extend the season. Cherry tomatoes and early hybrids such as Early Girl supply the kitchen until the Big Boys finally come in. Fist-sized heirlooms such as Paul Robeson color slowly and sweeten as they bask in the heat. Eat them dead ripe, still warm from the vine, coagulated sunshine. I let the last flush of canning tomatoes hang until soft and nearly shriveled. They are noticeably sweeter, ideal for paste. The last green tomatoes of the fall are suited to canning projects such as chutney and chow-chow.

See Tomato Varieties (page 416) for a few recommended types. Every spring I swear to not plant so many tomatoes, and every summer I'm glad I got carried away. You can't have too many. Unlike surplus zucchini, tomatoes always find a taker. Or they can be canned for space-efficient storage—something like 6 pounds of tomatoes goes into 1 quart of sauce.

The tomato kin—tomatillos, peppers, and eggplants—are also started indoors. Sow 4 to 6 weeks before last frost and put out late, once the weather has settled. Protect with a cutworm sleeve. Plant tomatillos deeply, as with tomatoes; two or more plants are required for successful pollination. Stake peppers and tomatillos to support the somewhat brittle bushes and improve fruit quality. Eggplants don't necessarily need support, but without it they sometimes topple under the heavy burden of their goth fruit.

The tomato kin are susceptible to some of the same pests and diseases as tomatoes but, in practice, I find them more carefree. Cool climate growers should plant through black plastic mulch.

AUGUST 22: San Marzano

TOMATO VARIETIES

Each variety of tomato—each phenotypic expression of the genomic potential—meets a specific need in the garden or the kitchen. What a seed catalogue offers, in effect, is a set of viable responses to known issues. Whatever question is posed by your garden, there will be a varietal to answer it. The chart below samples a few proven favorites.

Tomato taxonomy is informal, and seed suppliers sometimes sort the varieties in contrary ways. There are more kinds of tomatoes than a sensible gardener could hope to grow. As with shopping for antiques, you look tirelessly for the thing you have in mind until you happen upon the perfect something else.

TYPE	DESCRIPTION	SAMPLE VARIETIES
Beefsteak	Large, meaty, juicy, heavy tomatoes for slicing and eating raw	Brandywine, Halladay's Mortgage Lifter, Cherokee Purple, German Pink, Gold Medal, Big Boy
Bush or patio	Determinate plants; compact growth; bloom and fruit in a single crop	Glacier, Celebrity Plus, Principe Borghese (a paste tomato)
Cherry (also currant, grape, plum, pear)	Small fruited, early, productive, multicolored	Sweet 100, Sungold, Tommy Toe, Yellow Pear
Cocktail or salad	Larger than cherry, smaller than slicers; grow in clusters	Black Prince, Juliette (a small paste tomato I like for salads)
Heirloom	Open-pollinated variety passed down by seed savers, could overlap with any other category, such as an heirloom slicer	Abe Lincoln, Striped German, Green Zebra, Paul Robeson, Costoluto Genovese, many others
Paste (also sauce, plum)	Less juicy; for canning and sauce	Roma, San Marzano, Amish Paste, Pink Fang
Slicer	Consistent shape, medium size, and solid texture for sandwiches and burgers	Early Girl, New Girl, Mountain Fresh, Carmello

TOMATOES AND TOMATILLOS—START INDOORS

SOW	SPACING	DEPTH	EMERGENCE	TRANSPLANT	SPACING	ROW SPACING
5 to 6 weeks before last frost	1 per cell	1/4 inch	5–7 days	After last frost	24–36 inches	4–6 feet

Note: Plant starts deeply. Remove lowest leaves and bury two-thirds of the stem, or "up to the neck." To sow a succession crop outdoors, wait until after all threat of frost is past. Put out 3 seeds per stake or cage, thinning to the strongest.

PEPPERS, CHILES—START INDOORS

SOW	SPACING	DEPTH	EMERGENCE	TRANSPLANT	SPACING	ROW SPACING
4 to 5 weeks before last frost	2 per cell, thin to 1 per cell	1/4 inch	8–25 days	After last frost	18 inches	2–3 feet

Note: Seeds are slow to germinate in cool conditions.

EGGPLANT—START INDOORS

SOW	SPACING	DEPTH	EMERGENCE	TRANSPLANT	SPACING	ROW SPACING
4 to 5 weeks before last frost	3 per cell, thin to 1 per cell	1/4 inch	5–17 days	After last frost	24 inches	2–3 feet

Note: Seeds are slow to germinate in cool conditions.

In the Kitchen

Because tomatoes, peppers, chiles, eggplants, and tomatillos are sun fruits, borne aloft on staked plants, they often grow clean enough to eat off the vine. Cherry tomatoes and ground cherries are irresistible garden snacks. Use a damp cloth to wipe away the fine grit that settles on big tomatoes during dry weather, as if cleaning knick-knacks. In rainy weather, rinse splatters and dry to a shine.

Pick tomatoes and their kin on a dry day after morning dew has burned off. Many have a harvest indicator—tomatoes and peppers signal red or yellow when ripe. (Even Green Zebra tomato takes on a yellowish cast.) Tomatillos split their calyx. With eggplants, it can be hard to know when to pick. Consult the seed packet for mature size. Harvest regularly to encourage further blossoming. Gently search through the foliage. Green chiles blend in until betrayed by their high polish.

Cherry tomatoes and plum tomatoes release from the stem with the least jiggle. Tomatillos also detach easily. Small chiles can be snapped off if you find yourself without a sharp knife. Preferably use clippers for larger tomatoes, eggplants, and bell peppers, taking an inch of stem. If barehanded, use one hand to steady the branch while you gently pull the fruit with the other. No yanking.

Pick tomatoes as soon as they are ripe, or even a day early, letting them reach peak in a sunny windowsill. They will hold better inside, less exposed to birds and mishap. If you come across damaged or overripe fruits, pick them for the compost pile. Keep the vines clean.

Tomatoes and especially cherry tomatoes are prone to cracking after heavy rain. Rain-split tomatoes can be used for sauce if gotten right away, but they quickly sour in the heat.

Don't refrigerate tomatoes. Instead, spread them in a single layer on a platter or sheet pan and check daily for soft spots and sour smells. Intact tomatoes will last a week at room temperature. Rot usually begins in a nick or crack, where microbial spores get through the watertight skin. Cut away minor blemishes to salvage the rest.

The culinary uses of tomatoes are limitless. To state the obvious, summer tomatoes are juicy and fresh, while canned tomatoes are cooked and concentrated. Consider the culinary virtues of each. Straight from the garden, a ripe tomato approaches perfection in its natural state, balancing as it does sweet, tangy, and savory qualities. Preserved tomatoes—sun-dried, canned, paste—intensify all the tomato characteristics, supporting entire wings of the grand palazzo of Italian cuisine. Turn to Tomato Basics (page 420) for a survey of tomato uses.

Waxy, hollow peppers and chiles don't spoil as quickly as tomatoes. Even so, it is best to bring them in from the garden as soon as they are ripe. They will keep at room temperature for a week or more. Don't wash until just before using; turn occasionally to check for soft spots. Red and yellow bell peppers bring sweet depths of flavor when cooked with other ingredients. They shine on their own. Dice roasted red peppers for pimento cheese or slice them into strips to strew over summer meals, where they flatter anything grilled. Horn-shaped sweet Italian frying peppers don't need peeling. Skip ahead to Pepper and Chile Basics (page 436) for handling tips.

Chiles cross nearly every culinary boundary these days. They are a hallmark of Blade Runner cooking, the exciting flat-world amalgam of European, Asia, and Latin America influences that once seemed futuristic but today represents the here-and-now of our multicultural culinary world.

Use fresh chiles to accent dishes, almost like herbs. Pickle the extra. String ripe cayenne peppers in ristras, or spread them on screens to dry. Fleshier red chiles can be grilled and peeled as you would red bell peppers. Or fermented for hot sauce.

Ripe eggplants hold on the plant for a week or more. Once picked, they will keep on the counter for 5 days until their glossy skin wrinkles.

AUGUST 12: Brandywine is a so-called potato-leaf variety.

See Eggplant Basics (page 442) for cooking suggestions.

Ripe tomatillos will keep on the counter for several days. Unripe ones, green and dense, will keep for 10 days in the refrigerator without much change. They go into salsa verde, which can be served raw as a bright green sauce or sizzled in fat for a richer, silkier salsa. Pork stewed in chile verde is a taqueria staple. Chicken in chile verde is lighter. Chile verde also makes a terrific vegetarian stew with summer squash, potatoes, carrots, and purslane (verdolagas). Whole tomatillos and tomatillo salsa be canned using the same methods as tomatoes.

TOMATO BASICS

In return for the extra care they require, tomatoes yield abundant culinary rewards. Pappaw said, "There's only two things that money can't buy, and that's true, true love and homegrown tomatoes." Sliced and eaten raw, tomatoes are uniquely delicious, and they are more versatile as a cooking ingredient than any other vegetable, apart from onions. The following overview of basic techniques is meant to jog your memory when faced with summer's welcome glut.

TO PEEL A TOMATO

Tomatoes are usually peeled before adding to sauce, stews, or braises, otherwise the skins detach and curl into unpalatable quills. Bring a large pot of unsalted water to a boil. Slash an X in the blossom end of the tomatoes. Working in batches of 4 to 6, blanch for about 45 seconds, or until the slashed skin peels back. Cooking time varies with ripeness. Remove the tomatoes with a slotted spoon and plunge into an ice-water bath. As soon as they are cool enough to handle, slip off the skins.

TO SEED A TOMATO

Seeding fresh tomatoes eliminates juiciness and thus reduces cooking times. Slice a peeled or unpeeled tomato in half through its "equator," midway between stem and blossom end, to reveal the seed cavities in cross section. Squeeze out the seeds and discard. Probe the cavities with a fingertip to get them all.

SLICED AND SEASONED TOMATOES

Tomatoes eaten raw should be sliced and salted 15 to 20 minutes before serving to draw juices. Thick slices, up to 1 inch, enhance their mouth-filling succulence. Or cleave off faceted chunks, mixing multiple varieties, and arrange on a platter with a few smaller tomatoes, cut into rounded halves, thrown in for visual contrast. Freshly ground black pepper or ground toasted coriander are equally good accents. A drizzle of the best extra-virgin olive oil is not quite a dressing, more of a flavor garnish. A *few drops* of balsamic vinegar or red wine vinegar can sharpen the flavor, if needed—usually it's not needed. Scatter basil over the top. If you like, nest burrata at the center of the platter, or tuck in sliced mozzarella.

TOMATO SANDWICH

Surely you've had one. If not, as a gardener, you surely will. The first of the summer is a milestone at my house. Large beefsteak tomatoes work best; Brandywine is supreme and Mr. Stripey, spectacular. As in the previous recipe, slice a large tomato thickly, salt well, and set aside for several minutes to draw the juices. Use white bread for the sandwich, either 2 slices of the squishy store-bought type or, for an open-faced sandwich, a single thick slice of good sourdough, lightly toasted. Slather the bread with mayonnaise and cover with a single layer of sliced tomato. Sprinkle with coarse finishing salt and quite a lot of freshly ground black pepper. No other embellishment is needed.

SEPTEMBER 5: Striped German grown in the Hudson Valley by Max Morningstar

PAN CON TOMATE

This is a Spaniard's drinking snack, or for eating outside while grilling. It works best with very ripe, medium-sized tomatoes. Seed the tomatoes, as described on page 420. Grill or toast thick slices of crusty artisan bread. While the toast is still hot, rub it lightly with a whole, peeled garlic clove. Immediately rub it with a cut tomato, in effect grating the tomato against the sharp edges of the grilled open crumb. You'll be left with an empty skin. Drizzle with extra-virgin olive oil and sprinkle with salt. Lay an anchovy on top if you like.

CHERRY TOMATO SALAD

A mixed bowl of cherry tomatoes will taste sweet and look darling; their somewhat simple flavor can be tuned up. Slice a pint of cherry tomatoes into halves and toss with a half-dozen whole basil leaves, a lightly crushed peeled clove of new garlic, a few drops of fish sauce or Worcestershire sauce, a fat pinch of salt, and freshly ground black pepper. Set aside for 30 minutes to draw the juices, mixing occasionally to lubricate. Taste and adjust the flavors. Drizzle with the best extra-virgin olive oil. Serve as a salad, as a side dish for grilled fish or chicken, or heaped on toast rounds for bruschetta.

FRESH TOMATO SAUCE

Making sauce will absorb large quantities of tomatoes. So-called paste tomatoes—including the Roma types, such as famous San Marzanos—are often recommended because they are less juicy, but any variety will work for sauce, including heirloom beefsteaks. You decide how much to cook the sauce, striking a balance between concentration (more cooking) and fresh flavor (less cooking). I tend to leave it loose.

Peel and seed 3 pounds of tomatoes. Roughly chop and drain in a colander for 10 minutes. Heat 2 tablespoons extra-virgin olive oil in a large nonreactive sauté pan. Sizzle 2 lightly crushed peeled cloves of garlic for 1 minute until fragrant. Add the tomatoes and salt generously. Bring to a boil, then reduce the heat to maintain a lively simmer. Tuck in a basil sprig with a half-dozen leaves. Cook, uncovered and stirring frequently, until the juices are reduced and the oil separates, about 25 minutes. Discard the garlic and spent basil. Swish through a fresh sprig of basil or a fresh tomato leaf to revive the flavor. Serve with pasta.

TEN-MINUTE CHERRY TOMATO SAUCE

This variation on the Fresh Tomato Sauce (opposite) is faster and has the special taste of fresh cherry tomatoes. A chunky, half-cooked sauce, it goes with nearly any summer meal. Toss it with linguine or pour over sautéed zucchini. Spoon it over crispy-skin salmon, broiled marinated chicken thighs, or grilled vegetables. Roll it in a soft omelet for lunch or load onto toast topped with a sliced hard-boiled egg for a messy but delicious kitchen-counter snack. Or put a bowl of it on the table as an all-purpose side dish/condiment.

Bring in 1 pint of cherry tomatoes from the garden, which is roughly the amount you can carry if you stretch the front of your T-shirt into a kangaroo pouch. Mixed varieties will give you an ideal balance of sweet, tart, and savory notes. Rinse and remove the caps, leaving whole. Heat 2 tablespoons extra-virgin oil in a large sauté pan until it shimmers. Sizzle 2 peeled, lightly crushed cloves of garlic for 1 minute, until fragrant but not colored. Add the tomatoes and salt. Cover and cook for 3 minutes, shaking frequently. Uncover. If the tomatoes haven't started bursting, lightly crush several with a wooden spoon to release their juice. Shake the pan or toss frequently. As more tomatoes burst their skins, the liquids in the pan should stay just ahead of evaporation so that the tomatoes are neither watery nor dry but constantly bathed in thickened sauce. When all the tomatoes have collapsed, about 10 minutes, remove from the heat. Taste and, if needed to balance the flavors, dribble over the least amount of red wine vinegar. Sprinkle with freshly ground black pepper and a scattering of basil chiffonade. Serve immediately or at room temperature.

MARCELLA HAZAN'S TOMATO SAUCE

Often republished, this iconic three-ingredient recipe deserves the acclaim. Begin with 1 pint canned tomatoes or 2 pounds fresh tomatoes that have been peeled, seeded, and chopped. Combine in a nonreactive pan with 5 tablespoons unsalted butter and 1 medium onion, peeled and halved. Cook over medium heat, crushing the tomatoes with a fork and stirring occasionally, until the sauce is thickened and the butter incorporated, 45 minutes or more. Remove the onion but certainly do not discard. Instead save it to season soup, add to a braise, or chop for an omelet filling or sandwich topping.

For a variation dairy-lovers will appreciate, stir in ¼ cup cream or crème fraîche at the end of cooking. The richer, smoother sauce makes a wintry pasta topping or a luxurious blanket for mild white fish.

PROVENÇAL TOMATOES

Small-to-medium tomatoes baked quickly emerge from the oven with concentrated flavors; the sprinkled bread crumbs turn crunchy. Seed 5 or 6 firm, small-to-medium tomatoes, such as Early Girl (see To Seed a Tomato, page 420). Arrange the halves in a single layer, cut-side up, in an oiled baking dish. Salt generously. In a bowl, combine a heaping ½ cup bread crumbs, 1 tablespoon extra-virgin olive oil, a minced garlic clove, and 1 tablespoon minced parsley. Using a light hand, sprinkle the mixture over the tomatoes. Scatter optional herbes de Provence over the top if you have it handy. Drizzle with olive oil. Bake at 400°F until the tomatoes are bubbling and the bread crumbs browned, about 25 minutes.

FRIED GREEN TOMATOES

End-of-season fruit may not ripen in September's waning sunlight. Hard and tart, green tomatoes were traditionally used for pickles and chutneys. Fried in a crunchy coating, they turn meaty and pleasantly tangy—a sort of vegetarian cutlet. Fried green tomato makes a great sandwich or country-style side dish.

In a shallow bowl, whisk to combine ½ cup all-purpose flour, ¼ cup cornmeal, and ¼ cup cornstarch. Season heavily with fine sea salt, freshly ground black pepper, and pinches of cayenne pepper. Slice green tomatoes ½ inch thick. Press both sides into the flour mixture to coat. Pour vegetable oil into a heavy skillet to ¼-inch depth and heat over high heat until nearly smoking. Working in batches (do not overcrowd the skillet), fry the tomatoes at a fast sizzle until golden, 5 minutes per side. Drain on a rack or paper towels. To serve, dress with hot sauce, goat cheese crumbles, or a drizzle of your best balsamic vinegar.

CANNED TOMATOES

Canned tomatoes are a pantry staple and an essential homegrown ingredient. Canning is a project, to be sure. But by another light, a day of hot work in August accrues benefits across dozens of meals throughout the year.

Tomatoes are "canned" in mason jars, which are processed in a boiling-water bath for safe long-term keeping. The procedure is a multistep process with important safety protocols. I describe canning tomatoes in detail in my cookbook *Saving the Season*. Reliable instructions can also be accessed online at the University of Georgia's National Center for Home Food Preservation website.

WINTRY BAKED TOMATOES

This recipe made with canned tomatoes is a family favorite. I cook it on New Year's Day to serve alongside my annual lunch of Hoppin' John and collards. It also perks up mild comfort food, such as chicken and rice or mac 'n' cheese. Serve leftovers on crackers as a sort of wintertime bruschetta. Fans of Marcella Hazan will pick up on the background inspiration.

Lightly butter a baking dish. Pour in 1 quart canned tomatoes. Tuck in the pieces of a small peeled onion, cut into quarters. Cut 4 tablespoons unsalted butter into bits and scatter over the top. Sprinkle with ½ teaspoon fine sea salt, ½ teaspoon sugar, ¼ teaspoon fennel seeds, ½ teaspoon dried savory or thyme, ¼ teaspoon red chile flakes, and freshly ground black pepper. Bake in a 375°F oven, uncovered and without stirring, until the juices are syrupy, 1 hour to 1½ hours.

TOMATO SAUCE AND TOMATO PASTE

Making homemade tomato sauce and tomato paste involves a similar process to canning tomatoes. (Refer to the same resources mentioned in Canned Tomatoes, above.) They are mighty culinary assets to have on hand. In fact, I consider tomato paste a superpower ingredient (see the headnote to Pungent-Herb Sauces and Condiments, page 252). It brings umami depth to sauces and braises. I'll fry a large spoonful of tomato paste in extra-virgin olive oil with onions and garlic, then add Romano beans with enough water to make a loose sauce. Cook gently until the vegetables are silky and the water nearly gone, 45 minutes to 1 hour. In a simpler vein, spread tomato paste on grilled bread for a snack.

TOMATOES • BELL PEPPER • ALLIUMS • HERBS | STARTER | SERVES 6 TO 8 | VEGAN, GLUTEN-FREE | SUMMER

GAZPACHO

There are as many gazpachos as there are mortars and pestles, writes Pepita Aris, whose beautiful *Recipes from a Spanish Village* was among my first cookbooks. This version—made with a blender and not a mortar and pestle, thank goodness—has willfully inauthentic California vibes. It skips the traditional thickener of soaked bread and picks up the globe-hopping flavors of avocado, lime, and shiso. The result is familiar, if not authentic gazpacho rojo de Sevilla. It's gazpacho-ish.

Use dead-ripe tomatoes of the most delicious types, ideally including a few Cherokee Purple or Black Krim. The blender reveals the beauty in "ugly" ones less suited to slicing and serving on a platter. Peeling the tomatoes and straining the puree for an extra-velvety texture are optional refinements. Or you could eliminate the extra steps and call it rustic gazpacho.

Small, dense cucumbers from the garden won't need to be peeled. Roasted red peppers have superb flavor—another optional refinement suggested by my kitchen collaborator Dash. Good Spanish sherry vinegar is gazpacho's signature.

As with any garden soup, replace ingredients you don't have or don't like. Alternative garnishes include basil and sliced cherry tomatoes or chopped scallions and cubed cucumber.

2 pounds ripe tomatoes, peeled, seeded, and roughly chopped
2 to 3 teaspoons fine sea salt
3 (4-inch) cucumbers, roughly chopped
1 red bell pepper, roasted, seeded, and chopped
1 jalapeño, seeded and chopped
1 shallot, diced
1 clove garlic, peeled
3 tablespoons sherry vinegar
¼ cup extra-virgin olive oil, plus more for drizzling
1 avocado, cubed
Chiffonade-cut shiso, basil, opal basil, or parsley, for garnish
Lime wedges, for squeezing

1. Place the chopped tomatoes in a bowl and sprinkle with 1 teaspoon of the salt. Set aside for 15 minutes while prepping the other ingredients.
2. Transfer the tomatoes and their juices to a blender. Add the cucumbers, roasted pepper, jalapeño, shallot, garlic, sherry vinegar, and 1 teaspoon of the salt. Blend for 2 minutes or until smooth. With the blender running, stream in the olive oil. Taste and add the remaining 1 teaspoon of salt, if needed. Although not essential, you can rub the puree through a fine-mesh sieve with a rubber spatula or wooden spoon to refine the texture. You should have 5 cups. Transfer to a glass pitcher and chill for several hours or overnight.
3. When ready to serve, stir in up to 1 cup ice water, ¼ cup at a time, to adjust the consistency. (Thicker to serve in bowls, looser for glasses.) Garnish each serving with the avocado, herb chiffonade, a drizzle of olive oil, and squirt of lime juice.

SLICING CUCUMBERS • HEIRLOOM TOMATOES • TWO BASILS | SIDE | SERVES 6 TO 8 | VEGAN, GLUTEN-FREE | SUMMER

CURRIED TOMATOES, PEACHES, AND CUCUMBERS

Some summer dishes are not so much cooked as cut up. This salad, for one. It's made from heirloom tomatoes, which ripen with Del's peaches, during the same peak season when green Genovese basil and dream-dark opal basil signal to each other from across the garden and trellised cucumbers drape like baroque festoons. Combine everything on a serving platter, and the result is a flashy first course for nearly any summer meal. I'd put it on the table alongside thin-crust onion pizza and juicy grilled skewers. Or I'd not cook at all and instead crowd the buffet with sliced prosciutto, a country terrine, olives, bread, whole-stem parsley, salad, a few naked lettuce greens, shaved summer squash, and bottles of rosé.

- 1½ teaspoons curry powder
- ½ teaspoon fine sea salt, plus more to taste
- 2 tablespoons fresh lemon juice
- ¼ cup extra-virgin olive oil
- 2 pounds medium cucumbers, cut into ¼-inch rounds
- 1 pound yellow peaches, peeled and cut into chunky wedges
- ½ cup lightly packed fresh basil leaves, mixed green and purple
- 1 pound Brandywine or other heirloom tomatoes, in large chunks

1. In a small bowl, whisk together the curry powder, salt, and lemon juice. Stream in the olive oil, whisking constantly.
2. In a large bowl, combine the cucumbers, peaches, and basil. Pour in the dressing and toss to coat. Transfer to a serving platter.
3. Place the tomatoes in the same bowl and swirl them around until they are coated with the residual juices. Lightly season with more salt to taste. Tuck the tomatoes around the edge of the serving platter.

TOMATOES • WAXY POTATOES • CORN • OKRA • SUMMER SQUASH • GREEN BEANS • ALLIUMS

STARTER, MAIN

SERVES 6 TO 8

GLUTEN-FREE, VEGETARIAN OPTION, VEGAN OPTION

FOUR SEASONS (FROM THE PANTRY AND FREEZER)

NANNY'S SUMMER VEGETABLE SOUP

This is my grandmother's stone soup. It's made from bits and bobs—a zucchini and a handful of beans and an ear of corn and a few potatoes, all stewed together with tomatoes. Nanny made it when my mother and uncles were small, at a time they depended on the garden to eat. She continued to make it all her life, and my mother made it all year long. In winter, the ingredients come out of the freezer, and the soup is no less good.

My version didn't taste right until my mother showed me exactly how to do it. For one, she didn't start by sautéing the onions in the usual way. Everything went straight into the pot, along with a tablespoon of sugar. The sequence of ingredients mattered. Nanny cut up everything with whatever knife she grabbed and dropped the ingredients into the pot as she worked, letting the irregular chunks fall from her hands into bubbling tomato-rich broth. Prepping and cooking was a single continuous activity.

The resulting soup has a precise all-American flavor and is delicious if made with prime garden produce. The core of it is tomatoes—preferably home-canned, although you could stew down fresh peeled tomatoes if that's what you have. Okra gives the broth body. Adjust the other ingredients to suit what's on hand. No herbs, no olive oil, just a lot of freshly cracked black pepper at the end—pure country cooking.

1 quart canned tomatoes, or 2½ pounds fresh tomatoes, peeled and stewed
1 quart light chicken or vegetable stock
2 bouillon cubes or 1 teaspoon fine sea salt
1 tablespoon sugar
½ large sweet onion, diced
¾ pound waxy potatoes, such as Pinto Gold, peeled and cubed
1 cup roughly chopped peeled carrots
½ pound okra, trimmed and cut into ½-inch pieces
1 medium zucchini, cut into ½-inch chunks
1 medium yellow squash, cut into ½-inch chunks
⅓ pound green beans, topped and snapped
1 cup cut corn
Fine sea salt and freshly ground black pepper

1. In a large soup pot, combine the tomatoes, stock, 1 quart water, bouillon cubes or salt, and sugar. Bring to a boil, uncovered, over high heat. Add the onion, potatoes, carrots, and okra. Return to a boil, then reduce the heat to maintain a low boil for 20 minutes.

2. Add the zucchini, squash, and green beans. Bring back to a boil, then adjust the heat to cook at a low boil, stirring occasionally, until the potatoes and carrots are tender, 20 to 25 minutes.

3. Add the corn and cook for 5 minutes longer. Taste and adjust the salt, if needed. Finish with a generous amount of freshly ground black pepper.

TOMATILLOS • ZUCCHINI • POTATOES • ALLIUMS • HERBS | MAIN | SERVES 8 TO 10 | OMNIVORE, GLUTEN-FREE, VEGAN OPTION | FOUR SEASONS

CHILE VERDE STEW

Tomatillos are tangier than tomatoes. Stripped of their papery husks and pulsed in a blender with a little onion, garlic, cilantro, and as many green chiles as you like, they become kicky salsa verde in roughly the time it takes to stir up a margarita. Have chips handy.

The same ingredients are transformed by heat into a silky, rich, tangy base for stews. Pork in chile verde is inescapable at Mexican restaurants, but, even so, I like to make it at home with local pastured pork. I add more vegetables—summer squash, small potatoes, strips of roasted jalapeño, and purslane, if I have any growing. You could also replace the cubed pork shoulder with a whole chicken cut into 10 pieces. Or do away with the meat altogether. Vegetables in chile verde make a deeply satisfying vegan stew. In winter I use home-canned tomatillos to stew winter squash, carrots, hearty greens, tiny turnips, whatever vegetables I have. See the notes on the next page for adjusting the recipe.

Serve chile verde with a stack of warm corn tortillas, a bowl of rice, and a dish of sour cream loosened with buttermilk. Throw cilantro confetti over top.

SALSA VERDE

2 pounds tomatillos, husked and halved
4 jalapeños, stemmed
1 small onion, peeled and quartered
4 cloves garlic, peeled but whole
1 tablespoon grapeseed oil
1/2 teaspoon fine sea salt
4 large romaine leaves
A dozen cilantro sprigs
1 tablespoon cumin seeds, toasted and ground
1 teaspoon dried Mexican oregano or oregano
1 cup stock or water

STEW

2 1/2 pounds boneless pork shoulder, cut into 1-inch cubes
1 tablespoon fine sea salt
2 tablespoons grapeseed oil
1 large onion, diced
3 cups chicken stock or water
1 pound zucchini, cut into 3/4-inch chunks
3/4 pound small whole potatoes, such as Upstate Abundance

FOR SERVING

Steamed rice
Corn tortillas, warmed
Sour cream (loosened with buttermilk)
Chopped fresh cilantro
Radish coins or matchsticks, for garnish
Nasturtium flowers, for garnish
Purslane rosettes, for garnish

(CONTINUED)

1. **Make the salsa verde:** Preheat the broiler.
2. In a bowl, toss together the tomatillos, jalapeños, onion, garlic, grapeseed oil, and ¼ teaspoon of the salt. Spread on a sheet pan and broil until scorched, about 15 minutes, turning once. Remove from the oven and turn the oven temperature to 300°F.
3. Reserve 2 jalapeños and transfer the remaining broiled vegetables to a blender. Add the romaine leaves, cilantro, cumin seeds, oregano, the remaining ¼ teaspoon of salt, and the stock and pulse until smooth, adding more liquid if necessary for a milkshake consistency.
4. **Make the stew:** Season the pork with the salt. In a large Dutch oven, heat the oil over high heat until nearly smoking. Add the pork and brown on all sides, about 15 minutes total.
5. Add the diced onion and stir until translucent, 3 to 5 minutes. Add the salsa verde and stock to cover the pork and return to a boil. Partially cover the pot and transfer to the oven and bake for 1½ hours.
6. Remove from the oven and add the zucchini and potatoes. Partially cover and return to the oven and bake until the potatoes are cooked through and the pork shreds easily, 1 to 1½ hours.
7. Meanwhile, peel and seed the 2 reserved jalapeños. Cut into strips. Prep the garnishes and accompaniments.
8. When the pork is done, skim the grease without being too particular about it. If the stew is too soupy, bring to a boil on the stovetop and reduce, uncovered, for a few minutes.
9. To serve, ladle into individual bowls over rice with tortillas on the side. Top with a jalapeño strip. Place sour cream and other garnishes on the table.

To adapt for chicken stew: Make the salsa verde as directed. Cut a 3½-pound chicken into 10 pieces and salt liberally. Brown the chicken in oil and soften the onion as directed. Before pouring the salsa into the Dutch oven, add the zucchini and potatoes, as well as a roughly chopped carrot or two. Add enough water or stock to three-quarters cover the chicken and vegetables. Bring to a boil. Cover, leaving it ajar, and moderate the heat to maintain a lively simmer for 1 hour, or until the chicken is done. Stir occasionally.

To adapt for all-vegetable stew: Make the salsa verde as directed. Select about 3 pounds of mixed vegetables such as summer squash, carrots, winter squash, turnips, et cetera, plus 1 pound of small potatoes. Cut the vegetables into hearty chunks and sweat in oil for 5 to 7 minutes, without browning, then add the onion. Be quite liberal with oil throughout and replace the grapeseed with more flavorful avocado oil or olive oil. When you pour in the salsa, add the potatoes and a handful of hearty greens, stripped of their midribs and cut into ribbons. Bring to a boil, uncovered, and moderate the heat to maintain a lively simmer for about 1 hour. If you like, add 2 cups of cooked chickpeas or other favorite bean.

TOMATOES • ALLIUMS • HERBS | MAIN | SERVES 2 (OR 4 TO 6 IF USING FILLETS) | GLUTEN-FREE | SUMMER

FISH ROASTED ON A BED OF ONIONS AND TOMATOES

The tomato crop always includes some that are blemished or misshapen. In addition to gazpacho and sauce, the uglies make a flavorful bed for roasted fish in this failsafe one-dish recipe. The cooking is accomplished in two stages. First, let the tomatoes and aromatics cook down while you wander back outside to pick a handful of salad greens. Then you roast the fish over the concentrated tomato-onion reduction. Whole branzino makes a picturesque presentation. Fillets of halibut or bluefish go further if cooking for several people. For a splurge, use meaty tuna or swordfish steaks, adjusting the cooking time as needed. Throw in a handful of green or ripe olives near the end to pretend you're in Provence.

¼ cup extra-virgin olive oil, plus more for the baking dish and fish
2 medium onions, thinly sliced
2 cloves garlic, sliced
¼ teaspoon red chile flakes
1¼ teaspoons fine sea salt
2 pounds tomatoes, seeded
1 cup dry white wine
1 bay leaf
2 whole fish (1½ pounds each), or 2 pounds thick fish fillets
1 lemon, thinly sliced
10 sprigs thyme

1. Preheat the oven to 450°F. Lightly oil a 9 × 13-inch baking dish.
2. Add the onions, garlic, chile flakes, and the ¼ cup of olive oil to the baking dish. Toss to coat and sprinkle with ¼ teaspoon of the salt. Bake for 10 minutes.
3. Remove the baking dish from the oven. Add the tomatoes, wine, and bay leaf and season with ½ teaspoon of salt. Return to the oven and bake until the liquids are reduced to a syrup and the tomatoes are blackened at the edges, 30 to 40 minutes.
4. Meanwhile lightly oil the fish and sprinkle with the remaining ½ teaspoon of salt. Stuff with the lemon slices and thyme sprigs. If using fillets, make a bed of the herbs and lemon on a plate and place the fish on top.
5. Lay the fish (or slide the bedded fillets) onto the tomato base. Reduce the oven temperature to 400°F and bake, uncovered, until the fish is flaky and barely translucent at the center. The Canadian cooking method for fish provides a rule of thumb: 10 minutes of cooking per 1 inch of thickness at the thickest part of the fish.
6. Bring to the table in the baking dish.

PEPPER AND CHILE BASICS

With their juicy snap, crisp texture, and grassy flavor, green peppers and small green chiles are best appreciated raw, whether as diced jalapeños on tacos or thin bell pepper rounds thrown into a brimming salad bowl like a carnival ring toss.

Ripe peppers and chiles—red, yellow, or orange—are sweet and richly flavored. They get along fraternally with tomatoes, as in red salsa and gazpacho, or can be cooked down until supple.

Peppers and chiles love fire. Scorch them, blacken them, grill them—they only improve. Restaurant chefs turned blistered shishito peppers into a craze, and roasted red peppers marinated in olive oil are both fruity and meaty—one of the best antipasti. Smoke turns red jalapeños into chipotles, culinary alchemy.

The popularity of dried chiles approaches that of black pepper (the dried fruit of an unrelated tropical vine, *Piper nigrum*), making it the world's other universal spice.

TO ROAST PEPPERS AND CHILES

"Roasting" peppers and chiles is a euphemism. The technique is to scorch them until completely blackened. Cooking over intense heat concentrates flavor and loosens the skin, which, like tomato skin, is inoffensive when raw but unpalatable once cooked. Place peppers directly over a gas burner, on a grill, or under a broiler. Turn occasionally with tongs when the skin chars and pulls away from the flesh. Continue methodically, charring all over, about 15 minutes total cooking time. Transfer to a covered bowl and set aside for 15 minutes to steam. (Or wrap in aluminum foil.) When cool enough to handle, rub off the loosened skins. Work at tough spots by meticulously scraping with the dull edge of a knife. Split open the cleaned peppers and scrape out the seeds and connective membranes (discard them and the stem). Never rinse peppers because water strips their flavor, but you can wipe them down with a paper towel. Slice or dice, as needed. Roasted peppers will keep overnight in a sealed container. Covered with olive oil, they will keep for several days.

Green and red chiles are prepared the same way. Protect your hands with rubber gloves.

Note: A Turkish kitchen implement sold online by Milk Street, the közmatik, is useful for charring vegetables on the stovetop. It is a thin vented pan—a stovetop grill—that you place over the gas burner to diffuse the heat and contain the vegetables while they char. It will hold 3 medium eggplants or 6 bell peppers.

MARINATED ROASTED PEPPERS

Roast and peel several red and yellow bell peppers, as described above. Slice into ¼-inch-wide strips. Transfer to a pretty dish and season lightly with fine sea salt and freshly ground black pepper. Tuck in a peeled, lightly crushed garlic clove and pour over a glug of your best extra-virgin olive oil. Stir to coat and set aside for 30 minutes. Sharpen the flavor with a light splash of red wine vinegar. Drape a strip or two over grilled bread spread with soft cheese, and serve alongside flatbread and hummus, or include as a part of any appetizer board.

BLISTERED OR GRILLED PEPPERS

Some smaller peppers don't need peeling. Bite-sized shishitos can be eaten off the stem whole, seeds and all. They are mostly mild apart from the stray spicy one. Ditto Spanish pimientos de Padrón, about which the Galicians say, *algúns pican e outros non,* "some sting and others don't." When you blister peppers for an appetizer, you cook them fast in a hot skillet until they are leopard-spotted and half-limp. Coat the inside of a cast-iron skillet with a scant tablespoon of grapeseed oil. Heat over high heat to the smoking point. Working in batches to prevent crowding, throw in a handful of peppers. Cook on the first side for about 1 minute, then shake the pan to turn the peppers. Cook another minute, shaking a time or two, until the peppers are blistered in places. Serve hot, sprinkled with finishing salt.

Sweet "frying" peppers, such as the legendary Jimmy Nardello, the best tasting of all peppers, can be prepared the same way, or grilled over hot coals. Serve them whole, with or without drizzled extra-virgin olive oil and sprinkled finishing salt.

PICKLED PEPPERS

Peter Piper aside, all peppers—and especially green chiles—make excellent pickles. To preserve their crisp texture and fiery flavor, make refrigerator pickles, which are not canned, or processed in a boiling-water bath. Vinegar-based pickling brine prevents spoilage. Fill a 1-quart mason jar with sliced jalapeños or other chiles, green or red. Pack it full and gently press to squeeze in as many as possible without crushing. Add 2 crushed cloves garlic, 12 whole peppercorns, a bay leaf, and 2 fat pinches of dried oregano. In a small saucepan, combine 1 cup white wine vinegar or apple cider vinegar, 1 cup water, and ¾ teaspoon fine sea salt. Heat until the salt dissolves. Taste. The pickling brine should be sharp, strong, and salty. Add more salt if needed. You can temper the vinegar with a teaspoon of sugar or honey, if you like. Once seasoned to your taste, heat the liquid to scalding, and pour over the peppers, filling the jar to the brim. Top up with straight vinegar if needed. Seal the jar. Store in the refrigerator for 3 days to marinate before using. The pickles will last several months.

TO DRY CHILES

The best chiles for air-drying are small and thin, like cayenne. Even in humid regions, they can be successfully dried by stringing loosely on a thread and hanging in a well-ventilated spot out of direct sun. In arid regions, larger chiles such as New Mexico and Anaheim can be strung into ristras. An alternative method is to scatter chiles on a screen to dry. Provide good ventilation and turn every few days. Fully dried chiles will be leathery. They will keep in a sealed container for a year.

RED BELL PEPPERS • NEW GARLIC | STARTER | MAKES 2 CUPS | VEGETARIAN | SUMMER, FALL

FAMILY SECRET PIMENTO CHEESE

Pimento cheese, the ubiquitous Southern spread, seemed new again when I first made it with roasted red bell peppers rather than jarred pimentos. Over time, I've embroidered upon the basic framework—diced peppers, sharp cheddar, mayonnaise—to boost flavor to family-secret intensity. Pickling brine from sauerkraut or kosher dill pickles adds the salty, savory, mouthwatering qualities of fermentation. A hint of grated new garlic flickers just below the surface, barely there, like minnows in a creek. Pimentón has a subtly rounding effect, while the modest amounts of hot sauce and cayenne pepper make it not-quite-spicy. For more zip, stir in a diced roasted red jalapeño.

Heart-shaped pimento peppers, thick-walled Lipstick peppers, and sweet Corno di Toro are extra-tasty. Let the peppers become dead ripe before picking.

- 2 red bell peppers, roasted (see page 437), peeled, and diced
- 1 clove garlic, grated
- 3/4 cup mayonnaise
- 1 teaspoon Fermented Hot Sauce (page 440)
- 2 tablespoons sauerkraut brine or kosher dill pickle brine
- 1/4 teaspoon pimentón (Spanish smoked paprika)
- 1/4 teaspoon cayenne pepper
- 1/2 teaspoon fine sea salt
- 1/4 teaspoon freshly ground black pepper
- 8 ounces extra sharp cheddar cheese, coarsely grated (2 cups packed)
- 1 ounce Parmigiano-Reggiano cheese, finely grated (1/3 cup packed)

1. In a medium bowl, whisk together the diced roasted pepper, garlic, mayonnaise, hot sauce, brine, pimentón, cayenne, salt, and black pepper. Taste and adjust spiciness, if needed. It should be highly seasoned.
2. Add the cheddar and Parmigiano and stir until blended. Pack into a wide-mouthed jar and refrigerate for several hours before use. Leftovers will keep for 1 week.

FERMENTED HOT SAUCE

My first cookbook, *Saving the Season: A Cook's Guide to Home Canning, Pickling, and Preserving,* deals with the problem of abundance. What do you do with all that seasonal produce? You find a way to keep some of it for later, when you know there will be none. You "put it up" or "put it by." A bushel of peaches in July = a shelf full of jam and chutney in January. Preserving garden vegetables, from asparagus to zucchini, gives you pickles, sauerkraut, kimchi, relish, onion "jam," canned tomatoes, and this hot sauce. All save the season.

Many preserving techniques create delicious new flavors. Consider kosher dill pickles. Whatever is lost in the transformation—a cucumber's garden freshness—is offset by a pickle's salty, yeasty, savory qualities.

Fermented hot sauce walks the same fence. It preserves the fire of fresh chiles and even something of their fruitiness, while introducing umami complexity. Fermentation, also known as lacto-fermenting or brining, is a pickling technique in which the brine sours on its own, thanks to the action of beneficial bacteria within a flourishing microbial wilderness. If you've made kosher dills, you already know how to ferment chiles. If you haven't, it couldn't be simpler. You fill a 1-quart jar with mixed fresh chiles, cover them with brine, and wait. Nature does the rest. The seeds of the ferment, the necessary microbial spores, are carried on the chiles themselves. Your work is to tend the jar, which mostly amounts to weeding unwanted growth, which appears as the occasional harmless speck of mold.

The finished hot sauce can be sharpened with vinegar, if you like, for a condiment that works on nearly anything, from mild eggs to soulful greens to robust stews. As a probiotic, it even moderates the worst effect of beans.

Fermented hot sauce has a considerable shelf life. I used jars of one large batch for 3 years, by which time it had developed an agreeable whiff of old leather.

1¼ pounds mixed red chiles, such as jalapeño and cayenne, stemmed and split lengthwise
500 grams unchlorinated water, such as bottled spring water (2 overflowing cups)
20 grams noniodized salt (3½ teaspoons fine sea salt, or 5 to 6 teaspoons kosher salt)
6 black peppercorns
1 to 2 tablespoons Champagne vinegar (optional)

1. Wash a 1-quart mason jar and scald it by pouring a kettle of boiling water over it. Drain and pack with the chiles.
2. In a large bowl, stir together the water and salt to dissolve. Pour over the chiles. Add the peppercorns. If needed, add more unchlorinated water to fill the jar to the brim. Settle the lid on top but do not tighten, so that gasses produced during fermentation can escape—otherwise, kaboom. Put the jar in a cool spot, ideally 70°F or less, but do not refrigerate. Place a plate or saucer underneath, in case of spillage.

3. Every few days, push the chiles down into the brine. At some point they will waterlog and settle to the bottom. A delicate veil will bloom across the surface of the brine. Skim it and discard—there's no better tool than your own clean hand. If a speck of mold appears, pick it out. Nothing to worry about. The brine will sour within 2 weeks, or less in warmer temperatures.
4. When it tastes suitably sour, put the jar in the refrigerator for 2 weeks. The very slow cold fermentation will add another layer of flavor. You can leave the chiles for as long as you like and rob spoonfuls of brine to season deviled eggs and bloody Marys.
5. To make the sauce, dump the chiles and about half of the brine into a blender. Pulse until liquefied. Thin with more brine, to taste. Add the vinegar, if you like. Refrigerate the hot sauce in a stoppered bottle or a loosely sealed mason jar. It will keep for many months.

EGGPLANT BASICS

For enthusiasts, eggplant is a staple of the summer garden. For others, it might raise the kohlrabi question—what do you do with it? When in doubt, roast eggplants whole. The custardy pulp will be smoky and delicious, with that special eggplant tang, a tingle along the sides of the tongue and the back of the soft palate. Thick eggplant steaks have meaty substance when grilled, broiled, baked, or roasted. The small Asian varieties adapt to stir-fries and curries. And there's always eggplant Parmesan. Among the more outré possibilities, fried sliced eggplant dusted with powdered sugar is a signature appetizer at Galatoire's in New Orleans. See the following for practical suggestions.

Much is made about salting sliced eggplants before cooking to remove bitterness. Perhaps bitterness was a trait of antique varieties, but it's not a problem I've noticed with freshly picked eggplant. Deborah Madison sensibly suggests that salting will extract water and thereby reduce the cooking time needed to concentrate eggplant's flavor and achieve the ideal pudding consistency. If I have the foresight, I'll salt sliced eggplant, up to 1 hour before cooking, and place it in a colander to drain, then pat dry. Just as often I skip this step.

TO ROAST EGGPLANT

Prick an unblemished whole Italian eggplant with a fork in several places. Place directly over a gas burner, on a hot grill, or under a broiler. (See the Note on page 437 about a stovetop grill called the közmatik.) Turn occasionally until charred all over and collapsing, 20 to 40 minutes total cooking time, depending on size. Transfer to a covered bowl or wrap with aluminum foil. (Alternatively, wrap whole eggplants in aluminum foil and bake on the top rack of a 400°F oven until very soft, 30 to 45 minutes.)

When the cooked eggplant is cool enough to handle, rub off the charred skin. Scrape stuck bits with the dull edge of a knife, or wipe with paper towels, but don't rinse. Drain pulp in a colander for 30 minutes. It doesn't look very promising at this stage, but the pulp is ready to use. Look ahead to Smoky Eggplant Dip (page 444).

EGGPLANT STEAKS

Eggplant loses a lot of water in cooking, so cut thickly, at least ⅝ inch. Slice an Italian eggplant crosswise into steaks or, for Asian eggplants, cut in half lengthwise. Rub the cut surfaces with oil. Be generous but not extravagant: Raw eggplant will absorb an excess of oil. Sprinkle with salt and pepper. Steaks can be grilled over hot coals, broiled, or roasted in a 400°F oven until browned and soft. Turn once during cooking. Serve with tomato sauce or white beans, as below.

EGGPLANT AND WHITE BEAN SALAD

Cook eggplant steaks (as above) and cut into 1-inch pieces. Combine in a large bowl with cooked and drained white beans, cucumber chunks, hunks of tomato, fresh chèvre crumbles, a handful of whole parsley leaves, and a pinch of minced rosemary. Add several glugs of good extra-virgin olive oil, a bit of red wine vinegar, salt, and pepper. Toss gently to combine. Serve on a bed of lettuces. Scatter more chèvre over top.

SIMPLIFIED EGGPLANT PARMESAN

Slice 2 large Italian eggplants into steaks. Salt and drain in a colander up to 1 hour before cooking, then rinse and pat dry. Heat ¼ inch of vegetable oil in a large cast-iron skillet over high heat until nearly smoking. Working in batches, fry the eggplant slices in a single layer until golden, turning once, about 5 minutes. Transfer to paper towels to drain. Cover the bottom of a baking dish with a thin layer of Fresh Tomato Sauce (page 422). Cover the sauce with a single layer of fried eggplant slices. Scatter over handfuls of grated mozzarella and grated Parmesan, and drop spoonsful of fresh ricotta. Add additional layers of tomato sauce and eggplant. Top with more tomato sauce, and sprinkle with grated mozzarella and grated Parmesan. Bake at 350°F for 45 minutes, or until the sauce is bubbling and the cheese browned.

EGGPLANT • NEW GARLIC • MIXED SMALL VEGETABLES AND HERBS | STARTER | MAKES 2½ CUPS | VEGAN | HIGH SUMMER

SMOKY EGGPLANT DIP

Smoke is an adaptable ingredient. It can boldly frame the fatty grandeur of whole-hog BBQ or subtly waft through other flavors and disappear. I once had a dish of raw fish at Jua, a wood-fired Korean restaurant in New York, so delicately smoked that it seemed the chef had merely thought of fire. Wherever there is smoke, the elemental cooking technique always flickers in the background, a culinary echo of the fatted calf and the hecatombs of the Greeks, who sent up meaty incense to their hungry gods.

In the vegetable kingdom, eggplant is exalted through fire. Cook one whole until the thing looks ruined, burnt beyond eating, and the pulp will absorb fire's essence. Olive oil and tahini, the soul of baba ghanoush, establish a direction. The hidden charm is a grated clove of new garlic—especially the Rocambole varieties, such as exquisite Spanish Roja, or the purple-stripe types, including Persian Star and Chesnok Red.

Serve eggplant dip with grilled bread or an array of sliced raw vegetables, as in the photograph opposite, along with marinated sumac onions and whole-leaf herbs. Fold the same ingredients into a hot pita with sliced pickled beets and salty feta for a superb garden sandwich.

2 cups roasted eggplant pulp (see page 442), from 2 (1-pound) eggplants
1 clove garlic, grated on a Microplane
1 tablespoon minced fresh parsley
¼ teaspoon Aleppo pepper flakes
2 tablespoons tahini
½ teaspoon fine sea salt
2 tablespoons fresh lemon juice
¼ cup extra-virgin olive oil, plus more for drizzling
Grilled bread rubbed with garlic or sliced raw vegetables, for serving

1. Mash the eggplant pulp with a fork. Add the garlic, parsley, Aleppo, tahini, salt, and lemon juice and beat vigorously, as if beating eggs, until smooth. (Alternatively, pulse the ingredients in a blender.) Whisk in the olive oil in a stream until the mixture is fluffy. Taste and adjust the seasonings.

2. Transfer to a serving dish. Drizzle with olive oil and sprinkle with Aleppo. Serve with grilled bread or raw vegetables. Leftovers will keep in the refrigerator for 3 days.

ITALIAN EGGPLANT • ZUCCHINI • BELL PEPPERS • TOMATOES • ALLIUMS AND HERBS | SIDE | SERVES 12 TO 16 | VEGETARIAN, GLUTEN-FREE | HIGH SUMMER

RATATOUILLE

In the opening chapter, I pointed to ratatouille as an example of traditional cooking that doesn't just express a season of the garden—in this case, high summer—but seems to be a culmination of it. Created in the countryside outside Nice and made famous by a Pixar rat, ratatouille is a stew of eggplant, zucchini, bell pepper, and tomatoes, all of which peak as the new crop of onions and garlic mature. The funny name mixes two French verbs, *tatatouiller* and *ratouiller,* which, as erudite English cookbook author Jane Grigson notes, are "expressive forms of *touiller,* an old verb from the Latin *tudiculare,* meaning to stir and crush." The word dates to the eighteenth century but formerly described indeterminate meat stews; the meaning as it comes down to us today didn't settle until circa 1930. Be that as it may, who can doubt that summer stews resembling ratatouille have been prepared for a very long time wherever gardeners cultivated eggplant, zucchini, peppers, and tomatoes together.

Ratatouille can be made entirely on the stovetop—and even in a single pan—but I prefer to sauté the vegetables separately then layer them in a baking dish. They go in the oven long enough for the flavors to blend and the juices to reduce. The final consistency is jammy, but the vegetables retain their shape because—ironically, perhaps—you do not "stir and crush." My reference is a memory of the first ratatouille I ever ate. It was made by Monique Dubois, the mother of my high school best friend, Laurent, when I joined them for family vacation at the beach. We had dinner outside at a table carefully set with good plates, the adults drank wine, and we all had ratatouille omelets the next day for lunch. Meals with the Dubois family were an awakening.

The recipe is quite flexible and easy to make. The quantity can be halved or doubled. The only fuss is in the prep, and two steps should not be considered optional. Peeling and seeding the tomatoes prevents watery ratatouille. Salting and squeezing the eggplant also purges liquids and, no less important, prevents the eggplant from absorbing excessive oil when sautéed.

2 medium Italian eggplants (about 2 pounds)
1 tablespoon fine sea salt
1 cup extra-virgin olive oil
4 medium zucchini, sliced in 1/4-inch slices (about 2 pounds)
1/2 pound pear tomatoes, halved, or plum tomatoes sliced in 1/4-inch slices
2 medium onions, diced
3 sprigs thyme
1 sprig winter savory
1 dried red chile, minced
Freshly ground black pepper
2 medium red bell peppers, destemmed, seeded, and sliced into 1/4-inch strips
4 cloves garlic, chopped
4 medium tomatoes (about 2 pounds), peeled, seeded, and coarsely chopped
1/4 cup minced parsley
8 whole basil leaves, roughly chopped
Basil chiffonade, for strewing

1. Preheat the oven to 350°F. Lightly oil a 9½ × 13½-inch baking dish.
2. Cut the eggplants into ¾-inch pieces. Sprinkle with 1 teaspoon of the salt and place in a colander in the sink to drain for 30 minutes. Squeeze firmly, one handful at a time, to extract excess moisture.
3. Heat 3 tablespoons of the oil in a 12-inch sauté pan over high heat until it shimmers. Add half the eggplant and salt generously, about ¼ teaspoon. Cook until lightly brown, about 5 minutes. Shake the pan a time or two, but don't stir excessively. Turn into a large mixing bowl. Add 2 tablespoons of oil to the pan. When it shimmers, add the remaining eggplant and ¼ teaspoon salt. Cook until lightly browned, about 5 minutes. Transfer to the mixing bowl.
4. Using the same pan, heat 2 tablespoons of oil until it shimmers. Add half the zucchini and about ¼ teaspoon salt. Sauté until light gold, about 5 minutes. Add to the mixing bowl. Add 2 tablespoons of oil to the pan, then add the remaining zucchini and ¼ teaspoon salt. Sauté for 5 minutes, but do not transfer to the mixing bowl yet. Add the pear tomatoes and a fat pinch of salt. Sauté for 3 minutes, or until the tomatoes start to soften. Transfer the mixture to the mixing bowl. Gently combine the eggplant and zucchini by turning 2 or 3 times, but don't stir or mash. Rinse and dry the pan.
5. Heat 2 tablespoons of oil over high heat until it shimmers. Add the onions, thyme, savory, chile, ½ teaspoon salt, and ¼ teaspoon freshly ground black pepper. Sauté until the onions are translucent, about 3 minutes. Add the bell peppers and a fat pinch of salt. Sauté until the peppers soften, about 5 minutes. Add the garlic and sauté for 2 minutes. Add the chopped tomatoes and ¼ teaspoon salt, and sauté for 5 minutes, or until the juices are furiously bubbling and slightly reduced. Add the parsley and basil and toss a few times to combine. Turn into the prepared baking dish and spread the mixture in an even layer.
6. Pour the eggplant-zucchini mixture over the top and spread evenly. Drizzle with the remaining olive oil.
7. Bake, uncovered, for 45 minutes or until lightly browned. Allow to cool for at least 15 minutes. To serve, strew with basil chiffonade. Ratatouille is equally good warm or at room temperature. Leftovers will improve overnight in the refrigerator.

Note: For easier measuring, pour the oil into a small pitcher before you start and put the salt into a small dish. Then dole out measures as you go or, as I often do, eyeball quantities based on the amount left in the container or dish.

ACKNOWLEDGMENTS

A gardener learns from more experienced gardeners. This book is a compendium of what I've learned through the generosity of wise stewards. Farmers and gardeners preserve an agricultural legacy as old as civilization and as intimate as family lore. That they feed the world in return for modest worldly gain proves their outstanding wisdom. This book attempts to repay my debt of gratitude.

First among my garden teachers was my mother, Carol Martin Satterfield, who showed me what it means to "preserve the legacy." This book belongs to *her* legacy, and I dedicate it to her memory. The book also honors Nanny, my grandmother with a green thumb. We used to say that Nanny could stick a two-by-four in the ground and it would grow. And, as always, memories of Gran and Pappaw suffused every page I wrote.

It was a privilege to meet professionals who invited me to visit their farms and gardens even during the busy growing season. My memories of this project will be of outdoor conversations among the plants. Lasting thanks to:

Pat Brodowski at Monticello

Eliot Coleman of Four Season Farm

John Coykendall at Blackberry Farm

Steve Cunningham of Berkshire Bounty Farm

Barbara Damrosch of Four Season Farm

Tracy Hayhurst at Husky Meadows Farm

Elizabeth Keen of Indian Line Farm

Jan Johnson and Peter Chapin of Mill River Farm

Max Morningstar of M|X Morningstar Farm

Keith Nevison at Monticello's Thomas Jefferson Center for Historic Plants

Jeana Park and Yong Yuk of Et Cetera Farm

Lauren Piotrowski at Hancock Shaker Village

Barbra and Andy Rothschild of Spruce Ridge Farm

Jen and Pete Salinetti of Woven Roots Farm

Bill and Barbara Spencer of Windrose Farm

Ria Ibrahim Taylor at Soul Fire Farm

Talea and Doug Taylor of Montgomery Place Orchards and their daughter Caroline Olivia

Evan Thaler-Null of Abode Farm

Ira Wallace of Southern Exposure Seed Exchange

Loving thanks to the lifelong gardeners in my family who continue to teach me: my father, Fred West, and stepfather Bob Sharp, and Barbara and David West, who embody Gran and Pappaw's shining legacy.

A deeply felt thank-you goes next to my editor, Lexy Bloom, although perhaps it should have come first because she initially approached me with the humbling opportunity to create a modern gardening handbook in the tradition of *The Victory Garden*. We discussed whether it should be a cookbook for gardeners or a gardening book for cooks, and in the end, she gave me the extra years needed to write both—the two halves of this single volume. I am deeply grateful to her and the entire Knopf team for their patience and commitment to this project as it grew. I was told its final length is nearly double that of a typical cookbook, so I offer double thanks to the tireless production team, especially Kate Slate, Kelly Blair, Shubhani Sarkar, production editor Andrea Monagle, and managing editors Meredith Dros and Anne Achenbaum, as well as editorial assistants Morgan Hamilton and Isa Connolly and senior editor Tom Pold. Bountiful thanks as well to Sarah New and to Sara Eagle, the wonderful publicity and marketing team.

David Kuhn and his Aevitas Creative Management colleagues Nate Muscato and Helen Hicks were, throughout the process, my best advocates and counselors.

I wouldn't have had a garden to cook from without my neighbors Christine and Del Martin, who generously allow me to garden on their land. Their friendship is inseparable from my life in the Berkshires, and I thank them for the many ways they sustain me.

Further fond thanks go to those who worked alongside me on the book. Rachel Oberg's cheerful and efficient office help got the project underway. A special shout-out is reserved for kitchen collaborator Dashiell Nathanson, who visited the Berkshires twice and poured boundless creativity into the project, improving it in a hundred ways. It was also a joy to have my cousin Jack Carey come to the Berkshires to pull weeds and shell peas. And no one put in more hours than Chen Li and Deeva Gupta, who, during their annual break from Deep Springs College, labored mightily to water through drought and then brought keen intellect to the dinner table. I am grateful to Tom Hudgens, my Deep Springs classmate, for his scrupulous review of the manuscript and his many generous suggestions, which I eagerly accepted. Stokes Young was another ideal first reader whose feedback improved the book.

Wes Whitsell was generous as a chef and more so as a friend. David Tanis was always with me in spirit in the kitchen. Talking shop while hiking with Sara Franklin—another generous first reader—solved many editorial dilemmas and lifted my spirits. Early in the project, a conversation with Samin Nosrat got me through a logjam. I am incredibly lucky to know such talented friends, and I thank each of them personally.

I would like to thank, as well, Dr. Amber Kerr and poets Kateri Kosek and Naomi Shihab Nye for kind permission to reprint their work. Leanne McQueen's pottery appears in a number of photographs because it is part of nearly every meal I serve.

More friends who got me through are Maryetta Anschutz, Jamey Collins, Christophe Farber, Claire Hoffman, Alex Pincus, Wild Pincus, Ian and Jane Purkayastha, the Rothschilds of Chatham, and Tina Rathbone. Stephanie Brummett has a unique place in my heart.

Finally, I'd like to thank my loving stepfather Don Satterfield for the years he shared a dinner table with my mother and the many times he patiently endured rattling pots and pans as I developed recipes in their kitchen.

INDEX

(Page references in *italics* refer to illustrations.)

C

D

H

M

N

O

P

Q

R

S

T

U

V

W

Y

Z